P9-BYQ-995

ARCHIMEDES TO HAWKING

SELECTED BOOKS BY CLIFFORD A. PICKOVER

The Alien IQ Test
A Beginner's Guide to Immortality
Black Holes: A Traveler's Guide
Calculus and Pizza
Chaos and Fractals
Chaos in Wonderland
Computers, Pattern, Chaos, and Beauty
Computers and the Imagination
Cryptorunes: Codes and Secret Writing
Dreaming the Future
Future Health
Fractal Horizons: The Future Use of Fractals
Frontiers of Scientific Visualization
The Girl Who Gave Birth to Rabbits
The Heaven Virus
Keys to Infinity
The Lobotomy Club
The Loom of God
The Mathematics of Oz
Mazes for the Mind: Computers and the Unexpected
The Möbius Strip
The Paradox of God and the Science of Omniscience
A Passion for Mathematics
The Pattern Book: Fractals, Art, and Nature
The Science of Aliens
Sex, Drugs, Einstein, and Elves
Spider Legs (with Piers Anthony)
Spiral Symmetry (with Istvan Hargittai)
Strange Brains and Genius
Sushi Never Sleeps
The Stars of Heaven
Surfing Through Hyperspace
Time: A Traveler's Guide
Visions of the Future
Visualizing Biological Information
Wonders of Numbers
The Zen of Magic Squares, Circles, and Stars

CLIFFORD A.
PICKOVER

ARCHIMEDES TO HAWKING

Laws of Science and the Great Minds Behind Them

OXFORD
UNIVERSITY PRESS
2008

OXFORD
UNIVERSITY PRESS

Oxford University Press, Inc., publishes works that further
Oxford University's objective of excellence
in research, scholarship, and education.

Oxford New York
Auckland Cape Town Dar es Salaam Hong Kong Karachi
Kuala Lumpur Madrid Melbourne Mexico City Nairobi
New Delhi Shanghai Taipei Toronto

With offices in
Argentina Austria Brazil Chile Czech Republic France Greece
Guatemala Hungary Italy Japan Poland Portugal Singapore
South Korea Switzerland Thailand Turkey Ukraine Vietnam

Published by Oxford University Press, Inc.
198 Madison Avenue, New York, NY 10016

ISBN 978–0–19–533611–5

Printed in the United States of America

Truly the gods have not from the beginning revealed all things to mortals, but by long seeking, mortals make progress in discovery.

—Xenophanes of Colophon (c. 500 B.C.)

I canna change the laws of physics, Captain!

—"Scotty" Montgomery Scott to Captain Kirk, in "The Naked Time," *Star Trek* TV series

It is indeed a surprising and fortunate fact that nature can be expressed by relatively low-order mathematical functions.—Rudolf Carnap, classroom lecture

Perhaps an angel of the Lord surveyed an endless sea of chaos, then troubled it gently with his finger. In this tiny and temporary swirl of equations, our cosmos took shape.—Martin Gardner, *Order and Surprise*

The great equations of modern physics are a permanent part of scientific knowledge, which may outlast even the beautiful cathedrals of earlier ages.

—Steven Weinberg, in Graham Farmelo's *It Must Be Beautiful*

ACKNOWLEDGMENTS

> When Newton worked out the force of gravity, he helped to set into motion the industrial revolution. When Faraday worked out electricity and magnetism, he set into motion the electric age. When Einstein wrote down $E = mc^2$, he unleashed the nuclear age. Now, we are on the verge of a theory of all forces which may, one day, determine the fate of the human species.
>
> —Michio Kaku, "BBC Interview on Parallel Universes"

I thank Teja Krašek, Steve Blattnig, Graham Cleverley, Mark Nandor, Paul Moskowitz, Pete Barnes, and Dennis Gordon for their comments and suggestions.

While researching the laws in this book, I studied a wide array of wonderful reference works and Web sites, many of which are listed in the reference section at the end of this book. These references include the *McGraw-Hill Encyclopedia of Science and Technology*, *Van Nostrand's Scientific Encyclopedia*, *Encyclopædia Britannica*, Jennifer Bothamley's *Dictionary of Theories*, the *MacTutor History of Mathematics Archive* (www-history.mcs.st-and.ac.uk/), *Wikipedia: The Free Encyclopedia* (en.wikipedia.org), "Eric Weisstein's World of Physics" (scienceworld.wolfram.com/physics/), and "HyperPhysics Concepts" (hyperphysics.phy-astr.gsu.edu/hbase/hph.html).

The *Dictionary of Scientific Biography*, edited by Charles Coulston Gillispie and published by Charles Scribner's Sons, was a particularly invaluable resource. For readers who are interested in learning more about this multivolume reference set, I recommend the fascinating paper "Publishing the *Dictionary of Scientific Biography*" by Charles Scribner, Jr., published in *Proceedings of the American Philosophical Society*, 124(5): 320–322, October 10, 1980.

I also reread many dusty and crumbling physics textbooks that I had saved in my basement since my college days, despite my family's occasional suggestion that they might be discarded to create a less cluttered bookshelf. Related useful works include Robert Krebs's *Scientific Laws, Principles, and Theories*, Michael Hart's *The 100: A Ranking of the Most Influential Persons in History*, Lawrence Krauss's *Fear of Physics*, Frederick Bueche's *Introduction to Physics for Scientists and Engineers*, Arnold Arons's *Development of Concepts of Physics*, Martin Gardner's *Order and Surprise*, Michael Guillen's *Five Equations That Changed the World*, Graham Farmelo's *It Must Be Beautiful: Great Equations of Modern Science*, Richard Feynman's *The Character of Physical Law*, and John Casti's *Paradigms Lost*.

The frontispiece illustration (accompanied by Xenophanes' quotation) and illustration at the end of this book are from Gerogius Agricola's *De re metallica*, originally published in 1556. *De re metallica* was the first book on mining and metallurgy to be based on field research and careful observations. The book is available today from Dover Publications. The astronomer with compass on the initial quotation page is by Albrecht Dürer, from the title page of *Messahalah, De scientia motus orbis* (1504).

A NOTE ON TERMINOLOGY AND SYMBOLS

At the end of each entry in this book, under "Further Reading," I list references that are targeted to specific laws. While many entries mention primary sources, I have often explicitly listed excellent secondary references that most readers can obtain more easily than older primary sources. Readers interested in pursuing any subject can use the references as a useful starting point.

The text within gray boxes denotes historical events that occurred when a law was discovered. The large symbols used when introducing a law (atom, flask, telescope, and π symbols) denote the subject areas of physics, chemistry, astronomy, and mathematics, respectively.

Mathematical variables or constants that assume values are italicized. Subscripts for variables that do not assume values are typeset in a nonitalic font. For example, the T in T_{L} is italicized because it assumes a value for temperature; however, the subscript L is not italicized because it stands for the word "low."

The scientific literature appears to be divided when referring to Einstein's theory as either the "General Theory of Relativity" or the "Theory of General Relativity." Similarly, I found many instances of the "Special Theory of Relativity" and the "Theory of Special Relativity." I have decided to use the phrases "General Theory of Relativity" and "Special Theory of Relativity," which Einstein used to title the main sections of his book *Relativity: The Special and General Theory*, first published in 1916.

> In 1676, Isaac Newton explained his accomplishments through a simple metaphor. "If I have seen farther it is by standing on the shoulders of giants," he wrote. The image wasn't original to him, but in using it Newton reinforced a way of thinking about scientific progress that remains popular: We learn about the world through the vision of a few colossal figures.
>
> —Peter Dizikes, "Twilight of the Idols," *New York Times Book Review*, November 5, 2006

CONTENTS

> I do not agree with the view that the universe is a mystery. . . . I feel that this view does not do justice to the scientific revolution that was started almost four hundred years ago by Galileo and carried on by Newton. They showed that at least some areas of the universe . . . are governed by precise mathematical laws. Over the years since then, we have extended the work of Galileo and Newton. . . . We now have mathematical laws that govern everything we normally experience.
>
> —Stephen Hawking, *Black Holes and Baby Universes and Other Essays*

INTRODUCTION AND BACKDROP

Which discusses the definition of eponymous laws, the lives and afflictions of the lawgivers, science and religiosity, the difference between laws and theories, and the geographical and temporal distribution of the lawgivers.

INTRODUCTION AND BACKDROP

> Isaac Newton was born into a world of darkness, obscurity, and magic... veered at least once to the brink of madness... and yet discovered more of the essential core of human knowledge than anyone before or after. He was chief architect of the modern world.... He made knowledge a thing of substance: quantitative and exact. He established principles, and they are called his laws.
>
> —James Gleick, *Isaac Newton*

> At every major step, physics has required, and frequently stimulated, the introduction of new mathematical tools and concepts. Our present understanding of the laws of physics, with their extreme precision and universality, is only possible in mathematical terms.
>
> —Sir Michael Atiyah, "Pulling the Strings," *Nature*

THE LAWS OF NATURE

> It is now generally accepted that the universe evolves according to well-defined laws. The laws may have been ordained by God, but it seems that He does not intervene in the universe to break the laws.
>
> —Stephen Hawking, *Black Holes and Baby Universes*

Albert Einstein once remarked that "the most incomprehensible thing about the world is that it is comprehensible." Indeed, we appear to live in a cosmos that can be described or approximated by compact mathematical expressions and physical laws.

In this book, I discuss landmark laws of nature that were discovered over several centuries and whose ramifications have profoundly altered our everyday lives and understanding of the universe. These laws provide elegant ways for characterizing natural phenomena under a variety of circumstances. For example, as you'll learn in greater detail, Bernoulli's Law of Hydrodynamics, $v^2/2 + gz + p/\rho = C$, has numerous applications in the fields of aerodynamics, where it is considered when studying flows over airplane wings, propeller blades, and ship rudders. Fick's Second Law of Diffusion, $(\partial c/\partial t)_x = D(\partial c^2/\partial x^2)_t$, can be used to explain insect communication through pheromones, the migration of ancient humans, or diffusion in soils contaminated with petroleum hydrocarbons. Forensic police sometimes use Newton's Law of Cooling,

$T(t) = T_{env} + [T(0) - T_{env}]e^{-k}$, to determine the time of death of corpses discovered in seedy motel rooms.

The laws enable humanity to create and destroy—and sometimes they change the very way we look at reality itself. In the 1940s, Graham's Law $R_1/R_2 = (M_2/M_1)^{1/2}$ helped scientists make the atomic bomb that was dropped on Japan. The various laws dealing with electromagnetism enabled technologists to unite the world through both wired and wireless communications. For many scientists today, the Heisenberg Uncertainty Principle, $\Delta x \Delta p \geq \hbar/2$, means that the physical universe literally does not exist in a deterministic form but is rather a collection of probabilities. All of these relatively simple expressions impress us with their brevity and utility.

I side with Martin Gardner and others who seem to suggest that nature is usually describable by simple formulas and laws—not because we have *invented* mathematics and laws, but because nature has some hidden mathematical aspect. For example, Gardner writes in his classic 1950 essay "Order and Surprise":

> If the cosmos were suddenly frozen, and all movement ceased, a survey of its structure would not reveal a random distribution of parts. Simple geometrical patterns, for example, would be found in profusion—from the spirals of galaxies to the hexagonal shapes of snow crystals. Set the clockwork going, and its parts move rhythmically to laws that often can be expressed by equations of surprising simplicity. And there is no logical or a priori reason why these things should be so.

Here Gardner suggests that simple mathematics governs nature from molecular to galactic scales. Isaac Newton likened our quest to discover the fundamental laws of science to a child looking for a pretty pebble on an infinite beach. Albert Einstein felt he was like a child entering an infinite library with books in foreign languages. The child cannot understand most of the books but senses or suspects a mysterious order to the arrangement of books.

Similarly, theoretical physicist Paul Steinhardt writes in John Brockman's *What We Believe but Cannot Prove*:

> Recent observations and experiments suggest that our universe is simple. The distribution of matter and energy is remarkably uniform. The hierarchy of complex structures, ranging from galaxy clusters to subnuclear particles, can be described in terms of a few dozen elementary constituents and less than a handful of forces, all related by simple symmetries. A simple universe demands a simple explanation.

Centuries ago, most of the lawgivers saw God's hand in nature's laws. For example, British scientist Sir Isaac Newton (1642–1727) and many of his contemporaries believed that the laws of the universe were established by the will of God, who acted in a logical manner. In the preface to the second edition of Newton's *Principia* (1713), English mathematician Roger Cotes wrote:

> Without all doubt, this world, so diversified with that variety of forms and motions we find in it, could arise from nothing but the perfectly free will of God directing and presiding over all.
>
> From this fountain . . . the laws of Nature have flowed, in which there appear many traces indeed of the most wise contrivance, but not the least shadow of necessity. These, therefore, we must not seek from uncertain conjectures, but learn them from observations and experiments.

George Stokes (1819–1903), famous for Stokes's Law of Viscosity, wrote in his book *Natural Theology*:

> Admit the existence of God, of a personal God, and the possibility of miracles follows at once. If the laws of nature are carried out in accordance with His will, He who willed them may will their suspension. And if any difficulty should be felt as to their suspension, we are not even obliged to suppose that they have been suspended.

Because the laws of nature provide a framework in which to explore the nature of reality, and because laws allow scientists to make predictions about the universe, the discoveries of the laws are among humanity's greatest *noetic*, or intellectual, achievements. In order to formulate the laws of nature, scientists usually needed to perform significant observations, invent creative experimental designs, or show vast insight—and for these reasons, the lawgivers in this book were often among the most capable scientists in their fields.

At first glance, this book may seem like a long catalogue of isolated laws with little connection between them. But as you read, I think you'll begin to see many linkages. Obviously, the final goal of scientists and mathematicians is not simply the accumulation of facts and lists of formulas; rather, they seek to understand the patterns, organizing principles, and relationships between these facts to form laws.

For this book, I selected a number of scientific laws from a larger possible set, with an eye toward those laws that strongly influenced the world. Candidates for this collection usually satisfy the following criteria:

- They are laws, rules, and principles that have broad explanatory power to account for facts, observations, or phenomena—and they are widely accepted in a particular discipline.
- The laws, rules, and principles are named after a person, which usually means that a particular scientist was instrumental either in the discovery of that law or in bringing the law to wide scientific attention.

Just imagine the amazing adventures these passionate people had as they sought, discovered, and tested elegant formulations of phenomena at the heart of reality. In the end, these lawgivers changed the way we imagine and categorize our universe.

THE LAWGIVERS

> The scientist's religious feeling takes the form of a rapturous amazement at the harmony of natural law, which reveals an intelligence of such superiority that, compared with it, all the systematic thinking and acting of human beings is an utterly insignificant reflection. This feeling is the guiding principle of his life and work.... It is beyond question closely akin to that which has possessed the religious geniuses of all ages.
>
> —Albert Einstein, *Mein Weltbild*, 1934

According to the American sociologist Robert K. Merton (1910–2003), the practice of *eponymy*—the naming of laws, theories, and discoveries after their discoverers—dates back to the time of Galileo. Science often places a premium on rewarding those scientists who are first to make a discovery, propose a natural law, or just happen to be at the right place at the right time with respect to an experimental finding.

Because this book focuses on *eponymous laws*, be sure to use the index when hunting for a favorite law or equation, which may be discussed in entries for laws that you might not have expected. I have not confined myself to laws that have affected the present status of science—influence on past generations is taken into account, as well. In the interest of space, some of the most famous "equations" of science—which are usually not referred to as "laws" for historical or other reasons—may be found in the "Final Comments" section at the end of this book.

I remind the reader that my focus on eponymous laws is not meant to suggest that eponymous laws cover all the important findings in physics—surely many of the laws and principles discovered in modern times, although named in a more general descriptive fashion, are of paramount importance for understanding the nature of the universe. In fact, the relative paucity of eponymous laws after 1900 is discussed further in the "Final Comments" section, where I give examples of powerful physics concepts that do not have main entries in this book. Nevertheless, I hope that my focus on eponymy will help introduce the general reader to a bit of the history behind important laws and the colorful characters who took part in their discovery over the course of several centuries.

The scientists who gave their names to these laws in this book are a fascinating, diverse, and sometimes eccentric group of people. Many were extremely versatile polymaths—human dynamos with a seemingly infinite supply of curiosity and energy and who worked in many different areas in science. For example, French physicist Jean-Baptiste Biot (1774–1862) made advances in applied mathematics, astronomy, elasticity, electricity, magnetism, optics, and mineralogy. Not only is a law of magnetic force named after him, but so is the shiny mineral biotite. His partner Félix Savart (1791–1841) embarked on a medical career, but during those times when business was slow, he conducted experiments on the violin and devoted himself to the study of the acoustics of air, bird songs, and vibrating solids.

Many individuals had nonconventional educations. For example, the Swiss-German physicist Johann Lambert (1728–1777) made discoveries concerning the mathematical constant π and on the laws of light reflection and absorption, yet he was almost entirely self-taught. The German physicist Georg Ohm (1787–1854) was able to fulfill his intellectual potential mostly from his private and personal studies, reading the texts of the leading French mathematicians. The British physicist Michael Faraday (1791–1867) had almost no formal education. He later wrote, "My education was of the most ordinary description, consisting of little more than the rudiments of reading, writing, and arithmetic at a common day school." At the age of thirteen, when he could barely read or write, he quit school to find a job. French physicist Pierre Curie (1859–1906) considered himself to have a feeble mind and never went to elementary school. He later shared the Nobel Prize with his wife, Marie.

Despite varied educational backgrounds, many lawgivers displayed their unusual talents from an early age. The German mathematician and physicist Carl Friedrich Gauss (1777–1855), for example, was a childhood prodigy and learned to calculate before he could talk. At age three, he corrected his father's wage calculations when they had errors. By age 10½, French chemist Aléxis Petit (1791–1820) had already completed the entrance requirements of the École Polytechnique in Paris. Petit had surpassed the entrance exam scores of all other candidates at the time. The Irish mathematician Sir William Rowan Hamilton (1805–1865), mentioned in the "Great Contenders" section at the end of this book, spoke Hebrew by age 7, and by 13 he had mastered many classical and modern European languages such as Farsi, Arabic, Hindustani, Sanskrit, and Malay. French physicist André-Marie Ampère (1775–1836) was reported to have worked out long arithmetical sums by means of pebbles and biscuit crumbs before he was familiar with numbers and their names.

When describing the lives of these creative individuals, I did not attempt to provide a comprehensive biography, due to space limitations. In many cases, I mention curious aspects of their lives that may also give readers a better feel for the times in which these scientists lived. For example, I describe the incident in which German astronomer Johannes Kepler (1571–1630) had to defend his mother from accusations of witchcraft. I discuss how German mathematical physicist Rudolf Clausius (1822–1888) was wounded while leading a student ambulance core in the Franco-Prussian War, which, along with his wife's death during childbirth, hampered his scientific progress in the later years of his career.

In fact, a significant number of the lawgivers in this book had wives who died much earlier then they did. For example, the French chemist Alexis Petit's wife became ill six months after he married her and died shortly thereafter in 1817, and he died before he was thirty years old. Gauss, Ampère, and French physicist Pierre Weiss (1865–1940) all had wives who died young. Kepler's wife Barbara died from typhus in 1611. British physicist James Joule's (1818–1889) wife, Alice, died in 1854, leaving him with two children to raise. In 1869, German physicist Gustav Kirchhoff's (1824–1887) wife Clara died, leaving Kirchhoff to raise his four children alone. The German physicist Max Planck's (1858–1947) wife Marie died in 1909, leaving him with four children.

As already mentioned, religion also played a role in many of the lawgivers' lives. The Irish natural philosopher and chemist Robert Boyle (1626–1691) was quite devout and loved the Bible. His constant desire to understand God drove his interest in discovering laws of nature. Boyle specified that, after his death, his money should be used to found the Boyle lectures that were intended to refute atheism and religions that competed with Christianity. Ampère believed that he had proven the existence of the soul and of God. When Gauss proved a theorem, he sometimes said that the insight did not come from "painful effort but, so to speak, by the grace of God."

Not all lawgivers were traditionally religious. For example, British physicist William Henry Bragg (1862–1942) said that the Bible brought him years of misery and fear. He wrote, "From religion comes a man's purpose; from science his power to achieve it."

Men like Faraday, William Thomson (Lord Kelvin), James Clerk Maxwell, and Joule were particularly religious and motivated by their

Christian faith. For them, God was glorified whenever humans discovered scientific laws that God had established. Joule wrote in 1873, in notes for a lecture he never gave due to poor health:

> After the knowledge of, and obedience to, the will of God, the next aim must be to know something of His attributes of wisdom, power, and goodness as evidenced by His handiwork.... It is evident that an acquaintance with natural laws means no less than an acquaintance with the mind of God therein expressed.

Kepler was also motivated by his faith in God. He, too, sought to discover God's plan for the universe and to read the mind of God. For Kepler, mathematics was the language of God. Because humans were made in the image of God, humans were capable of understanding the universe that God had created. In *Conversation with Galileo's Sidereal Messenger*, Kepler wrote, "Geometry is unique and eternal, and it shines in the mind of God. The share of it which has been granted to man is one of the reasons why he is the image of God." He also explained, "I had the intention of becoming a theologian... but now I see how God is, by my endeavors, also glorified in astronomy, for 'the heavens declare the glory of God.'"

Even modern-day physicists muse about the possible role of a god in establishing the laws of the universe. According to British astrophysicist Stephen Hawking (b. 1942), all the laws of physics would hold even at the precise instant the universe formed. If God exists, Hawking suggests, He may not have had the freedom to choose the initial conditions for the universe. However, God would "still have had the freedom to choose the laws that the universe obeyed." Hawking writes in *Black Holes and Baby Universes*:

> However, this may not have been much of a choice. There may only be a small number of laws, which are self-consistent and which lead to complicated beings like ourselves.... And even if there is only one unique set of possible laws, it is only a set of equations. What is it that breathes fire into the equations and makes a universe for them to govern? Is the ultimate unified theory so compelling that it brings about its own existence?

In 1623, Italian physicist Galileo Galilei (1564–1642) echoed a belief of many in this book that the universe could be understood using mathematics, writing, "Nature's great book is written in mathematical symbols." Newton supposed that the planets were originally thrown into orbit by God, but even after God decreed the Law of Gravitation, the planets required continual adjustments to their orbits. According to *New York*

Times columnist Edward Rothstein, "The conviction that there is an order to things, that the mind can comprehend that order and that this order is not infinitely malleable, those scientific beliefs may include elements of faith."

Many of the laws in this book that excite me the most deal with electrical discoveries of highly religious people. Today, we often hear our age referred to as the "information age" as a result of the rising importance of computers and the Internet. But underlying the information age is our use of electricity. Many scientists have contributed to our knowledge of electricity, including Ampère, Charles Augustine de Coulomb, Count Alessandro Volta, Hans Christian Ørsted, Faraday, and Maxwell. Many of these "electric thinkers" of the eighteenth and nineteenth centuries were religious Christians whose discoveries led to the building of the first electric dynamos and eventually to modern electric generators that power our cities. Michael Guillen writes in *Five Equations That Changed the World*:

> Long before Christians had come to believe in the Father, Son, and Holy Ghost, natural philosophers had stumbled on their own trinity: electricity, magnetism and gravitational force. These three forces alone had governed the universe, they believed.... Given the forces' disparate behaviors, it was no wonder that philosophers very early on were left scratching their heads: Were these three forces completely different? Or were they, like the Christian Trinity, three aspects of a single phenomena?

In the area of electricity and magnetism, I discuss Frenchman Charles-Augustin de Coulomb (1736–1806), who hung bar magnets from strings and who discovered that the force between them diminished as the square of their separation. He also found that electrically charged objects suspended on strings followed the "inverse square law"—similar to the law Newton discovered for gravity. I also discuss Faraday's discovery that moving magnets actually produce electricity. Inspired by the electrical experiments of Italian experimentalist Luigi Galvani (1737–1798), Faraday, a Christian theologian, declared "electricity is the soul of the universe."

Some of the scientists in this book experienced resistance to their ideas, causing significant personal anguish. For example, Ohm's Law was so poorly received and his emotions scraped so raw that he resigned his post at Jesuit's College of Cologne, where he was professor of mathematics. His work was ignored, and he lived in poverty for much of his life. One critic said of Ohm's physics book that its "sole effort is to detract from the dignity of nature." The German minister of education said that

Ohm was "a professor who preached such heresies was unworthy to teach science."

Newton was so distressed by criticism from his colleague British physicist Robert Hooke (1635–1703) that Newton decided to withhold publication of one of his greatest works, *Opticks*, until after Hooke died. Newton also went nearly mad in another argument about his theory of colors with several English Jesuits who had criticized Newton's experiments. The correspondence between Newton and these critics lasted until Newton finally had a nervous breakdown. Faraday's scientific fame spread to such an extent that his previous mentor, Humphry Davy, began to despise him and campaigned that Faraday not be elected to the Royal Society. Like Newton, Faraday also suffered a nervous breakdown.

Austrian physicist Ludwig Boltzmann (1844–1906) is best remembered for his work in thermodynamics, heat, and disorder. He used the concept of the atom to explain how heat was a statistical property of the motions of many atoms. However, several of his contemporaries, such as Ernst Mach and Wilhelm Ostwald, argued so forcefully against Boltzmann's position that Boltzmann's depression worsened, and he killed himself in 1906. Boltzmann appeared to have bipolar disorder, and his low emotional periods were only exacerbated by his failing eyesight and arguments with colleagues. All we know for certain about his suicide is that he hanged himself while on a holiday with his wife and daughter. Other depressed lawgivers include Petit, Newton, and hypochondriac Gauss.

Many cutting-edge scientific geniuses in addition to those discussed in this book had to persevere despite resistance. For example, the revolutionary discoveries concerning antibiotics by Scottish biologist Alexander Fleming (1881–1995) were met with apathy from his colleagues. Many surgeons initially resisted English surgeon Joseph Lister's (1827–1912) advocacy of antisepsis. American inventor Chester Carlson (1906–1968), inventor of the Xerox® machine, was rejected by more than twenty companies before he finally sold the concept. German scientist Alfred Wegner's (1880–1930) theory of continental drift was ridiculed by the geologists of his time.

A number of scientific lawgivers included in this book had to overcome the strong resistance of their own parents. For example, Coulomb's mother wanted him to be a medical doctor, but her son insisted on studying a more quantitative subject like engineering or mathematics. The disagreements became heated, and his mother virtually disowned him. Similarly, physicist and mathematician Daniel Bernoulli (1700–1782) and the polymath Biot both rebelled against their fathers, who insisted that their sons pursue careers in business. The father of Scottish chemist Thomas Graham (1805–1869) had always wanted Graham to become a minister in the Church of Scotland and opposed Graham's growing interest in chemistry. Luckily for Graham, his mother and sister were supportive of his interest in science, which helped Graham achieve his scientific dreams.

The lawgivers in this book often had afflictions of one form or another. British chemist, physicist, and meteorologist John Dalton (1766–1844) was color blind. Kirchhoff needed crutches to walk. French mathematician Joseph Fourier (1768–1830) had an ailment that made him perpetually cold—he rarely went outside without an overcoat and a servant bearing another in reserve, even in the middle of summer. Hooke was a sickly child with constant headaches, and he was not expected to reach adulthood. Kepler was bow-legged, often afflicted with large boils, and suffered from poor vision. British chemist William Henry's (1775–1836) childhood injury caused him incredible pain throughout his life, and he eventually killed himself. A number of the lawgivers suffered as a result of their research. For example, French chemist Pierre Dulong (1785–1838) blew off his fingers and lost an eye during a chemistry experiment.

Chronic physical ailments may have given some individuals the desire to compensate for their shortcomings, or to leave a mark on the world and achieve immortality through creative excellence. Perhaps the peculiarities, or even physical defects exhibited by some geniuses, have caused these individuals to overcompensate through constant creative activity. For example, when conducting research for my book *Strange Brains and Genius*, I found that many creative geniuses have had a sense of physical vulnerability because they felt that at any moment they could be sick and without a means of income. Perhaps this unease keeps some individuals on edge and serves as a source of creative tension.

A surprisingly large number of the scientists included in this book thought deeply about the existence of extraterrestrial life, and their religiosity convinced them that the universe literally teemed with life. As one example of the pervasiveness of this thinking, physicist Johann Lambert believed that all planets, comets, and moons were likely to contain life. In his *Cosmologische Briefe*, Lambert asserted, "The Creator is much too efficient not to imprint life, forces and activity on each speck of dust. . . . All possible varieties which are permitted by general laws ought to be realized. . . ." Similarly, astronomer Johann Bode (1747–1826) believed that all significant objects in space—the Sun, stars, planets, moons, and comets—were inhabited by intelligent beings. Bode remarked that habitability was "the most important goal of creation" and that alien life forms throughout the universe "are ready to recognize the author of their existence and to praise his goodness."

Physicist David Brewster (1781–1868), in his book *More Worlds Than One*, gave Biblical reasons as to why every star has a planetary system similar to ours, and believed that every planet, sun, and moon was inhabited by life forms. The astronomer Kepler even wrote a science-fiction tale, *Somnium*, in which the inhabitants of the Moon resembled large serpents with a spongy, porous skin.

A number of the "lawgivers of reality" in this book grew up in families that almost seemed to have physics in their genes! For example, father and son Braggs—William Henry Bragg (1862–1942) and William Lawrence Bragg (1890–1971)—were awarded the Nobel Prize in physics in 1915 for their pioneering studies in determining crystal structures. German physicist Friedrich Wilhelm Kohlrausch (1840–1910) collaborated with his brother, physicist Wilhelm Friedrich Kohlrausch, in the field of electrochemistry, and their father, Rudolph Kohlrausch, was also a famous physicist working in related areas of science. The Bernoullis, an extraordinary Swiss family, contained eight outstanding mathematicians within three generations, and in the seventeenth and eighteenth centuries they made great contributions to hydrodynamics, differential calculus, probability theory, geometry, mechanics, ballistics, thermodynamics, optics, magnetism, elasticity, and astronomy.

As another example of familial brilliance, the genius sons of German physicist Gustav Wiedemann (1826–1899) had an extremely intellectual and distinguished pedigree. Their father was famous for the Wiedemann-Franz Law and was professor of physical chemistry at Leipzig. Their maternal grandfather was Eilhard Mitscherlich (1794–1863), famous for his work on chemical isomorphism and similarity of crystal structures. Their mother, Clara, helped translate into German the Irish natural philosopher John Tyndall's (1820–1893) *Heat as a Mode of Motion*. The elder of Gustav's sons, Eilhard, became a physicist and historian of science and was the first individual to use the term "luminescence." The younger son, Alfred, became a famous egyptologist.

For two final examples of smart families, consider Adolf Fick (1829–1901), the German physiologist famous for his laws of diffusion. Fick had a brother who became a professor of anatomy and another who became a professor of law. Also consider the physicists Pierre and Marie Curie, who received the Nobel Prize for Physics in 1903 for their research involving radiation. Marie received another Nobel Prize in Chemistry in 1911 for, among other things, discovery of the elements radium and polonium. She was the first person to win or share two Nobel Prizes. Their elder daughter, Irène (1897–1956), married French physicist Jean Frédéric Joliot (1900–1958), and the husband and wife team received the Nobel Prize for Chemistry in 1935.

Chance sometimes affected the lawgivers' lives in important ways. For example, in 1812, the English chemist Humphry Davy (1778–1829) was temporarily blinded by a chemical explosion, and as a result, Faraday became Davy's assistant, which became an important stepping stone in Faraday's career. In fact, Faraday's interest in electricity was kindled earlier by his serendipitous encounter with the 127-page entry "Electricity" in the *Encyclopaedia Britannica*, which he happened to be rebinding for a client.

Author Sherwin B. Nuland suggests that the quirks and personalities of scientists are valid and important areas of study when trying to understand the evolution of scientific ideas. In his essay "The Man or the Moment?" Nuland suggests that historians of science should not write only about the effect of prevailing *social* forces on scientific discoveries "because part of the process is the distinctive *personality* of the discoverer." To understand scientific progress, I believe that we should examine the lives of the people who made the discoveries, at least to understand more about the kinds of scientists who may readily incubate the ideas, attitudes, and talents that foster the discovery of nature's laws.

IS IT FAIR TO NAME A LAW AFTER A PERSON?

> Good theories [don't necessarily] convey ultimate truth, or [imply] that there "really are" little hard particles rattling around against each other inside the atom. Such truth as there is in any of this work lies in the mathematics; the particle concept is simply a crutch ordinary mortals can use to help them towards an understanding of the mathematical laws.
>
> —John Gribbin, *The Search of Superstrings, Symmetry, and the Theory of Everything*

In hindsight, we can usually see that if one scientist did not discover a particular law, some other person would have done so within a few months or years of the discovery. Most scientists, as Newton said, stood on the shoulders of giants to see the world just a bit farther along the horizon. In fact, I give a few examples in this book in which more than one individual discovered a law within a few years of one another, but for various reasons, including sheer luck, history sometimes remembers only the more famous discoverer. Readers may enjoy noting the frequency with which this happens in the history of science.

From a more general perspective, it is fascinating the degree to which simultaneous discoveries appear in great works of science and mathematics. As just one example, French chemist and physicist Joseph Louis Gay-Lussac's (1778–1850) first publication of his gas law was notable not only because of its scientific value but also because almost identical research was carried out simultaneously and independently by Dalton. In their near-simultaneous publication in 1802 of research on the thermal expansion of gases, Dalton and Gay-Lussac both concluded that all gases expand by the same proportion for a particular temperature rise at constant pressure.

As I mention in my book *The Möbius Strip*, in 1858 the German mathematician August Möbius (1790–1868) simultaneously and independently discovered the Möbius strip along with a contemporary scholar, the German mathematician Johann Benedict Listing (1808–1882). This simultaneous discovery of the Möbius band by Möbius and Listing, just like that of calculus by Newton and German mathematician Gottfried Wilhelm Leibniz (1646–1716), may make us wonder why so many discoveries in science were made at the same time by people working independently. For another example, British naturalists Charles Darwin (1809–1882) and Alfred Wallace (1823–1913) both developed the theory of evolution independently and simultaneously. Similarly, Hungarian mathematician János Bolyai (1802–1860) and Russian mathematician Nikolai Lobachevsky (1793–1856) seemed to have developed hyperbolic geometry independently and at the same time. (Some legends suggest that both mathematicians may have learned about this geometry indirectly from Gauss, who worked in this area.)

The history of materials science is replete with simultaneous discoveries. For example, in 1886, the electrolytic process for refining aluminum, using the mineral cryolite, was discovered simultaneously and independently by American Charles Martin Hall (1863–1914) and Frenchman Paul Héroult (1863–1914). Their inexpensive method for isolating pure aluminum from compounds had an enormous effect on industry.

Most likely, such simultaneous discoveries have occurred because the time was "ripe" for such discoveries, given humanity's accumulated knowledge at the time the discoveries were made. Sometimes, two scientists are stimulated by reading the same preliminary research of one of their contemporaries. On the other hand, mystics have suggested that a deeper meaning exists to such coincidences. Austrian biologist Paul Kammerer (1880–1926) wrote, "We thus arrive at the image of a world-mosaic or cosmic kaleidoscope, which, in spite of constant shufflings and rearrangements, also takes care of bringing like and like together." He compared events in our world to the tops of ocean waves that seem isolated and unrelated. According to his controversial theory, we notice the tops of the waves, but beneath the surface there may be some kind of synchronistic mechanism that mysteriously connects events in our world and causes them to cluster.

Despite simultaneity in science and the quirks of naming laws, the laws in this book were actually formulated or promoted by the lawgivers for whom they were named, and we should give them substantial credit for their work. Thus, every lawgiver in this volume, in my opinion, is a truly remarkable figure in the history of science—even if a particular scientist built on ideas of others and even if the discovery was, in some sense, the result of group intelligence.

Some innovative scientists, such as Einstein and Newton, did not always discover their laws and theories through their own experiments and observations, but rather they pondered the implications of other scientists' observations. When discussing technological inventions, Leslie Berlin writes in her book *The Man Behind the Microchip*, "If nearly any invention is examined closely enough, it almost immediately becomes apparent that the innovation was not the product of a single mind, even if it is attributed to one. Invention is best understood as a team effort." Although Berlin's idea may have validity for most inventions, this group effort is perhaps less pronounced for the discovery of laws, even when scientists start by studying the work of others. In fact, many *laws* of science and mathematics, expressed as a single equation, do derive largely from the work of a single individual working long hours in relative isolation or who has a "eureka" moment. Unlike the invention of increasingly complicated devices such as computers and cars, in the past the "production" of laws has usually not required the same kind of collaborative team effort in which several contributors, working in a lab, supply subsets of a technology invention. While basic science today is often performed in large teams, the expression of a natural law in terms of a simple formula may still continue to be the work of specialized individuals with a spark of insight.

If we move our attention from laws to other kinds of scientific discoveries, some historians of science may hesitate about naming such discoveries after people, and some writers have gone as far as to suggest that scientists *always* pick the wrong scientists after whom to name a discovery. Jim Holt writes in "Mistaken Identity Theory: Why Scientists Always Pick the Wrong Man":

> Stigler's Law of Eponymy, which in its simplest form states that "no scientific discovery is named after its original discoverer," was so dubbed by Stephen Stigler in his recent book *Statistics on the Table* (Harvard).... If Stigler's law is true, its very name implies that Stigler himself did not discover it. By explaining that the credit

belongs instead to the great sociologist of science Robert K. Merton, Stigler not only wins marks for humility; he makes the law to which he has lent his name self-confirming.

Robert Merton suggested that "all scientific discoveries are in principle 'multiples.'" In other words, when a scientific discovery is made, it is made by more than one person, but for some reason, the discovery is named after the "wrong" one of its multiple discoverers. For example, sometimes a discovery is named after the person who *develops* the discovery rather than the original discoverer. Stigler has suggested that eponyms are inaccurately assigned because "eponyms are only awarded after long time lags or at great distances" by scientists who are not trained in history and who are more interested in "recognizing general merit than an isolated achievement."

Whether or not Stigler's Law is valid, his law tends to focus on other kinds of observations and discoveries, and not on scientific laws. For example, Halley's Comet, named after English astronomer Edmond Halley (1656–1742), was not first discovered by Halley because it had been actually seen by countless observers even before the time of Jesus. (But let's not downgrade Halley, whose calculations enabled earlier references to the comet's appearance to be found in the historical record!)

As I have already suggested, when it comes to natural laws, I have found that lawgivers for whom the laws are named did discover the law or contribute to the refinement and promotion of the law's application for practical scientific purposes. However, some fascinating instances exist of several people discovering the same law in different forms through the centuries. As one example, consider Snell's Law that in 1621 accurately described the refraction (bending) of light through glass. The law is named after Dutch mathematician Willebrord van Roijen Snell (1580–1626). However, perhaps the first person to understand the basic relationship expressed by Snell's Law was the Arabian mathematician Ibn Sahl in the year 984. In 1602, English astronomer and mathematician Thomas Harriot also discovered the law, but he did not publish his work. Though a quirk of fate, we call it Snell's Law today because, in 1662, Dutch scholar and manuscript collector Isaac Vossius discovered Snell's writings, and Dutch physicist Christian Huygens referred to the writings in his *Dioptrica* published in 1703. Note, however, that the French refer to Snell's Law as Descartes's Law, because René Descartes was the first person to publish the law in terms of sine functions in his 1637 *Discourse on Method*, but without experimental verification. Huygens and others actually accused Descartes of plagiarism, given that Descartes was in Leiden during and after Snell's work, but little evidence exists to support the plagiarism assertion.

THEORIES AND LAWS

> A correct theory is one that can presumably be verified by experiment. And yet, in some cases, scientific intuition can be so accurate that a theory is convincing even before the relevant experiments are performed. Einstein—and many other physicists as well—remained convinced of the truth of special relativity even when... experiments seemed to contradict it.
>
> —Richard Morris, *Dismantling the Universe*

All of the main entries in this book focus on scientific laws and principles. On the other hand, "theories" are usually only discussed *within* the main entries. However, I admit that the line between a scientific theory and a law is sometimes thin. To my mind, a scientific law exhibits a significant degree of universality and invariability, and can usually be summarized by a simple formula. A law is a relatively secure, high-level, and succinct formulation. But what exactly is a theory? I could have a theory about why snails in a Mozambique rain forest went through a population surge as a result of local conditions at the time. This theory might be elevated to the status of a law if I claimed that those conditions would always produce a doubling per month for a wide class of animals under given conditions and if I could provide evidence for this doubling that a majority of scientists came to accept.

Theories are often used to describe *why* certain laws work. On the other hand, a law often shows that the universe works in a certain way, but does not explain the "why"—and the laws usually do not even explain the "how." Knowledge moves in an ever-expanding, upward-pointing funnel. From the rim, we look down and see previous knowledge from a new perspective as new theories are formed to explain the universe in which we live.

Famous scientific theories usually explain facts or behaviors that have been shown to be true in many independent experiments. However, sometimes a theory exists before it is tested or confirmed. Thus, today many physicists use the phrase "string theory" when referring to the fundamental composition of subatomic particles and not the "law of strings." A law requires substantial confirmation; that is, we usually do not call a mathematical formulation a scientific law until it has been tested many times and has not been falsified. Nevertheless, even the great scientific laws are not immutable, and laws may have to be revised centuries later in the light of new information.

Science progresses mainly because both existing theories and laws are never quite complete. For example, Newton's Law of Universal

Gravitation describes the attraction between two bodies as a force that depends on the mass of the bodies and the distances between them. This law predicts with remarkable accuracy the motion of the Moon around Earth. It allows us to predict the trajectories of bullets and cannon balls. The law describes gravity in terms of a few variables, but it does not illuminate *what* gravity is or the precise mechanism of its action. It also does not accurately predict the bending of light rays that pass Earth, which requires us to invoke Einstein's General Theory of Relativity. Einstein's formulation generalizes Newton law and treats gravity as a manifestation of curved space and time. Today, General Relativity itself is often regarded as limited and not useful for subatomic distances. Thus, scientific laws generally account for everything that humans know about a phenomenon at a point in time. Philosopher Karl Popper in his *Conjectures and Refutations* suggested that all scientific models and laws are only tentative, and philosopher David Hume in *An Enquiry Concerning Human Understanding* asserted that no amount of testing and observations can absolutely prove a model correct.

I usually think of these evolving laws as incomplete rather than wrong, because I put great weight on the law's utility in helping humanity make predictions at a particular stage of human knowledge. In fact, most of these old laws, like Newton's laws, continue to be crucial and help us predict the functioning of the universe. The newer laws that "replace" the older laws generally have more predictive power while retaining the successes of the previous laws and also addressing new experimental observations. Many laws remind me of oil paintings in which the crucial visual themes are elucidated early in humanity's quest for understanding, but the tiny brush strokes are still to be added in light of new knowledge. If there were no more brush strokes to come, science would be dead. Perhaps a better analogy likens the sharpening of laws to the calculation of the square root of 2. A simple approximation is 1.4. A better approximation is 1.4142135623. Neither is absolutely correct, but the earlier historical approximation of 1.4 is obviously less refined than the later value.

Given this brief background, can we articulate the precise relationship between a law and theory? One way in which the *Shorter Oxford English Dictionary* defines a theory is "a scheme or system of ideas or statements held as an explanation or account of a group of facts or phenomena." As just discussed, theories are scientifically backed explanations, including overarching conceptual schemes such as Einstein's General Theory of Relativity or Darwin's Theory of Natural Selection. Laws concerned with optical lenses, for example, which allow technologists to build optical instruments, can be derived from theories about how light propagates. Arnold Arons in *Development of Concepts of Physics* writes:

There is a kind of symbiotic relationship here between law and theory. A theory becomes more and more respected and powerful the more phenomena can be derived from it, and the law describing these phenomena becomes more meaningful and useful if it can be made part of a theory. Thus, Newton's theory of universal gravitation gained greatly in stature because it enabled one to derive the laws governing the motion of the moon, which had been known by empirical rules since the days of Babylonian observers.

A majority of scientists consider the major scientific theories and laws to be true, and scientists use both to make predictions. On the other hand, a theory is usually more complicated and dynamic than a law. A law may apply to a single broad observation, whereas a theory explains a set of related phenomena and can have various subcomponents.

Consider, for example, a rock and a catapult. Both can be used as weapons. The action of the rock is straightforward and can be expressed succinctly. The rock is like a law. On the other hand, a medieval catapult often had numerous plates, cross pieces, cords, wheels, spindle heads, arms, rollers, supports, and a cup to hold a massive stone. The catapult is like a theory. Through time, improvements are made to the components. New kinds of pieces are invented to replace less effective ones. Despite these enhancements, the catapult's function remains the same and is useful through time. Few scientists doubt the overall "truth" of the famous scientific theories, such as the theory of evolution or quantum theory, although scientists certainly clamor to refine and better understand the various components.

According to John Casti in *Paradigms Lost*, the logical structure of science can be represented by the following sequence:

Observations/Facts
↓
Hypothesis
↓
Experiment
↓
Laws
↓
Theory

Our observations give rise to hypotheses that are studied with experiments. Hypotheses that are supported by experiments may become empirical relationships, or laws. Laws may become part of an encompassing theory with wide explanatory power.

As an example, consider the observations of French physicist Jacques Charles (1746–1823), who in 1787 investigated the relationship between the volume of a gas and how it changes with temperature. After many careful experiments, he observed precisely how temperature affects the volume of a gas. Charles's Law states that the volume of a given amount of an ideal gas is directly proportional to the temperature, provided the amount of gas and the pressure remain fixed. This behavior was further quantified and published in 1808 by Gay-Lussac and can be represented by the succinct formula $V = kT$, where V is the volume of gas, k is a constant, and T is the temperature. This law enabled scientists around the world to summarize the results of many experiments in a statement that could be expressed in a formula less than an inch in length. Note that Charles's Law does not tell us *why* the volume and temperature are directly related. We need the Kinetic Theory of Gases to explain that the behavior results from molecules in random motion. The moving particles constantly collide with each other and with the walls of the container. Increasing the temperature of the gas increases the kinetic energy of the gas particles and thus the pressure on the walls of the container.

Laws that simply summarize some observed regularity might be called *empirical laws*. Examples include Gay-Lussac's Law of Combining Volumes, Bode's Law of Planetary Distances from the Sun, Kepler's Laws of Planetary Motion, Hooke's Law of Elasticity, Snell's Law of Refraction, Boyle's Law of Gases, and Ohm's Law of Electricity—all discussed in this book.

The laws that impress me the most are those that suggest a more general principle that represents the behavior of very different phenomena. These laws often require us to define fundamentally important or new concepts. As one example, consider Newton's Second Law of Motion: When a net force acts upon an object, the rate at which the momentum changes is proportional to the force applied. Today, we would express this as $\mathbf{F} = \mathrm{d}\mathbf{p}/\mathrm{d}t$, where the boldface letters indicate vector quantities that have both a magnitude and direction. Here, $\mathbf{p}$ is the momentum, which is equal to the mass times the velocity of the object. $\mathbf{F}$ is the applied force, and $\mathrm{d}\mathbf{p}/\mathrm{d}t$ is the rate of change of momentum. Both mass and force are important concepts that needed either defining or further elucidation before the law could make sense.

Other related laws represent a general conclusion derived from some theory. For example, Newton's Law of Universal Gravitational Attraction, expressed as $F = (Gm_1m_2)/r^2$, is a generalization that is in some sense built upon Kepler's laws, Newton's Laws of Motion, and the hypothesis of mass attraction between bodies, such as the Moon and Earth, with masses m_1 and m_2. This kind of law can be particularly interesting because it is, in part, *derived* from some underlying theory. Often, scientists try to turn empirical laws into derived laws over the span of many decades. With empirical laws, we have less of a sense of what fields they apply to, because we often do not have an underlying reasoning for them.

DO WE DISCOVER OR INVENT LAWS?

> Heisenberg once made the following remark to Einstein: "If nature leads us to mathematical forms of great simplicity and beauty... that no one has previously encountered, we cannot help thinking that they are 'true,' that they reveal a genuine feature of nature."
>
> —Paul Davies, *Superforce*

In my mind, we do not invent laws in mathematics and science, but rather we discover them. They have an existence independent from us. My viewpoint is not without controversy, and certainly other points of view exist. For example, I believe that mathematical laws transcend us and our physical reality. The statement "3 + 1 = 8" is false. Was the statement false before the discovery of integers? I believe it was. Numbers and mathematics exist whether humans know about them or not. Martin Gardner once stated this idea as: "If two dinosaurs met two other dinosaurs in a clearing, there would be four of them even though the animals would be too stupid to know that." In other words, four dinosaurs are now in the clearing, whether or not humans are around to appreciate this fact.

G. H. Hardy in his famous *A Mathematician's Apology* wrote, "I believe that mathematical reality lies outside us, that our function is to discover and *observe* it, and that the theorems which we prove, and which we describe grandiloquently as our 'creations,' are simply our notes of our observations." I think that when we write down laws of nature, we are taking notes on our discoveries. Law creators are like archeologists, uncovering treasures as they mine the cosmos for truths.

As discussed above, many scientists feel that "simplicity" is a requirement for all "laws." For example, authors David Halliday and Robert Resnick suggest in *Physics*:

> One criterion for declaring the program of mechanics to be successful would be the discovery that *simple* laws do indeed exist. This turns out to be the case, and this fact constitutes the essential reason that we "believe" the laws of classical mechanics. If the force laws had turned out to be very complicated, we would not be left with the feeling that we had gained much insight into the workings of nature.

This enduring interest in simplicity is further discussed in the following section.

SIMPLE MATHEMATICS AND REALITY

> The mathematical take-over of physics has its dangers, as it could tempt us into realms of thought which embody mathematical perfection but might be far removed, or even alien to, physical reality. Even at these dizzying heights we must ponder the same deep questions that troubled both Plato and Immanuel Kant. What is reality? Does it lie in our mind, expressed by mathematical formulae, or is it "out there"?
>
> —Sir Michael Atiyah, "Pulling the Strings," *Nature*

> Our mathematical models of physical reality are far from complete, but they provide us with schemes that model reality with great precision—a precision enormously exceeding that of any description that is free of mathematics.
>
> —Roger Penrose, "What Is Reality?" *New Scientist*

Marilyn vos Savant has been listed in the *Guinness Book of World Records* as having the highest IQ in the world—an awe-inspiring 228. She is author of several delightful books and wife of Robert Jarvik, M.D., inventor of the Jarvik 7 artificial heart. One of her readers once asked her, "Why does matter behave in a way that is describable by mathematics?" She replied, "The classical Greeks were convinced that nature is mathematically designed, but judging from the burgeoning of mathematical applications, I'm beginning to think simply that mathematics can be invented to describe anything, and matter is no exception." Marilyn vos Savant's response is certainly one with which many people would agree. However, as mentioned throughout this Introduction, the fact that reality can be described or approximated by *simple* mathematical expressions suggests to me that nature has mathematics at its core.

The laws in this book impress us because of their compactness and predictive power. I am not suggesting that *all* phenomena, including subatomic phenomena, are described by simple-looking formulas; however, as scientists gain a more fundamental understanding, they hope to simplify many of the more unwieldy formulas. James Trefil in *The Nature of Science* writes of this simplicity:

> The laws of nature are the skeleton of the universe. They support it, give it shape, tie it together.... They tell us that the universe is a place we can know, understand, and approach with the power of human reason. In an age that seems to be losing confidence in its ability to manage things, [the laws of nature] remind us that even the most complex systems around us operate according to simple laws, laws easily accessible to the average person.

To best understand some of the laws, consider the first great question of physics: "How do things move?" Imagine a universe called Madness in which Kepler looks up into the heavens and finds that most planetary orbits can never be approximated by ellipses but rather by bizarre geometrical shapes that defy his mathematical description. Imagine Newton dropping an apple whose path requires a 200-term equation to describe. Luckily for us, we do not live in Madness. Newton's apple is a symbol of both nature and simple arithmetic from which reality may naturally evolve.

American theoretical physicist Richard Feynman (1918–1988) in *The Character of Physical Law* also suggests that laws should be simple, and he uses Newton's Law of Universal Gravitation as an example that has all the essential ingredients of a natural law:

> First, it is mathematical in its expression.... Second, it is not exact; Einstein had to modify it.... There is always an edge of mystery, always a place where we have some fiddling around to do yet.... But the most impressive fact is that gravity is simple.... It is simple, and therefore it is beautiful.... Finally, comes the universality of the gravitational law and that fact that it extends over such enormous distances....

Theoretical physicist Michio Kaku, in "Parallel Universes, the Matrix, and Superintelligence," echoes the belief that scientific laws or formulas that underlie the basic principles of the universe should be succinct: "Professionally, I work on something called Superstring theory, or now called M-theory, and the goal is to find an equation, perhaps no more than one inch long, which will allow us to "read the mind of God," as Einstein used to say." To Kaku, brevity is key when describing reality, and he believes that brevity is possible. Einstein went a step further and thought that we should be able to simply state all physical theories, irrespective of the math involved. In fact, he once wrote, "All physical theories, their mathematical expressions notwithstanding, ought to lend themselves to so simple a description that even a child could understand them." Similarly, American physicist Leon Lederman once said, "If the basic idea is too complicated to fit on a T-shirt, it's probably wrong." His ambition was to "live to see all of physics reduced to a formula so elegant and simple that it will fit easily on the front of a T-shirt."

Max Tegmark, professor of physics at the Massachusetts Institute of Technology, wrote, "In 2056, I think you'll be able to buy a T-shirt on which are printed equations describing the unified physical laws of our universe. All the laws we have discovered so far will be derivable from these equations." New Zealand born nuclear physicist Ernest Rutherford (1871–1937) echoed many of these beliefs when he said, "If a piece of physics cannot be explained to a barmaid, then it is not a good piece of physics."

WHAT IS REALITY REALLY?

> They came again, so many of them but this time I only smiled and I didn't open my eyes. You can come, you aren't going to make me jump and wake up. No, you can come, even if there are so many of you there are no numbers for you. You come from the place where there are no numbers.
>
> —Anne Rice, *Christ the Lord*

We often say that scientific laws describe reality, but what is reality really? Various schools of thought exist among scientists—such as realism, instrumentalism, and relativism—that look at reality from varying perspectives. *Realists* believe that reality exists independent of us, and it can be discovered and understood using the tools of science. We describe this reality with our equations. On the other hand, according to John Casti, the *instrumentalists* "cling to the belief that theories are neither true nor false, but have the status only of instruments or calculating devices for predicting the results of measurements." For *relativists*, truth is "not a relationship between a theory and an independent reality" but changes according to individual perspectives and thus changes from time to time. American historian of science Thomas Kuhn (1922–1996) suggested in his 1962 book *The Structure of Scientific Revolutions* that scientists are not really getting any closer to a "scientific truth" as they study nature and discover scientific laws. This means that we cannot even measure scientific progress by the degree with which it appears to get closer to describing reality. Kuhn says that as scientists study the universe, they are learning about a "different universe" each time that their methodology and observations advance.

Hawking holds views that are close to those of the instrumentalists. In his books *Black Holes and Baby Universes* and *The Nature of Space and Time*, coauthored with Roger Penrose, he writes,

> I don't demand that a theory correspond to reality because I don't know what [reality] is. . . . I take the positivist viewpoint that a physical theory is just a mathematical model and that it is meaningless to ask whether it corresponds to reality. All that one can ask is that its predictions should be in agreement with observations. . . . All I'm concerned with is that the theory should predict the results of measurements. . . .

The physicist Victor J. Stenger expresses similar views in his paper "A Scenario for a Natural Origin of Our Universe":

> What is it that science does when it "explains" some phenomenon? At least in the case of the physical sciences, it builds a mathematical model to describe the empirical data associated with the phenomenon. When that model works well in fitting the data, has passed a number risky tests that might have falsified it, and is at least not inconsistent with other established knowledge, then it can be said to successfully explain the phenomenon.

Stenger continues by musing on the relationships of mathematical models to reality:

> Further discussion on what the model implies about "truth" or "ultimate reality" falls into the area of metaphysics rather than physics, since there is nothing further the scientist can say based on the data. What is more, nothing further is needed for any practical applications. For example, not knowing whether or not electromagnetic fields are real does not prevent us from utilizing the theory of electromagnetic fields.

BOOK ORGANIZATION AND PURPOSE

> I tell them that if they will occupy themselves with the study of mathematics they will find in it the best remedy against the lusts of the flesh.
>
> —Thomas Mann, *The Magic Mountain*

The main entries for the laws in this book are divided into two sections. The first section provides a brief introduction to the law and the equation used to represent the law. The second section provides biographical information for the lawgiver. Also included is a brief "curiosity file" that contains fascinating, lesser-known tidbits relating to the law or lawgiver.

Historical events that occurred at the time when a law was discovered are highlighted. Additionally, a short "Cross Reference" section indicates connections to other key laws, equations, or people. Each entry concludes with an section entitled Interlude: Conversation Starters, which contains thought-provoking quotations that are related to scientific laws and the mathematics of reality. For several entries, I have included simple numerical examples and solved problems so that readers may have a more direct understanding of the application of a law.

My goal in writing *Archimedes to Hawking: Laws of Science and the Great Minds Behind Them* is to provide a wide audience with a brief guide to important scientific ideas and thinkers, with entries short enough to digest in a few minutes. Most laws are accompanied by an equation. Readers need not dwell on these formulas to understand the gist of an idea. Sometimes, simply seeing *how* the laws are expressed in a mathematical notation gives readers an indication of the compactness of a law. While a Ph.D. may be required to master some of these laws, only a few pages are needed to state their essence. Because we are aware of just a few fundamental laws that appear to rule the cosmos or shape modern science, this book need not stretch over many volumes in order to give an overview.

The equation-based entries were chosen in consultation with scientific colleagues. Not all eponymous laws of science are included in this book, but I believe that I have included a majority of those with historical significance and that have had the greatest influence on science and human thought. Only "laws of nature," based on observations of the physical universe, are discussed in this book. Thus, in the interest of brevity, laws of economics, psychology, biology, geology, or pure mathematics are not included as main entries. Similarly, important computing technology laws, such as Moore's Law, Amdahl's Law, or Gustafson's Law, are not discussed.

Most laws of nature in this book come from the field of physics. Lawrence Krauss in *Fear of Physics* reiterates some of the ideas discussed above under "Simple Mathematics and Reality," namely, that physicists are often able to follow the amazing complexity and depth of modern physics because the concepts are largely based on the same "handful" of fundamental ideas and laws. He writes, "Any phenomenon described by one physicist is generally accessible to any other through the use of perhaps a dozen basic concepts. No other realm of human knowledge is either so extensive or so simply framed."

Several excellent books have catalogued guiding principles of science—including Trefil's *The Nature of Science*, Jennifer Bothamley's *Dictionary of Theories*, and Robert Krebs's *Scientific Laws, Principles and Theories*. Because these books do not focus exclusively on eponymous scientific laws, they must be extremely brief in their coverage due to space limitations. For example, Bothamley's interesting book usually describes each principle in two or three sentences. Although I promised to be brief in this volume, I attempt to delve into more detail than past compendiums when discussing both the laws and the lives of the lawgivers.

Archimedes to Hawking reflects my own intellectual shortcomings, and while I try to study as many areas of science as I can, it is difficult to become fluent in all aspects. Thus I am sure that this book reflects my own personal interests, strengths, and weaknesses. I am responsible for the choice of laws included in this book and, of course, for any errors and infelicities. This is not a comprehensive or scholarly dissertation, but rather is intended as recreational reading for students of science and interested lay people. The entries have different lengths, in part because limited biographical material is available for some of the lawgivers. I welcome feedback and suggestions for improvement from readers, as I consider this an ongoing project and a labor of love.

This book is organized chronologically, according to the year that a law was discovered. In some cases, the literature may report slightly different dates for the discovery of a law because some sources give the publication date as the discovery date of a law, while other sources give the actual date that a law was discovered, regardless of the fact that the publication date is sometimes a year or more later. In many cases, if I am uncertain of a precise earlier date of discovery, I have used the publication date for a law.

In the case of lawgivers such as Newton, a number of years may intervene between the discovery date and the publication date. For example, Newton claimed that in 1666, while his university was closed due to the plague and he was isolated at home, he discovered the chromatic composition of light and discovered the inverse-square principle of the Law of Universal Gravitation. However, upon critical analysis, the discovery date that he gave may be a little early because it was in his self-interest to place his discoveries as early as possible. Some of his ideas on mechanics may not have fully gelled until around 1685–1687, when he was actually composing his great book, the *Philosophiae Naturalis Principia Mathematica* (now usually referred to as the *Principia*). Thus, I have listed the date for some of his laws as 1687, the year the *Principia* was published. In any case, despite a very few discrepancies in the dates of laws in the literature, the reader should get a general sense of the progression of discoveries from the ordering of laws in this book, even if the precise date may subject to discussion.

Some readers may wonder why no eponymous laws for women are listed in this book. Several reasons for this absence exist. Until the twentieth century, very few women received much education, and the path to more advanced studies was usually blocked. Many women mathematicians had to go against the wishes of their families if they wanted to learn. Some were even forced to assume false identities, study in terrible conditions, and work in intellectual isolation. Consequently, very few women contributed to mathematics and the scientific laws featured in this book.

In the "Final Comments" and "Great Contenders" sections at the end of this book, I catalogue a number of additional eponymous laws or equations. These two final sections are meant to prompt questions from readers. In particular, have any deserving laws been left out of the main section? Should, for example, Schrödinger's and Maxwell's equations or $E = mc^2$ be considered "laws" and highlighted in the main section of the book rather than relegated to the final sections? I welcome comments from readers for a follow-up book that will focus on great equations of science.

DISTRIBUTION OF LAW DISCOVERIES THROUGH TIME

> Einstein's fundamental insights of space/matter relations came out of philosophical musings about the nature of the universe, not from rational analysis of observational data—the logical analysis, prediction, and testing coming only after the formation of the creative hypotheses.
>
> —R. H. Davis, *Skeptical Inquirer*

In conclusion, I must admit a personal reason for my interest in physical laws, and I think my reason resonates with a wide audience. Laws give us a feeling of triumph, understanding, and even being in control of our destinies. Most of our daily lives are fraught with challenges that have no clear solutions. Some of our problems have no resolutions at all. We go through life doing the best we can. However, when discovering laws, the human race can have a feeling of purpose and pride. With each scientific law that we uncover and express mathematically, we have a sense that we have encapsulated the cosmos in neat little packages and wrapped them with sparkling bows. Will each century bring us new laws?

Table 1 indicates the historical distribution of the laws in the main section of this book. Notice that a significant number of the laws were discovered in the nineteenth century. Very few physical principles after 1900 are

TABLE 1 Time Distribution of Main-Entry Laws

Time Period	Number of Laws
250 B.C.–1700 A.D.	10 (20%)
1700–1800	6 (12%)
1800–1900	30 (60%)
1900–2000	4 (8%)

See the "Great Contenders" section at the end of this book for a similar breakdown of a large number of additional scientific laws.

commonly called laws. Scientists demand that fundamental physical principles be defined in terms of experiments, but this demand became a growing challenge as scientists treaded in the realms of quantum mechanics, particle theory, and relativity. Of course, with advancing technology, scientists can and do perform many experiments, but today the experiments are often more difficult and expensive to perform than those conducted in previous centuries. (Some degree of subjectivity was involved when creating Table 1, and sometimes I counted several laws as one when they were closely related. For example, both of Kirchhoff's electrical circuit laws counted toward one law, and I counted only Planck's Law of Radiation when considering all the various closely related blackbody radiation laws in Planck's time.)

The "Final Comments" section of this book presents additional musings as to why eponymous laws fade away after 1900. Perhaps this diminution is caused by the fact that modern science has become increasingly organized around large, collaborative research projects. Peter Dizikes writes in "Twilight of the Idols" that "today's insights are not so much perceived from the shoulders of giants as glimpsed from a mountain of jointly authored papers announcing results from large labs, and rapidly circulated through journals, conferences and the Internet." James Gleick writes in his biography of Richard Feynman, "The world has grown too vast and multifarious for the towering genius of the old kind."

To place the dates in Table 1 in perspective, consider the scientific revolution that occurred roughly during the period between 1543 and 1687. In 1543, Copernicus published his heliocentric theory of planetary motion. In 1609 and 1618, Kepler established his three laws that described the paths of the planets about the Sun, and in 1687, Newton published his fundamental laws of motion and gravity.

A second scientific revolution occurred between 1850 and 1865, when scientists introduced and refined various concepts concerning energy and entropy. Fields of study—such as thermodynamics, statistical mechanics, and the kinetic theory of gases—began to blossom. In the twentieth century, Quantum Theory and Special and General Relativity were among the most important insights in science to change our views of reality.

The ideas of quantum mechanics flourished after 1925. According to Paul Quincey, a physicist at the United Kingdom's National Physical Laboratory, quantum mechanics was not only useful for accounting for the properties of atoms, but the ideas

> were absolutely central to explaining why atoms did not collapse, how solids can be rigid, and how different atoms combine together in what we call chemistry and biology.... But this triumph of quantum mechanics came with an unexpected problem—when you stepped outside of the mathematics and tried to explain what was going on, it didn't seem to make any sense.

Physics had finally entered the age of the nonintuitive science, or as Feynman said, "I think I can safely say that nobody understands quantum mechanics." Danish physicist Niels Bohr (1885–1962) wrote, "There is no quantum world. There is only an abstract physical description. It is wrong to think that the task of physics is to find out how nature *is*. Physics concerns what we can *say* about nature." Theoretical physicist Jim Al-Khalili writes in *Quantum: A Guide for the Perplexed*:

> When it comes to the world of the quantum, we really are crossing into a quite extraordinary domain... where it seems we are free to choose any one of a number of explanations for what is observed, each of which is in its way so astonishingly strange that it even makes tales of alien abduction sound perfectly reasonable.

Nonetheless, the mathematical framework of quantum mechanics is precise, and it accurately predicts the behavior of particles at the atomic and subatomic levels.

WHERE THE LAWGIVERS LIVED

> During the Renaissance... the interaction among different European cultures stimulated creativity through new ways of thinking and new paradigms for the observation of nature.... The foundation of scientific academies,

> notably the Accademia dei Lincei, the Royal Society and the Académie des Sciences, and the establishment of universities throughout Western Europe, contributed to scientific progress....
>
> —Maurizio Iaccarino, "Science and Culture"

We may also catalogue the laws in this book by the lawgiver's countries of birth, or primary country affiliation (Table 2). Germany, France, and Great Britain are clearly the most significant contributors.

Historians of science acknowledge that Europe was special for many reasons when it comes to the discoveries of scientific laws. Richard Koch and Chris Smith, authors of "The Fall of Reason," note that

> [s]ome time between the 13th and 15th centuries, Europe pulled well ahead of the rest of the world in science and technology, a lead consolidated in the following 200 years. Then in 1687, Isaac

TABLE 2 Country Distribution of Main-Entry Laws

Country	Number of Laws
Germany	14
France	12
Britain	10
Ireland	2
Netherlands	2
Italy	1
Switzerland	1
United States	1
Hungary	1
Greece	1

See the "Great Contenders" section at the end of this book for a similar breakdown of additional eponymous scientific laws.

> Newton—foreshadowed by Copernicus, Kepler, and others—had his glorious insight that the universe is governed by a few physical, mechanical, and mathematical laws. This instilled tremendous confidence that everything made sense, everything fitted together, and everything could be improved by science.

According to Koch and Smith, many scholars who lived in medieval and early modern Europe began to feel that the mysteries of the universe could be solved because God had provided laws of naturc that were logical and that made sense. Koch and Smith write, "The full emergence of science required belief in one all-powerful God, whose perfect creation" could be understood by rational thought. They go further to suggest the controversial idea that this condition was special to Christianity of the time, especially compared with other religions that had a less consistently rational creator and for religions in which "the universe is inexplicable, unpredictable." Catholic philosopher Thomas Aquinas (1225–1274) thought so much about the laws of the universe that he finally decided to make various categories of laws—such as eternal, natural, human, and divine laws.

According to Jan Wojcik, author of *Robert Boyle and the Limits of Reason*, many influential scientist Christians of the seventeenth and eighteenth centuries believed that reason could be used to address theological questions. Although reason alone could not be used to *discover* the mysteries of Christianity, "reason was considered competent to aid the believer in *understanding* the content of what has been revealed." Wojcik writes:

> After the Christian had come to believe that something was true simply because God had revealed it, the believer could analyze that truth philosophically in an attempt to determine *how* it could be true. Further, reason was seen as playing an important role in convincing atheists of the truth of Christianity by showing that the mysteries of Christianity are rationally possible.

Of course, the Christian religion in Europe was not the only motivator for many of the scientific lawgivers. Newton, for example, was thoroughly immersed in alchemical and magical thinking, which also contributed to his interest in the laws of nature. Additionally, starting around 750 A.D., science and mathematics flourished under the Abbasid caliphs of Baghdad, and the knowledge learned under Arab-Islamic patronage spread throughout Europe and central Asia. Going further back in time, pagan Greek culture produced thinkers ranging from Aristotle and Pythagoras to Archimedes.

Maurizio Iaccarino, a scientist at the International Institute of Genetics and Biophysics of the National Research Council in Naples, Italy, writes in *EMBO Reports*:

> The Muslims were the leading scholars between the seventh and fifteenth centuries, and were the heirs of the scientific traditions of Greece, India and Persia.... Participants were Arabs, Persians, Central Asians, Christians and Jews, and later included Indians and Turks. The transfer of the knowledge of Islamic science to the West... paved the way for the Renaissance, and for the scientific revolution in Europe.

Additionally, while Christianity could be conducive to science and serve as a motivator for many of the lawgivers in this book, the Church could also have supreme failings in this regard. Sam Harris writes with passion in "The Language of Ignorance":

> Lest we forget: Galileo, the greatest scientist of his time, was forced to his knees under threat of torture and death, obliged to recant his understanding of the Earth's motion, and placed under house arrest for the rest of his life by steely-eyed religious maniacs. He worked at a time when every European intellectual lived in the grip of a Church that thought nothing of burning scholars alive for merely speculating about the nature of the stars.... This is the same Church that did not absolve Galileo of heresy for 350 years (in 1992).

Nevertheless, Europe's economic expansion after 1000 A.D., which resulted from a growing number of interacting and free cities, led to the development of simple industries and their related sciences.

WHEN WILL THE LAST LAW BE DISCOVERED?

> Eternity is a child playing checkers.
> —Heraclitus (535–475 B.C.)

Western "faith" in the rationality and logic of the universe faded slightly in the twentieth century, with the gradual demise of the Newtonian universe, leading the way to new physical theories that suggested an inscrutable and uncertain universe. Perhaps the rise of quantum theory is one reason among many that fewer eponymous laws are found in the twentieth century than in the nineteenth century. As I have already mentioned, experiments that could lend support to new laws are becoming very difficult to perform. At a Royal Society meeting, Nobel laureate Stephen Weinberg focused on this challenge, saying, "Quantum gravitation seems inaccessible to any experiment we can devise. In fact, physics in general is moving into an era where the fundamental questions can no longer be illuminated by conceivable experiments. It's a very disquieting position to be in."

I conclude this introduction by asking readers, Do you feel that we will have all the laws of nature in our hands in the next fifty years? Personally, I think that there will always be more laws for us to uncover. Isaac Asimov had the right idea about the future of knowledge: "I believe that scientific knowledge has fractal properties; that no matter how much we learn, whatever is left, however small it may seem, is just as infinitely complex as the whole was to start with. That, I think, is the secret of the Universe." Hawking has said that he believes that the search for the ultimate laws will soon come to an end. He notes in his books *A Brief History of Time* and *Black Holes and Baby Universes* that in the early 1900s, many scientists thought that the universe and its laws could be explained in terms of "the properties of continuous matter, such as elasticity and heat conduction." However, the discovery of subatomic structure and Heisenberg's Uncertainty Principle led humanity to the next level of understanding. In 1928, physicist and Nobel Prize winner Max Born (1882–1970) told a group of visitors to Göttingen University, "Physics, as we know it, will be over in six months." Born's proclamation took place immediately after British theoretical physicist Paul Dirac (1902–1984) formulated the Dirac Equation, which characterizes the behavior of an electron. Hawking explains both Born's confidence and folly in *Black Holes and Baby Universes*:

> It was expected that a similar equation [to Dirac's] would govern the proton, the only other supposedly elementary particle at that time. However, the discovery of the neutron and of nuclear forces disappointed those hopes.... Nevertheless, we have made a lot of progress in recent years, and... there are some grounds for cautious optimism that we may see a complete theory within the lifetime of some of those reading these pages.

REFERENCES

Al-Khalili, Jim, *Quantum: A Guide for the Perplexed* (London: Weidenfeld & Nicolson, 2004).

Arons, Arnold, *Development of Concepts of Physics* (Reading, Mass.: Addison-Wesley, 1965).

Asimov, Isaac, *I. Asimov: A Memoir* (New York: Bantam, 1995).

Atiyah, Michael, "Pulling the Strings," *Nature*, 438: 1081–1082, December 22, 2005.

Berlin, Leslie, *The Man Behind the Microchip* (New York: Oxford University Press, 2005).

Blank, Brian, "*The Road to Reality: A Complete Guide to the Laws of the Universe*" [book review], *Notices of the American Mathematical Society*, 53(6): 661–666, June/July 2006.

Bothamley, Jennifer, *Dictionary of Theories* (Washington, D.C.: Gale Research International Ltd, 1993).

Brewster, David, *More Worlds Than One* (London: John Murray, 1854).

Brockman, John, *What We Believe but Cannot Prove: Today's Leading Thinkers on Science in the Age of Certainty* (New York: Harper Perennial, 2006).

Casti, John, *Paradigms Lost* (New York: William Morrow & Company, 1989).

Dizikes, Peter, "Twilight of the Idols," *New York Times Book Review*, Section 7, p. 31, November 5, 2006.

Durbin, Paul, *Dictionary of Concepts in the Philosophy of Science* (New York: Greenwood Press, 1988).

Feynman, Richard, *The Character of Physical Law* (New York: Modern Library, 1994).

Gardner, Martin, *Order and Surprise* (Amherst, N.Y.: Prometheus Books, 1985); reprints Gardner's 1950 essay "Order and Surprise."

Gardner, Martin, "Interview with Martin Gardner," *Notices of the American Mathematical Society*, 52(6): 602–611, June/July 2005.

Gleick, James, *Genius: The Life and Science of Richard Feynman* (New York: Vintage, 1992).

Guillen, Michael, *Five Equations That Changed the World* (New York: Hyperion, 1995).

Halliday, David, and Robert Resnick, *Physics* (New York: John Wiley & Sons, 1966).

Hardy, G. H., *A Mathematician's Apology* (Cambridge, U.K.: Cambridge University Press, 1940).

Harris, Sam, "The Language of Ignorance," *Truthdig*, August 15, 2006; see www.truthdig.com/report/item/20060815_sam_harris_language_ignorance/.

Hawking, Stephen, *A Brief History of Time* (New York: Bantam, 1988).

Hawking, Stephen, *Black Holes and Baby Universes* (New York: Bantam, 1993).

Hawking, Stephen, and Roger Penrose, *The Nature of Space and Time* (Princeton, N.J.: Princeton University Press, 2000).

Holt, Jim, "Mistaken Identity Theory: Why Scientists Always Pick the Wrong Man," *Lingua Franca Online*, 10(2), March 2000; see linguafranca.mirror.theinfo.org/0003/hypo.html; this article discusses Stigler's Law of Eponymy.

Hume, David, *An Enquiry Concerning Human Understanding*, edited by Tom L. Beauchamp (New York: Oxford University Press, 1999).

Iaccarino, Maurizio, "Science and Culture," *EMBO Reports*, 4(3): 220–223, 2003.

Jennings, Byron K., "On the Nature of Science," July 27, 2006; see xxx.lanl.gov/abs/physics/0607241.

Joule, James, "Remarks on God," 1873. (These are notes from an address Joule was to deliver at a meeting as president of the British Association for the Advancement of Science. Joule never delivered the address due to poor health. The quotation can be found in J. G. Crowther, *British Scientists of the 19th Century* [London: K. Paul, Trench, Trubner & Co., Ltd., 1935], pp. 138–140.)

Kaku, Michio, "Parallel Universes, the Matrix, and Superintelligence"; see www.kurzweilai.net/meme/frame.html?main=/articles/art0585.html.

Kepler, Johannes, *Conversation with Galileo's Sidereal Messenger,* translated by Edward Rosen. (New York: Johnson Reprint Corp., 1965).

Koch, Richard, and Chris Smith, "The Fall of Reason," *New Scientist*, 190(2557): 25, June 24, 2006.

Krauss, Lawrence, *Fear of Physics* (New York: Basic Books, 1993).

Krebs, Robert, *Scientific Laws, Principles, and Theories* (Westport, Conn.: Greenwood Press, 2001).

Nuland, Sherwin, "The Man or the Moment?" *The American Scholar*, 73: 129–132, 2004; reprinted in *The Best American Science and Nature Writing 2005*, Tim Folger, editor (Boston: Houghton Mifflin, 2005).

Pickover, Clifford, *The Loom of God* (New York: Plenum, 1997).

Pickover, Clifford, *The Möbius Strip* (New York: Thunder's Mouth Press, 2006).

Pickover, Clifford, *A Passion for Mathematics* (Hoboken, N.J.: Wiley, 2005).

Pickover, Clifford, *Strange Brains and Genius* (New York: Quill, 1999).

Penrose, Roger, "What Is Reality?" *New Scientist*, 192(2578): 39, November 18, 2006.

Popper, Karl, *Conjectures and Refutations: The Growth of Scientific Knowledge* (London: Routledge, 1963).

Quincey, Paul, "Why Quantum Mechanics Is Not So Weird after All," *Skeptical Inquirer*, 30(4): 39–43, July/August 2006.

Rothstein, Edward, "Reason and Faith, Eternally Bound," *New York Times*, Section B, p. 7, December 20, 2003.

Stenger, Victor, "A Scenario for a Natural Origin of Our Universe," *Philo* 9(2): 93–102, 2006.

Stokes, George, *Natural Theology* (London: Adams and Charles Black, 1891).

Tegmark, Max, "Max Tegmark Forecasts the Future," *New Scientist*, 192(2578): 37, November 18, 2006.

Trefil, James, *The Nature of Science* (New York: Houghton Mifflin Company, 2003).

Wojcik, Jan, *Robert Boyle and the Limits of Reason* (Cambridge, U.K.: Cambridge University Press, 2002).

INTERLUDE: CONVERSATION STARTERS

> As measured by the millions of those who speak it fluently..., mathematics is arguably the most successful global language ever spoken.... Equations are like

poetry: They speak truths with a unique precision, convey volumes of information in rather brief terms.... And just as conventional poetry helps us to see deep *within* ourselves, mathematical poetry helps us to see far *beyond* ourselves.

—Michael Guillen, *Five Equations That Changed the World*

There is no reason why the most fundamental aspects of the laws of nature should be within the grasp of human minds ... or why those laws should have testable consequences at the moderate energies and temperatures [of life-bearing planets].... As we probe deeper into ... the nature of reality, we can expect to find more deep results which limit what can be known. Ultimately, we may even find that their totality characterizes the universe more precisely than the catalogue of those things that we can know.

—John Barrow, *Boundaries and Barriers: On the Limits of Scientific Knowledge*

It is the most persistent and greatest adventure in human history, this search to understand the universe, how it works and where it came from. It is difficult to imagine that a handful of residents of a small planet circling an insignificant star in a small galaxy have as their aim a complete understanding of the entire universe, a small speck of creation truly believing it is capable of comprehending the whole.

—Murray Gell-Mann, in John Boslough's *Stephen Hawking's Universe*

Scientists are remarkably sloppy about their use of the word "law." It would be nice, for example, if something that had been verified a thousand times was called an "effect," something verified a million times a "principle," and something verified 10 million times a "law" ... but the use of these terms is based entirely on historical precedent and has nothing to do with the confidence scientists place in a particular finding.

—James Trefil, *The Nature of Science*

The chessboard is the world, the pieces are the phenomena of the universe, the rules of the game are what we call the laws of Nature.

—Thomas Huxley, "A Liberal Education" in *Autobiography and Selected Essays*

Time was when all scientists were outsiders. Self-funded or backed by a rich benefactor, they pursued their often wild ideas in home-built labs with no one to answer to but themselves. From Nicolaus Copernicus to Charles Darwin, they were so successful that it's hard to imagine what modern science would be like without them. Their isolated, largely unaccountable ways now seem the antithesis of modern science, with consensus and peer review at its very heart.

—Editors of *New Scientist*, "It Pays to Keep a Little Craziness," *New Scientist*, December 9, 2006

Explaining the simultaneity of invention by different people in different places at the same time, Mark Twain said, "When it's steamboat time, you steam."

—*Automobile Magazine,* September, 2006

The contributions of Muslim scientists typically occurred in spite of Islam rather than because of it. Orthodox Islamic scholars absolutely rejected any conception of the universe that involved consistent physical laws, because the absolute autonomy of Allah could not be restricted by natural laws.... Catholicism admits the possibility of miracles and acknowledges the role of the supernatural, but the very idea of a miracle suggests that the event in question is *unusual*, and of course it is only against the backdrop of an orderly natural world that a miracle can be recognized in the first place.

—Thomas E. Woods, Jr., *How the Catholic Church Built Western Civilization*

Galileo had championed a view of the universe, Copernicus's [Sun-centered universe], that seemed not only new but shocking. Many churchmen who had never even heard of Copernicus now learned that he had fathered these disturbing ideas. An Italian bishop wanted Copernicus throw in jail and was surprised to learn that he had been dead for seventy years.

—James C. Davis, *The Human Story*

250 B.C.—1700 A.D.

Our experience hitherto justifies us in believing that nature is the realization of the simplest conceivable mathematical ideas. I am convinced that we can discover by purely mathematical constructions the concepts and the laws connecting [mathematics and physical reality] with each other, which furnish the key to the understanding of natural phenomena.... In a certain sense, therefore, I hold it true that pure thought can grasp reality, as the ancients dreamed.

—Albert Einstein, "On the Methods of Theoretical Physics," 1933

Even stranger things have happened; and perhaps the strangest of all is the marvel that mathematics should be possible to a race akin to the apes.

—Eric T. Bell, *The Development of Mathematics*

As far as the laws of mathematics refer to reality, they are not certain, and as far as they are certain, they do not refer to reality.

—Albert Einstein, "Geometry and Experience," Address to the Prussian Academy of Sciences, 1921

The essential fact is simply that all the pictures which science now draws of nature... are mathematical pictures.... It can hardly be disputed that nature and our conscious mathematical minds work according to the same laws.

—Sir James Jeans, *The Mysterious Universe*

The most important fundamental laws and facts of physical science have all been discovered, and these are now so firmly established that the possibility of their ever being supplemented in consequence of new discoveries is exceedingly remote.

—Albert Michelson, 1894 dedication address, Ryerson Physical Laboratory, University of Chicago

ARCHIMEDES' PRINCIPLE OF BUOYANCY

The Greek city-state of Syracuse, Sicily, c. 250 B.C. The vertical force of buoyancy on a submerged object is equal to the weight of fluid the object displaces.

CROSS REFERENCE: JOHANNES KEPLER AND GALILEO GALILEI.

Close to the time when Archimedes discovered his Principle of Buoyancy, the Septuagint Greek version of the Old Testament was being written, the La Tène Iron Age people invaded Britain, the first Roman prison Tullianum was erected, and the Carthaginian general Hannibal was born.

Imagine that you are weighing an object, like a carrot, that is submerged in a kitchen sink. If you weigh the carrot by hanging it from a scale, the carrot would weigh less while in the water then when the carrot is lifted out of the sink and weighed. The water exerts an upward force that partially supports the weight of the carrot. This force is more obvious if we perform the same experiment with an object of lower density, such as a cube made out of cork, which floats while being partially submerged in the water.

The force exerted by the water on the cork is called a *buoyant force*, and for the cork, the upward force is greater than its weight. This buoyant force depends on the density of the liquid and the volume of the object, but not on the shape of the object or the material with which the object is composed. Thus, in our experiment, it does not matter if the carrot is shaped like a sphere or a cube. One cubic centimeter of carrot or wood would experience the same buoyant force in water.

According to Archimedes' principle, a body wholly or partially submerged in liquid is buoyed up by a force equal to the weight of displaced liquid. Physicists write this with the compact expression

$$\boxed{B = w_f,}$$

where B is the buoyant force and w_f is the weight of the fluid that the object displaces.

As another example, consider a small pellet of lead placed in a bathtub. The pellet weighs more than the tiny weight of water w_f it displaces, and the pellet sinks. A wooden rowboat is buoyed up by the large weight of water that it displaces, and hence the rowboat floats. Archimedes' principle helps us understand how flotation works and is one of the founding principles of hydrostatics.

Another form of Archimedes' Principle of Buoyancy can be stated as

$$B = \rho_{liquid} G V_{solid}$$

where ρ_{liquid} is the density of the liquid, G is the constant of gravitational acceleration, and V_{solid} is the volume of the solid. (The volume of the displaced fluid is equal to the volume of an object either fully submerged or to that fraction of the volume below the surface for an object that is partially beneath the water.)

The buoyant force occurs because the pressure of the liquid at the bottom of the object is greater than that at the top. If the density of the object—for instance, the carrot in our example—is greater than that of the liquid, the object's weight will be greater than the buoyant force, and the object will sink if not supported. If the density of a submerged object is less than the density of the fluid, the object will accelerate upward to the liquid's surface. Part of the object will usually rise above the surface so that the weight of the displaced liquid equals the weight of the object.

A submarine floating in the sea displaces a volume of water that has a weight that is precisely equal to the submarine's weight. In other words, the average weight of the submarine—which includes the people, the metal hull, and the enclosed air—equals the weight of displaced seawater. When plesiosaurs (extinct reptiles) floated in the middle of the sea, their average weights also equaled the weights of the water they displaced. Gastroliths (stomach stones) have been discovered in the stomach region of skeletons of plesiosaurs, and these stones may have helped maintain useful buoyancy for the creatures.

A rowboat lowered into a pond sinks into the pond until the weight of the water it displaces is equal to its own weight. The buoyant force may be thought of as acting through the centroid, or center of gravity, of the displaced fluid volume. The stability of a floating body, like the rowboat, depends on this location and its relationship to the center of gravity of the body.

Perhaps it is not intuitive that objects with equal volumes experience equal buoyant forces when held beneath the water. Imagine that we have equal-sized cubes of cork, aluminum, and lead that we hold beneath the water. The buoyant force would be the same on each object because of the equal amounts of water they displace; however, the behavior of the three cubes would certainly be different when we release them. The cork cube would rush upward, the aluminum would sink, and the lead would sink more rapidly. The various behaviors arise from the different ratios of buoyant force to object weights.

Archimedes' principle has various applications; for example, the principle can be used to determine the pressure of a liquid as a function of depth.

Archimedes of Syracuse (287–212 B.C.), Greek mathematician and inventor famous for his geometric and hydrostatic studies, as well as the Archimedean screw, still used today to move water.

> Archimedes will be remembered when Aeschylus is forgotten, because languages die and mathematical ideas do not. "Immortality" may be a silly word, but probably a mathematician has the best chance of whatever it may mean.
>
> —G. H. Hardy, *A Mathematician's Apology*, 1941

> Give me a place to stand on, and I will move the earth.
>
> —Archimedes, upon discovering the principles of leverage (as told by Pappus of Alexandria)

> Eureka! Eureka!
>
> —Archimedes, when he realized how to determine if King Hieron's wreath was made of pure gold

---

CURIOSITY FILE: Archimedes was an inspiration to Sophie Germain (1776–1831), one of the greatest female mathematicians who ever lived. At the age of thirteen, Sophie read an account of Archimedes' murder while he was solving mathematical problems, and she was so moved by this story that she decided to become a mathematician. Sadly, her parents forbade her to study mathematics; thus, she had to hide beneath blankets to secretly study the works of Isaac Newton and mathematician Leonhard Euler. • Archimedes sometimes sent his colleagues false theorems in order to trap them when they stole his ideas.

Archimedes, the ancient Greek geometer, is often regarded as the greatest mathematician and scientist of antiquity and one of the four greatest mathematicians to have walked Earth—together with Isaac Newton, Leonhard Euler, and Carl Friedrich Gauss. Archimedes was the son of an astronomer named Phidias and spent most of his life in the Greek city-state of Syracuse, where he worked on numerous inventions, including weapons used against the Romans, Archimedean screws for conveying water upward, and model planetariums.

At some point during his life, Archimedes traveled to Egypt and studied at Alexandria, one of the intellectual centers of Greek scholarship during his time. Today, he is famous for his inventions, for his geometrical works that included formulas for the volume and surface area of a sphere, and for mathematical procedures that bordered on more modern methodologies that involve logarithms and calculus.

The famous legend of Archimedes' death describes his encounter with a Roman soldier around 212 B.C., during the capture of Syracuse in the Second Punic War. The soldier had come upon Archimedes, who was studying a mathematical diagram drawn in the sand. Archimedes was annoyed by the soldier's interruption and allegedly uttered his famous last words before being killed by the soldier: "Μη μου τους κύκλους τάραττε" ("Don't disturb my circles"). According to the Greek writer Plutarch, Archimedes had asked his friends to engrave a cylinder circumscribing a sphere on his grave, along with "the ratio by which the including solid exceeds the included."

A lunar crater with a diameter of 82 kilometers was named after Archimedes in 1935 by the International Astronomical Union General Assembly. The International Astronomical Union (IAU) is in charge of reviewing and approving names for most solar system objects, as well as their surface features. The IAU's General Assembly formally adopts the names that scientists and laypersons have submitted.

Although a genius mathematician, Archimedes is perhaps more famous for his mechanical inventions, including:

- The water snail or Archimedean screw to raise water to help irrigate crops (mentioned by Greek historian Diodorus Siculus)
- A long screw used to launch a ship (mentioned by Greek writer Athenaeus)
- The compound pulley used to help move heavy ships with minimal effort (mentioned by Greek writer Plutarch)
- Globelike planetariums (mentioned by Roman orator Cicero)
- Ballistic defensive devices used to repel the Romans (mentioned by Greek historian Polybius, Roman historian Livy, and Plutarch)
- Burning mirrors used to repel the Romans (some scholars are skeptical of the legend of this device)

Regarding Archimedes' burning mirrors, in 313 B.C. Archimedes was said to have made a "death ray" consisting of a set of mirrors that focused sunlight on Roman ships, setting the ships afire. Various individuals have tried to test the practical use of such mirrors and declared their use to have been unlikely. However, mechanical engineer David Wallace of MIT encouraged his students in 2005 to build an oak replica of a Roman warship and focus sunlight on it, using 127 mirrored tiles. Each mirror was about 12 square meters in area, and the ship was about 30 meters away. After ten minutes of exposure to the focused light, the warship burst into flames!

In 1973, a Greek engineer employed seventy flat mirrors (each about five feet by three feet in length) in order to focus sunlight onto a rowboat. In this experiment, the rowboat also quickly burst into flames. However, while it is possible to set a ship afire with mirrors, this task probably would have been very difficult for Archimedes if ships were moving.

As an interesting aside, Arthur C. Clark's short story "A Slight Case of Sunstroke" describes the fate of a disliked soccer referee. When the referee makes an unpopular decision, the spectators focus sunlight onto the referee using their shiny souvenir programs that they hold in their hands. The shiny surfaces act like Archimedes' mirror, and the poor man is burned to ashes.

According to Plutarch, Archimedes' ballistic weaponry was used effectively against the Romans in the siege of 212 B.C. For example, Plutarch wrote,

> When Archimedes began to ply his engines, he at once shot all sorts of missile weapons against the land forces, and immense masses of stone came down with incredible noise and violence, against which no man could stand for they knocked down those upon whom they fell in heaps.

Plutarch also writes that Archimedes was so obsessed with mathematics that his servants had to force him to take baths, and while they scrubbed him, he would continue to draw geometrical figures. "And while the servants were anointing of him with oils," Plutarch writes, "with his fingers he drew lines upon his naked body, so far was he taken from himself, and brought into ecstasy or trance, with the delight he had in the study of geometry."

Perhaps the most famous legend of Archimedes involves King Hieron, who needed to check the authenticity of a wreath-shaped crown allegedly made of pure gold, but which Hieron suspected had silver impurities. Archimedes used the displacement of water in a bathtub to uncover the scam, demonstrating that the wreath actually consisted of a mixture of silver and gold rather than pure gold. Roman architect and engineer Marcus Vitruvius writes of this incident in *De Architectura*:

> While Archimedes was turning the problem over, he chanced to come to the place of bathing, and there, as he was sitting down in the tub, he noticed that the amount of water which flowed over the tub was equal to the amount by which his body was immersed. This

showed him a means of solving the problem.... In his joy, he leapt out of the tub and, rushing naked towards his home, he cried out with a loud voice that he had found what he sought.

In principle, Archimedes could have performed his volume experiment using a bucket filled to the brim and then measured the volume of water that overflowed. He is said to have conducted an experiment with equal weights of gold and silver. Because gold has a greater density than silver, a gold cube would be smaller than the silver cube of equal weight, causing less water to spill out of the bucket. Once Archimedes was able to measure the volumes of water that represented the volumes of the gold, silver, and the crown, he could determine their relative densities and determine that the crown was not of pure gold.

The legend of Hieron and Archimedes concludes with the density of the crown measured to be between 10.5 and 19.3 grams per cubic centimeter, which are the densities of silver and gold, respectively. This meant that the wreath was not made out of pure gold, and the royal goldsmith was executed.

However, today, scholars suggest that this story of the gold and silver may have been embellished through time because it is unlikely that Archimedes had measuring equipment of sufficient accuracy to detect the rather small difference in displacement between a wreath made of pure gold and one fashioned from gold mixed with other metals.

As for Archimedes' invention of the Archimedean screw, his creation of the screw seems plausible, and Vitruvius gives a detailed description of its operation for lifting water, which required intertwined helical blades. For example, the bottom end of the screw is immersed in a pond. The act of rotating the screw raises water from the pond to a higher elevation. Archimedes may also have designed a related helical pump, a corkscrew-like device used to remove water from the bottom of a large ship. The Archimedean screw is still used today in societies without access to advanced technology. It works well even if the water is filled with debris. The Archimedean screw also tends to minimize damage to aquatic life. Modern Archimedean screwlike devices are used to pump sewage in water treatment plants.

As discussed above, Archimedes contributed to mathematics by proving theorems that concerned areas and volumes, determining the formulas for the volume and surface area of a sphere (he showed that the surface area of a sphere is four times the area of a circle that passes through the sphere's center), determining an approximation for the value of π, and discussing the physics of floating bodies. His scientific work on flotation is the first known work in the field of hydrostatics, the field of physics that deals with fluids at rest and under pressure.

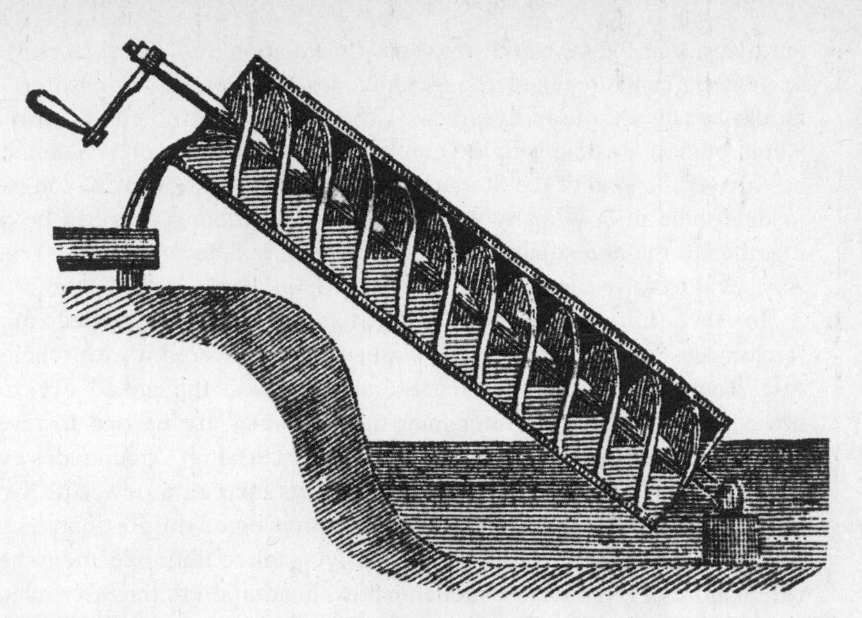

Archimedean screw, from *Chambers's Encyclopedia* (Philadelphia: J. B. Lippincott Company, 1875).

Archimedes' research on the value of π involved inscribing polygons on the inside and outside of a circle. The greater the number of sides of the polygon, the more closely they approached the actual edge of the circle. By calculating the areas of the internal and surrounding polygon, he determined a value for π that was between 3.140845 and 3.142857 (today, we know the correct value is 3.14159...). Beginning with a hexagon, he worked all the way up to a polygon with ninety-six sides!

Archimedes proved various theorems relating to levers and fulcrums. He also explained why it is easier to move an object up a long, sloping ramp than to move the same object along a shorter but steeper ramp to the same height. He discovered that the volume of a sphere is exactly two-thirds the volume of a cylinder that tightly encloses it. As I have mentioned, despite the fact that he lived more than eighteen centuries before Newton, Archimedes came close to formulating integral calculus but lacked a satisfactory system of mathematical notation to aid his thinking. In Archimedes' *The Sandreckoner*, he proposed a number system in which humans could express numbers up to 8×10^{63} in modern notation. According to Archimedes, this number is sufficiently large to count the number of grains of sand needed to fill the entire universe.

Two favorite puzzles from Archimedes concern geometry and numbers, and I discuss these in my book *A Passion for Mathematics*. The first of these puzzles is called the Stomachion of Archimedes. In 2003, math historians discovered long lost information on the object in an ancient parchment called the Archimedes Palimpsest, overwritten by monks nearly a thousand years ago. The puzzle involved combinatorics, a field of math dealing with the number of ways that a given problem can be solved. The goal of the Stomachion (pronounced sto-MOCK-yon) was to determine in how many ways fourteen flat puzzle pieces can be put together to make a square. In 2003, researchers determined that 17,152 ways exist to solve the puzzle.

In the thirteenth century, Christian monks had ripped apart Archimedes' original manuscript, washed it, and covered it with religious text. Today, we cannot see the Stomachion with the naked eye, and ultraviolet light and computer imaging techniques are needed to reveal the hidden mathematical gem. Scholars are uncertain if Archimedes ever correctly solved the problem. In 2006, experts focused a powerful X-ray beam from the Stanford Linear Accelerator Center on portions of the manuscript to cause the iron in the underlying ink to fluoresce and to help reveal more of Archimedes' hidden words. In addition to the Stomachion, the ancient palimpsest contains seven of Archimedes' treatises, including *Method of Mechanical Theorems* (which is known only through the Palimpsest document) and the only surviving copy of *On Floating Bodies* in the original Greek.

A second puzzle, called Archimedes' Cattle Problem, can be stated as follows:

> Oh stranger, compute the number of cattle of the Sun, who once upon a time grazed on the fields of the Thrinacian Isle of Sicily, divided into four herds of different colors—one milk white, another glossy black, the third yellow, and the fourth dappled. The number of white bulls was equal to (1/2 + 1/3) the number of black bulls plus the total number of yellow bulls. The number of black bulls was (1/4 + 1/5) the number of dappled bulls plus the total number of yellow bulls. The number of spotted bulls was (1/6 + 1/7) the number of white bulls, plus the total number of yellow bulls. The number of white cows was (1/3 + 1/4) the total number of the black herd. The number of black cows was (1/4 + 1/5) the total number of the dappled herd. The number of dappled cows was (1/5 + 1/6) the total number of the yellow herd. The number of yellow cows was (1/6 + 1/7) the total number of the white herd.
>
> If you can accurately tell, Oh stranger, the total number of cattle of the Sun, including the number of cows and bulls in each color, you would not be called unskilled or ignorant of numbers, but not

yet shalt thou be numbered among the wise. But understand also these conditions: [The white bulls could graze together with the black bulls in rows, such that the number of cattle in each row was equal and that number was equal to the total number of rows, thus forming a perfect square. And the yellow bulls could graze together with the dappled bulls, with a single bull in the first row, two in the second row, and continuing steadily to complete a perfect triangle.] If you are able, oh stranger, to find out all these things and gather them together in your mind, giving all the relations, you shall depart crowned with glory and knowing that you have been adjudged perfect in this species of wisdom.

The solution to the full problem is about 7.76×10^{206544} cattle. Not until 1880 did mathematicians find an approximate answer. The more precise value for the total number of cattle was first calculated in 1965 by Hugh C. Williams, R. A. German, and C. Robert Zarnke using an IBM 7040 computer. I am not aware of any other numerical problems that required twenty-two centuries to solve. Note that the Cattle Problem has at least two versions attributed to Archimedes, and the one just described is a more complex version. Archimedes obviously never solved this particular version of the problem. Author Heinrich Dörrie cites four scholars who suggest that that this version may not actually be due to Archimedes, but he also cites four authors who believe that the problem *should* be attributed to Archimedes.

Unlike the works of the Greek geometer Euclid, Archimedes' mathematical discoveries did not have a great influence on humanity until the Arabs rediscovered and revived them in the eighth century. In the 1500s, his works began to circulate widely and affected the work of Kepler and Galileo, the latter of whom used Archimedes' name numerous times in his writings. Kepler's method of finding areas was similar to that used by Archimedes to find the area of a circle or the value of π.

Mathematics has surely come a long way since the time of Archimedes, and I wonder how much mathematics humanity can ever know. The body of mathematics has generally increased from ancient times, although this has not always been true. Most mathematicians in Europe during the 1500s knew less than Greek mathematicians at the time of Archimedes. However, since the 1500s, humans have made tremendous journeys along the vast landscape of mathematics. Today, several hundred thousand mathematical theorems are proved each year.

In the early 1900s, a great mathematician was expected to know nearly all of the known mathematics. Mathematics was akin to a shallow pool that could be inspected by a single individual. Today the mathematical waters have grown so deep that a great mathematician can be intimate with only a small percentage of the entire ocean. What will the future of mathematics be like as specialized mathematicians know more and more about less and less until they know everything about nothing?

FURTHER READING

Clagett, Marshall, "Archimedes," in *Dictionary of Scientific Biography*, Charles Gillispie, editor-in-chief (New York: Charles Scribner's Sons, 1970).

Dörrie, Heinrich, *100 Great Problems of Elementary Mathematics: Their History and Solution* (New York: Dover, 1965).

Heath, Thomas (translator), *The Works of Archimedes* (New York: Dover, 2002).

Holden, Constance, "Death-Ray Test," *Science* 310(5747): 435.

Kolta, Gina, "In Archimedes' Puzzle, a New Eureka Moment," *New York Times*, 153(52,697): 1, December 14, 2003.

Pickover, Clifford, *A Passion for Mathematics* (New York: Wiley, 2005).

Rorres, Chris, "The Turn of the Screw: Optimal Design of an Archimedes Screw," *American Society of Civil Engineers Journal of Hydraulic Engineering* 126: 72–80, January 2000; available on the Web from the author at www.mcs.drexel.edu/~crorres/ screw/screw.pdf.

Simmons, John, *The Scientific 100: A Ranking of the Most Influential Scientists, Past and Present* (New York: Citadel Press, 1996).

Stein, Sherman, *Archimedes: What Did He Do Besides Cry Eureka?* (Washington, D.C.: Mathematical Association of America, 1999).

INTERLUDE: CONVERSATION STARTERS

> It is these connections that form the fabric of physics. It is the joy of the theoretical physicists to discover them, and of the experimentalist to test their strength.... In the end, what science does is change the way we think about the world and our place within it.... There is a universal joy in making new connections.
>
> —Lawrence Krauss, *Fear of Physics*

> Archimedes' principle can be understood in terms of kinetic theory.... When the fluid is displaced by the solid object, the molecules in the fluid will collide with the body, exerting the same pressure as they did before the object was placed there. For a completely submerged object...the molecules of the fluid will be hitting the

bottom of the object with a greater force than those hitting the top. This is the molecular origin of the upward buoyant force.

—James S. Trefil, *The Nature of Science: An A–Z Guide to the Laws and Principles Governing Our Universe*

Most of the papers which are submitted to the *Physical Review* are rejected, not because it is impossible to understand them, but because it is possible. Those which are impossible to understand are usually published.

—Freeman Dyson, *Innovation in Physics*

Isaac Newton discovered laws of motion that apply equally to a planet moving through space and to an apple falling earthward, revealing that the physics of the heavens and the earth are one. Two hundred years later, Michael Faraday and James Clerk Maxwell showed that electric currents produce magnetic fields, and moving magnets can produce electric currents, establishing that these two forces are as united as Midas' touch and gold.

—Brian Greene, "The Universe on a String," *The New York Times*, October 20, 2006

KEPLER'S LAWS OF PLANETARY MOTION

Germany, 1609, 1618. Three laws that describe the motions of planets about the Sun.

CROSS REFERENCE: ISAAC NEWTON'S LAWS, EINSTEIN'S GENERAL THEORY OF RELATIVITY, AND THE WORKS OF EUCLID, GALILEO GALILEI, AND TYCHO BRAHE.

> The same year that Kepler discovered his Law of Orbits, Henry Hudson discovered the Hudson Bay. For the first time, the Dutch East India Company shipped tea from China to Europe. Also in 1609, Galileo demonstrated his first telescope to Venetian lawmakers.

German astronomer Johannes Kepler, working with data collected by the Danish astronomer Tycho Brahe (1546–1601), discovered three laws that described the elliptical motion of the planets in space. Brahe had made all his measurements by eye, because the telescope had not yet been invented when he had studied the sky.

In order for Kepler to formulate his laws, he had to first abandon the prevailing notion that circles were the "perfect" curves for describing the cosmos and its planetary orbits. When Kepler first expressed his laws, he had no theoretical justification for them. They simply provided an elegant means by which to describe orbital paths obtained from experimental data. Roughly fifty years later, Newton showed that Kepler's laws are a direct consequence of motion under the influence of gravity.

KEPLER'S FIRST LAW (LAW OF ORBITS, 1609)

All of the planets in our Solar System move in elliptical orbits, with the Sun at one focus. Written mathematically, we may express this as

$$\boxed{R_{\text{aphelion}} = a(1+e),\ R_{\text{perihelion}} = a(1-e)}$$

where R_{aphelion} is the distance between the Sun and the most distant point of the elliptical orbit of the planet (the "aphelion"); $R_{\text{perihelion}}$ is the distance between the Sun and the closest point of the elliptical orbit (the "perihelion"); a is the length of the ellipse's semimajor axis, and e is the eccentricity of the ellipse. *Eccentricity* is a measure of the elongation of the ellipse and has a value of zero for a circle. Pluto, which was designated a "dwarf planet" in 2006, has an orbit with $e = 0.25$. The orbit of Venus, on the other hand, is very close to circular with $e = 0.0068$. The orbit of Earth has an eccentricity of $e = 0.017$. Today, we understand that the elliptical shape of the orbit arises from the inverse-square principle of Newton's Law of Universal Gravitation. Note that the Sun is at one focus of the ellipse, but no object resides at the other focus of the elliptical orbit.

We may also express the law in terms of the gravitational constant G ($G = 6.67 \times 10^{-11}$ N·m^2/kg^2) and the mass of the Sun M:

$$\boxed{r = \frac{l^2/GM}{1 + e\cos\theta}}$$

As in the previous formulation, e is the eccentricity of the ellipse. l is the satellite's specific angular momentum, that is, the relative angular momentum per unit mass. (Angular momentum increases with the angular velocity of rotating body.) The variable r is equal to the distance between the planet and the focus that is located near the center of the Sun. The angle θ sweeps from 0 to 360 degrees.

In 1610, only a year after Kepler published his First Law, Galileo discovered several satellites of Jupiter. These bodies also follow Kepler's First Law. In 1687, Newton showed that any satellite that orbits another body must move along a path described by a conic section (e.g., ellipse, parabola, or hyperbola), if one body, such as the Sun, attracts the satellite with a force that varies inversely as the square of the distance between the bodies. A comet with sufficiently high energy may have a parabolic or hyperbolic orbit and leave the solar system forever after it has entered. If the speed of Earth were suddenly increased by approximately a factor of 1.4, the orbital shape would change into a parabola, and Earth would leave the Solar System.

KEPLER'S SECOND LAW (LAW OF EQUAL AREAS, 1618)

Kepler observed that when a planet is far from the Sun, the planet moves more slowly than when it is close to the Sun. Thus, when far from the Sun, the planet travels a shorter path along the orbit in a given time than it does when close to the Sun. As the planet travels toward the Sun, it accelerates due to the gravity of the Sun. An imaginary line that connects a planet to the Sun sweeps out equal areas in equal intervals of time.

In order to visualize these equal areas, imagine that Earth takes one day to travel from points A to B on an ellipse. If we draw a line from A to the Sun and from B to the Sun, we create a sector of an ellipse (like cutting a wedge from an apple pie). The area of these sectors will be the same each day, regardless of where the planet is in its orbit. Of course, in order for the area being swept to remain constant, for a given interval of time, a planet must change its speed. (Imagine cars whizzing around a racetrack with the driver always going faster near one end of the track than the other.)

As was the case with Kepler's First Law, Newton understood the underpinnings of this Second Law and realized that the law is a consequence of the conservation of angular momentum.

Given Kepler's first two laws, he no longer needed more complicated shapes in order to explain orbital paths, such as the "epicycloids" that prior astronomers had used. Planetary orbits and positions could now be easily calculated and with an accuracy that matched observations.

KEPLER'S THIRD LAW (LAW OF PERIODS, 1618)

For any planet, the square of the period of its revolution about the Sun is proportional to the cube of the semimajor axis of its elliptical orbit. This law is sometimes known as the Harmonic Law. Today, given our knowledge of Newton's Law of Universal Gravitation, Kepler's Third Law is sometimes expressed as:

$$T^2 = \left[\frac{4\pi^2}{G(M+m)}\right] a^3$$

where T is the period of the orbiting body (the time required to complete one orbit), a is the length of semimajor axis of the orbit, G is the gravitational constant, M is the mass of the Sun, and m is the mass of a planet. For the example of our Solar System, m is so much smaller than M that it can be removed from this expression for many practical calculations. Thus, the quantity in brackets may be considered a constant for our Sun, such that T^2/a^3 has essentially the same value for all the planets of the Solar System, namely, 3.00×10^{-19}. When Kepler formulated his laws, the masses of the relevant objects were not in the equations.

In high-level terms, Kepler found that the square of the planet's "year" is a multiple of the cube of the planet's distance from the Sun. Thus, planets far from the Sun have very long years. For example, Pluto, the dwarf planet, has a year of 90,410 days, but Mercury, which is close to the Sun, has a year that is 88 days long. For both of these measurements, the "days" refer to the durations of Earth days.

KEPLER'S LAWS (GENERAL)

Kepler's laws were initially derived for planetary orbits around the Sun, but they have more general applications and later helped Newton formulate the Laws of Motion and Universal Gravitation, which depended strongly on Kepler's work. The laws also apply to artificial satellites, such as those launched by NASA. In the cases of modern satellites, Earth replaces the Sun as the main body around which the satellite orbits. Of course, if a satellite is orbiting too closely to our atmosphere, it will eventually heat up and be destroyed as its kinetic energy is converted to heat.

None of the three laws precisely describes the motions of orbiting planets and moons because the laws neglect any other gravitational interactions that may exist, such as the forces of attraction among planets. For example, when the Moon orbits Earth, it is affected by the gravitation of both Earth and the Sun. In the case of planetary orbits, because the Sun is so much more massive than the planets, the gravitational interactions between planets are very small and can be neglected for most purposes. However, when studying other systems, such as orbits that involve double stars (binary stars), masses of both stars must be considered.

Additionally, as I discuss in the entry on Newton's laws, we know that a planet does not orbit around a stationary Sun as Kepler believed. Instead, both the planet and the Sun orbit around the common center of mass located between the planet and the Sun, and sometimes this requires scientists to modify Kepler's Third Law to make it more accurate.

Kepler's Laws are among the earliest scientific laws to be established by humans, and his statements of the laws provided a stimulus to subsequent scientists who attempted to express the behavior of reality in terms of simple formulas. Although the laws are slightly modified by Einstein's General Theory of Relativity, Kepler's laws, along with Newton's Law of Universal Gravitation, provide the underpinning to practical celestial mechanics. Kepler believed that simple rules could be used to describe the motions of planets, and with his three simple laws, he summarized thousands of years of planetary observations.

Johannes Kepler (1571–1630), German astronomer and theologian-cosmologist, famous for his laws that describe the elliptical orbits of Earth and other planets around the Sun.

CURIOSITY FILE: Kepler never numbered his three laws. • He was fascinated by astrology and suggested that the sudden appearance in 1604 of a star, which we now call a "nova," was God's way to encourage the conversion of Native Americans to Christianity. He urged all sinners to repent. • By postulating the ray theory of light and showing how images are formed on the retina, Kepler is considered by some to be the founder of modern optics. Kepler was also first to explain why both eyes were required for depth perception. • Kepler wrote a story called "The Dream," which is one of the oldest examples of a "modern" science-fiction story.

> I wanted to become a theologian. For a long time I was restless. Now, however, behold how through my effort, God is being celebrated in astronomy.
>
> —Johannes Kepler, 1595 letter to German astronomer and mathematician Michael Maestlin

> Although Kepler is remembered today chiefly for his three laws of planetary motion, these were but three elements in his much broader search for cosmic harmonies.... He left [astronomy] with a unified and physically motivated heliocentric system nearly 100 times more accurate.
>
> —Owen Gingerich, "Kepler," in *Dictionary of Scientific Biography*

> It should no longer seem strange that man, the ape of his Creator, has finally discovered how to sing polyphonically, an art unknown to the ancients. With this symphony of voices, man can play through the eternity of time in less than an hour and can taste in small measure the delight of God the Supreme Artist by calling forth that very sweet pleasure of the music that imitates God.
>
> —Johannes Kepler, *Harmonice mundi* (Harmony of the World), 1619

Johannes Kepler was born in Weil der Stadt, a town now part of Germany. Max Caspar, author of *Kepler*, provides a vivid glimpse of the appearance of the city during Kepler's life: "The little streets, the spacious market place surrounded by high gabled houses, the towers and gates of the city wall... [are all situated] in a rolling landscape on the edge of the Black Forest, surrounded by gardens and meadows, fields and woods...." Kepler's staunch adherence to Lutheranism, and refusal to convert to Catholicism, forced him to relocate many times during his life, sometimes sacrificing his career and safety.

From what Kepler wrote, we know that his mother and father were poor parents. He described his father Heinrich as "criminally inclined." When Kepler was three years old, Heinrich became a mercenary, and in

1588, Heinrich abandoned young Kepler forever. Kepler called his mother Katharina "bad-tempered," but later in his life he came to her aid when she was about to be tortured: Katharina was tried for witchcraft in 1617, and the adult Kepler spent months helping to prepare a legal defense, which was eventually successful. During the ordeal, Katharina's ruthless questioners showed her torture devices in an attempt to make her confess that she was a witch. James A. Connor, author of *Kepler's Witch*, describes Kepler's state of mind during this difficult time:

> The executioner showed Katharina Kepler the instruments of torture, the pricking needles, the rack, the branding irons. Her son Johannes Kepler was nearby, fuming, praying for it to be over. He was forty-nine and, with Galileo Galilei, one of the greatest astronomers of the age—the emperor's mathematician, the genius who had calculated the true orbits of the planets and revealed the laws of optics to the world.... He never gave up trying [to stop the interrogation], and in that he was a good deal like his mother.

Katharina was finally freed in 1621, partly because of her steadfast refusal to confess under threat of torture.

As a child, Kepler was frequently bullied, and he believed himself to be ugly. Most if not all of his brothers and sisters suffered from severe mental and physical handicaps. Kepler himself was bow-legged, often afflicted with large boils, and suffered from poor vision.

Kepler studied astronomy at the University of Tübingen, where he was virtually a "straight A" student. In 1591, he received his master's degree, and his plan was to study theology and become a clergyman. However, the local authorities in Graz, Austria, were able to coax Kepler to teach mathematics at a Lutheran high school, and he soon moved to Austria, where he taught mathematics and the works of Virgil. While in Graz, he spent some of his time making futuristic predictions that included topics ranging from the local weather to politics. His predictions often turned out to be correct, which elevated his status in the eyes of the townspeople.

Although Kepler appeared to be quite interested in astrology, he also could be skeptical, having written in *De fundamentis astrologiae certioribus* (1601), "If astrologers sometimes do tell the truth, it ought to be attributed to luck." In 1610, he seemed to both praise and condemn astrology in *Tertius interveniens*: "No one should consider unbelievable that here should come out of astrological foolishness and godlessness also cleverness and holiness... out of evil-smelling dung, a golden corn scraped for by an industrious hen."

Throughout his life, he attributed his scientific ideas and motivation to his quest for understanding the mind of God. For example, in his work *Mysterium cosmographicum* (The Sacred Mystery of the Cosmos, 1596), he wrote, "I believe that Divine Providence intervened so that by chance I found what I could never obtain by my own efforts. I believe this all the more because I have constantly prayed to God that I might succeed."

Kepler's initial vision of the universe rested upon his studies of symmetrical, three-dimensional objects known as Platonic solids. Centuries before Kepler, the Greek mathematician Euclid (325–265 B.C.) showed that only five such solids exist with identical faces: the cube, dodecahedron, icosahedron, octahedron, and tetrahedron. Although Kepler's theory in the 1500s seems strange to us today, he attempted to show that the distances from the planets to the Sun could be found by studying spheres inside these regular polyhedra, which he drew nested in one another like layers of an onion. For example, the small orbit of Mercury is represented by the innermost sphere in his models. The other planets known at his time were Venus, Earth, Mars, Jupiter, and Saturn. Here is how Kepler explained his Platonic-solid model of the Solar System in *Mysterium cosmographicum*:

> We must first eliminate the irregular solids, because we are only concerned with orderly creation. There remain six bodies, the sphere and the five regular polyhedra. To the sphere corresponds the heaven. On the other hand, the dynamic world is represented by the flat-faces solids. Of these there are five: when viewed as boundaries, however, these five determine six distinct things: hence the six planets that revolve about the sun. This is also the reason why there are but six planets....
>
> I have further shown that the regular solids fall into two groups: three in one, and two in the other. To the larger group belongs, first of all, the Cube, then the Pyramid, and finally the Dodecahedron. To the second group belongs, first, the Octahedron, and second, the Icosahedron. That is why the most important portion of the universe, the Earth—where God's image is reflected in man—separates the two groups. For, as I have proved next, the solids of the first group must lie beyond the Earth's orbit, and those of the second group within....Thus, I was led to assign the Cube to Saturn, the Tetrahedron to Jupiter, the Dodecahedron to Mars, the Icosahedron to Venus, and the Octahedron to Mercury....

To explain his theory, Kepler published a diagram showing spheres within smaller and smaller Platonic solids. An outer sphere surrounds a cube. Inside the cube is a sphere, followed by a tetrahedron, followed by another

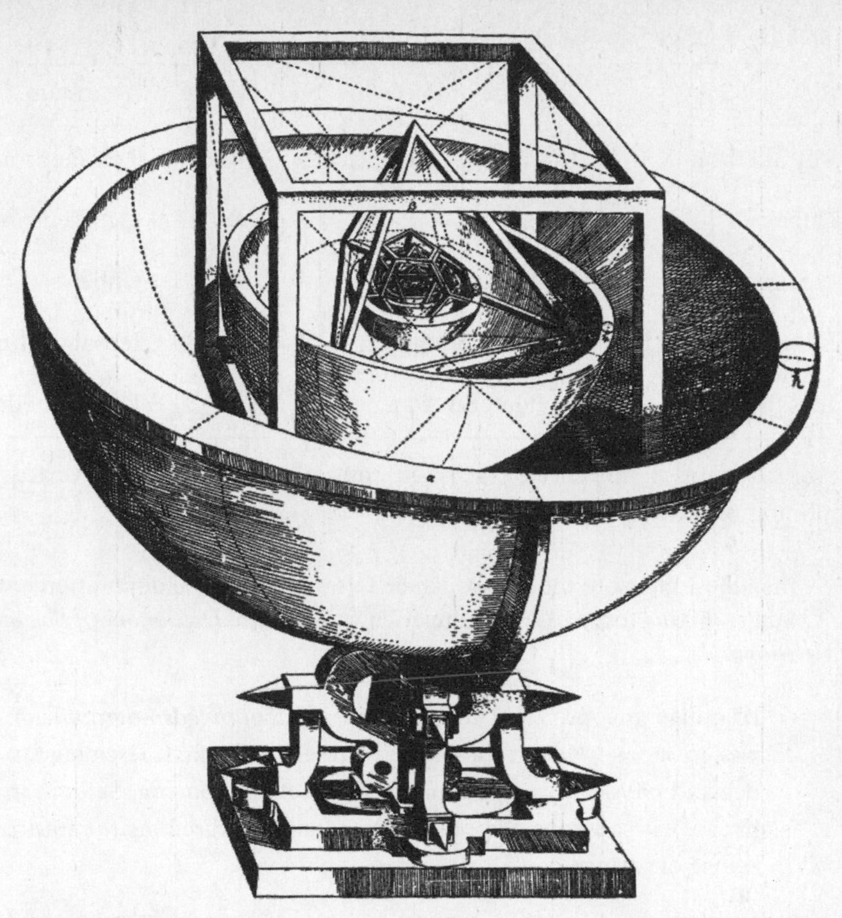

Johannes Kepler's Platonic solid model of the solar system, published in *Mysterium cosmographicum* (1596).

sphere, followed by a dodecahedron, followed by a sphere, an icosahedron, sphere, and finally a small inner octahedron. A planet may be imagined as being embedded in each sphere that defines an orbit of a planet.

As an example, the regular dodecahedron is situated between Mars and Earth, the regular icosahedron between Earth and Venus, and the regular octahedron between Venus and Mercury. For Kepler, the spheres explained the spacing of the planets. With a few subtle compromises, Kepler's scheme worked fairly well as a rough approximation to what was known about planetary orbits at the time.

Kepler supplemented this initial work with formulas that expressed relationships between planetary periods and their distances from the Sun. He understood that planets more distant from the Sun had longer periods

TABLE 3 Platonic Solids

Platonic Solid	F	Shape of Faces	NF	V	E	Dual
Tetrahedron	4	Equilateral triangle	3	4	6	Tetrahedron
Cube	6	Square	3	8	12	Octahedron
Octahedron	8	Equilateral triangle	4	6	12	Cube
Dodecahedron	12	Regular Pentagon	3	20	30	Icosahedron
Icosahedron	20	Equilateral triangle	5	12	30	Dodecahedron

Abbreviations: F, number of faces; NF, number of faces at each vertex; V, number of vertices; E, number of edges.

than those closer to the sun, which he felt was due to the diminution of the Sun's "driving force." Owen Gingerich writes in the *Dictionary of Scientific Biography*,

> Although the principal idea of the *Mysterium cosmographicum* was erroneous, Kepler established himself as the first . . . scientist to demand physical explanations for celestial phenomena. Seldom in history has so wrong a book been so seminal in directing the future course of science.

For those readers interested in the precise properties of Platonic solids, table 3 lists the name of each Platonic solid, the number of faces (F), face shape, the number of faces at each vertex (NF), number of vertices (V), number of edges (E), and the "dual," that is, the Platonic solid that can be inscribed within the outer solid by connecting the midpoints of the faces.

The Platonic solids were described by Plato in his *Timaeus*, circa 350 B.C. Pythagoras of Samos (582–507 B.C.), the famous Greek mathematician and mystic, who lived in the time of Buddha and Confucius, was aware of three of the five regular polyhedra (the cube, tetrahedron, and dodecahedron). The shapes of some of the Platonic solids have been discovered carved into ancient stones in Scotland and dated to approximately 2000 B.C.

In *Mysterium cosmographicum*, Kepler resurrected the ancient Greek idea that the heavens were ruled by simple, geometrical laws. In 1597, Kepler sent *Mysterium cosmographicum* to Galileo, but Galileo replied that he read only the Preface. He also sent a copy to Tycho Brahe, who said that Kepler's notion of nesting polyhedrons was clever speculation.

Let's return to a discussion of Kepler's personal life. In 1596, he married Barbara Müller, twice widowed before they had met. Gingerich says:

> The initial happiness of the marriage gradually dissolved as he realized that his wife understood nothing of his work—"fat, confused, and simple-minded" was Kepler's later description of her. The early death of his first two children grieved him deeply.

In 1600, Kepler started to work for Brahe as part of the research staff in Brahe's castle observatory near Prague. Here, Kepler studied the orbit of Mars. When Brahe died in 1601, Kepler was appointed to Brahe's position of imperial mathematician. Kepler continued to wonder how the Sun could exert an influence on the planets, and he came to think of the Sun as somehow producing a magnetic "emanation" that drove the planets in their orbits. For Kepler, the universe was like a clockwork in which all of the motions arose from a magnetic force, "just as in a clock, all motions arise from a very simple weight." Over the next few years, Kepler arrived at his famous three laws of planetary motion, never understanding the nature of gravity that Newton would further elucidate years later.

In 1605, Kepler published an article on a new star (a nova) that appeared in the sky. In his 1606 book *De stella nova*, he speculated on the astrological significance of the star, which he believed had appeared in order to (1) convert the Indians of America, (2) herald the return of Jesus, or (3) mark the downfall of Islam. In 1609, he published *Astronomia nova*, in which he described his theories on planetary orbits, which are now called Kepler's first two laws of planetary motion.

In 1611, his wife Barbara Müller died from typhus, and two years later, Kepler married the 24-year-old Susanna Reuttinger. Being a famous astronomer must have been quite an allure for women in Kepler's time. Gingerich writes of Kepler's letter to a nobleman:

> Kepler details his slate of eleven candidates for marriage and explains how God had led him back to [woman] number five, who had evidently been considered beneath him by his family and friends. The marriage was successful, far happier than the first; but of their seven children, five died in infancy or childhood.

In his five-part book *Harmonice mundi* (1618), Kepler established his grand cosmic vision that involved music (e.g., musical harmonies and intervals), geometry (e.g., polygons and polyhedrons), astronomy (e.g., Kepler's Third Law), and astrology (e.g., the position of planets had the power to influence the soul). In *Harmonice mundi*, he likened the ocean tides to the breathing of a gigantic organism. So excited was Kepler by his book that he said in *Harmonice mundi*, "I yield freely to the sacred frenzy." He wrote that the book was "to be read now or by posterity, it matters not. It can wait a century for a reader, as God himself has waited six thousand years for a witness."

In 1627, he published *Tabulae rudolphinae*, which contained tables of astronomical data that were used by astronomers for more than a hundred years. The book also included Brahe's catalogue of hundreds of star positions.

Kepler's science-fiction work was *Somnium*, which described a fantasy journey of a young Icelander named Duracotus to the Moon. Kepler had actually written the piece in 1609 and sent copies of the manuscript to colleagues. Because the book contained conversations with a demon, the witch-hunters made this known at his mother's witchcraft trial. In 1634, four years after Kepler's death, his son Ludwig Kepler finally published *Somnium*.

In *Somnium*, the seventeenth-century astronauts are made more comfortable during the process of liftoff from Earth by being put to sleep with opiates. Their limbs are arranged so as to minimize physical stress of acceleration. The inhabitants of the moon resemble large serpents with a spongy, porous skin. *Somnium* does not contain mathematics, but Kepler provided many notes that contained astronomical calculations to accompany the fiction.

Kepler died in Regensburg in 1630. His grave was destroyed two years later because of the Thirty Years War. Today, nothing remains of the tomb, but scientists have honored him by naming a lunar crater after him. The crater has a diameter of 31 kilometers, and its name was approved in 1935 by the International Astronomical Union General Assembly. A crater on Mars has also been named after Kepler.

Today we remember Kepler not only for his Three Laws of Planetary Motion but also for a number of famous equations. For example, Kepler's Orbital Equation can be expressed as $M = E - e \times \sin E$, where e is the eccentricity of an elliptical planetary orbit, M the mean angular motion about the sun (sometimes called the "mean anomaly"), and E is the auxiliary angle (sometimes called the "eccentric anomaly"). Kepler's equation gives the relation between the polar coordinates of a planet and the time elapsed from a given initial point. Kepler's Log Equation, related to logarithms that are used today, can be expressed in modern notation as

$$\log_{\text{Kepler}}(x) = \lim_{n\to\infty} 2^n \left[1 - \left(\frac{x}{10^5}\right)^{\frac{1}{2^n}}\right] \cdot 10^5$$

In *Matter in Motion*, Ernest Abers and Charles F. Kennel suggest that Kepler could not have achieved his scientific accomplishments without reliance on experimental data obtained by others; nevertheless, Kepler's laws had a profound effect on science and reinforced the growing belief that mathematics could be used to explain a vast yet orderly universe:

> One man had the imagination and extraordinary patience to extract from Tycho Brahe's mountain of observations a truly simple picture of the planets' motions. . . . Kepler worked with whatever portions of Brahe's data he could cajole out of the Imperial Mathematician. . . . Kepler would have remained merely an eccentric genius had his flights of speculative fancy not been confronted by Tycho's hard facts. From their brief and tempestuous encounter came a new cosmic order.

FURTHER READING

Abers, Ernest, and Charles F. Kennel, *Matter in Motion* (Boston, Massachusetts: Allyn & Bacon, 1977).

Caspar, Max, *Kepler* (New York: Dover, 1993).

Connor, James, *Kepler's Witch: An Astronomer's Discovery of Cosmic Order amid Religious War, Political Intrigue, and the Heresy Trial of His Mother* (San Francisco, California: HarperSanFrancisco, 2005).

Gingerich, Owen, "Kepler," in *Dictionary of Scientific Biography*, Charles Gillispie, editor-in-chief (New York: Charles Scribner's Sons, 1970).

Stephenson, Bruce, *Kepler's Physical Astronomy* (Princeton, New Jersey: Princeton University Press, 1994).

INTERLUDE: CONVERSATION STARTERS

> The Christians know that the mathematical principles, according to which the corporeal world was to be created, are coeternal with God. Geometry has supplied God with the models for the creation of the world. Within the image of God it has passed into man, and was certainly not received within through the eyes.
>
> —Johannes Kepler, *Harmonice mundi* (Harmony of the World), 1619

> The search [for physical laws and particles may] be over for now, placed on hold for the next civilization with the temerity to believe that people, pawns in the ultimate chess game, are smart enough to figure out the rules.
>
> —George Johnson, "Why Is Fundamental Physics So Messy?" *WIRED* magazine, February, 2007

Galileo loved a fight, and he took to calling his opponents "mental pygmies" and "hardly deserving to be called human beings." Two professors at his university hadn't even deigned to peer through his telescope. When one of them died a little later, Galileo wrote that he "did not choose to see my celestial trifles while he was on earth; perhaps he would do so now that he has gone to heaven."

—James C. Davis, *The Human Story*

Galileo refused to give Kepler one of his telescopes, although he gave them to many political heads of the world.... Galileo wrote his discoveries to Kepler only in anagrams, so that Kepler could not understand them, but Galileo later could prove that these were his discoveries. After this, Galileo broke off all further contact with Kepler. He totally ignored Kepler's famous book *Astronomia Nova* with the vital proposal of elliptical orbits....

—Thomas Schirrmacher, "The Galileo Affair: History or Heroic Hagiography?" *Technical Journal*, April, 2000

The system which Galileo advocated was the orthodox Copernican system ... nearly a century before Kepler threw out the epicycles.... Incapable of acknowledging that any of his contemporaries had a share in the progress of astronomy, Galileo blindly and indeed suicidally ignored Kepler's work to the end, persisting in the futile attempt to bludgeon the world into accepting a Ferris wheel with forty-eight epicycles as "rigorously demonstrated" physical reality.

—Arthur Koestler, *The Sleepwalkers: A History of Man's Changing Vision of the Universe*

SNELL'S LAW OF REFRACTION

Netherlands, 1621. The angle of refraction of light that travels between two media depends on the refractive indices of the media and is described quantitatively by Snell's Law.

CROSS REFERENCE: JOHANNES KEPLER AND MAXWELL'S EQUATIONS.

> During the year that Snell discovered his law, the English attempted to colonize Nova Scotia and Newfoundland, and potatoes were planted for the first time in Germany. The *Mayflower* sailed from Plymouth colony in North America on a return trip to England.

Light traveling through air bends, or *refracts*, when it passes into another material such as glass. When waves such as light waves are refracted, they experience a change in the direction of propagation due to a change in their velocities. Refraction typically occurs when a ray of light passes from one medium to another, and every known material slows light relative to its speed in a vacuum. In particular, the refraction of light occurs at the boundary between the media (e.g., between air and water), at which point the phase velocity of the wave changes, and the light changes direction of travel. Additionally, the wavelength of light changes at the interface between media, but the frequency of the light remains constant.

To understand the concept of *phase velocity*, imagine a sinusoidal wave made of a piece of wood and sliding to the right. The phase velocity is simply the ordinary speed with which the wooden wave is moving. Now imagine a wave in a pond, in which a leaf on the surface is oscillating vertically as the wave passes. In this case, the wave pattern moves to the right with phase velocity v_p, just as with the wooden wave, but the leaf may have no lateral motion at all.

I like to demonstrate refraction to young people by placing my finger into my aquarium filled with large fish. Because air has a refractive index of 1.0003, and water has a refractive index of 1.33, when my guests look at my straight finger that is partially submerged in the water, the finger appears to bend abruptly at the surface of the water. Before the fish bite my finger, I explain to my guests that this apparent bending is due to the bending of light rays as they move from the water to the air. I then scribble Snell's Law on a napkin,

$$\boxed{n_1 \sin(\theta_1) = n_2 \sin(\theta_2),}$$

and explain how the law is used to calculate the degree to which light is refracted when traveling from air to water. (You can demonstrate this more safely with a pencil in a glass of water.) Here, n_1 and n_2 are the refractive indices of media 1 and 2. The angle between the incident light and a line perpendicular to the interface is called the *angle of incidence*, θ_1. The light ray continues from medium 1 into medium 2 and leaves the boundary between the media with an angle θ_2 to a line that is perpendicular to the boundary. This second angle is known as the *angle of refraction*.

Recall that refraction refers to the bending of a wave when it enters a medium where its speed changes. When light passes from a "fast medium" to a "slow medium," refraction bends the light ray toward the *normal* (the imaginary perpendicular line) to the boundary between the two media. The amount of bending depends on n_1 and n_2 and is described quantitatively by Snell's Law.

The refractive index n of water, glass, air, or any material is the factor by which the phase velocity of electromagnetic radiation is slowed in that material, relative to its velocity in a vacuum. n depends on the wavelength of radiation under study. Some example values for n at a wavelength of 589.3 nm are vacuum, 1; air, 1.00029; liquid water, 1.333; glass, 1.5–1.9; and diamond, 2.419.

The law of reflection, which in simplified form states that the angle of reflection of a light ray from a surface is equal to the angle of incidence, and the law of refraction can be derived from James Clerk Maxwell's equations of electricity and magnetism and generally hold for a wide region of the electromagnetic spectrum. The laws can also be derived from simpler theories of light, such as those discussed by the Dutch physicist Christiaan Huygens (1629–1695) in 1678. Huygens used geometrical constructions to indicate where a given wavefront will be at any time.

Refraction has numerous applications today. For example, a convex lens makes use of refraction to cause parallel light rays to converge. Without the refraction of light by the lenses in our eyes, we could not see properly, and no traditional lenses for cameras would exist. Seismic waves—for example, the waves of energy caused by the sudden breaking of subterranean rock—change speed within Earth and are bent when they encounter interfaces between materials in accordance with Snell's Law. Geologists can investigate the layers within Earth by studying the behavior of refracted and reflected waves.

When a beam of light is transmitted from a material with high index of refraction to one of low index, the beam can, under some conditions, be totally reflected. This optical phenomenon is usually called *total internal reflection*, and it occurs when light is refracted at a medium boundary to such an extent that it is reflected back. Such would be the case in certain kinds of optical tubes in which the light enters at one end remains trapped inside until it emerges from the other end.

You can use Snell's Law to understand and calculate the condition necessary for total internal reflection. First, set the refracted angle θ_2 to 90°, and then calculate the incident angle. Because we cannot refract the light by more than 90°, under this condition, all of the light will reflect for angles of incidence greater than the angle that gives refraction at 90°.

The properties of a cut diamond provide an additional example of total internal reflection. Beams of light are reflected within a properly shaped diamond crystal so that the diamond sparkles and often emits light in the direction of the observer's eye. In other words, the cut of the diamond makes use of total internal reflections so that a large portion of the rays entering the diamond will internally reflect within the diamond until they leave at specific upper faces in order to give diamond its bright sparkle. A light ray will often undergo total internal reflection several times before finally refracting out of the diamond.

Total internal reflections makes it possible to confine light so that it travels along a fiber made of the appropriate material, and thus the fiber can pipe light around corners. For example, an optical fiber may resemble a glass "hair" that is so thin that once light enters one end, it can never strike the inside walls at less than the critical angle that causes total internal reflection. The light undergoes total internal reflection each time it strikes the surface of the fiber and finally exits at the end of the fiber. Fiber optic cables are used to transmit telephone and computer signals and have many advantages over traditional electrical wires. For example, the optical fibers can carry more information in a smaller cable faster than wires can, and the fibers are not sensitive to stray electromagnetic fields in the vicinity of the fibers. Optical fibers have facilitated an explosion in worldwide communications during the last twenty years and enabled the proliferating use of the Internet.

Fiber optics has also been used in medicine as a way to allow physicians to look inside the body with minimal invasion. Some endoscopes have made use of two fiber optic lines. An "image fiber" is surrounded by "light fibers" that carry light down to the end to illuminate the tissue of interest. In other words, two separate fiber bundles exist in certain flexible endoscopes, one for viewing and one for light transmission.

Total internal reflection also plays a role in mirages, such as those that may appear above asphalt or deserts on a hot day. In particular, an observer can sometimes see inverted images of the landscape and nearby trees as if they were reflected in a pool of water. Air close to the ground is hotter than the air farther away and has a lower refractive index than the lower temperature air. Light traveling downward from any point on an object may be refracted away from the normal, and as it passes near the ground it is totally reflected at a layer of lower refractive index. The refracted light enters the observer's eye and can give the appearance of an image that is below the surface of the ground.

Until the year 2001, all known materials had a positive index of refraction. However, in 2001, scientists from the University of California at San Diego described an unusual composite material that had a negative index, essentially reversing Snell's Law. This odd material is a mix of fiberglass, copper rings, and wires that is capable of focusing light in novel ways. Early tests revealed, for example, that microwaves emerged from the material in the exact opposite direction from that predicted by Snell's Law. Physicists Sheldon Schultz, David R. Smith, and Richard A. Shelby hypothesize that the new material is more than a physical curiosity because it may one day lead to the development of novel antennas and other electromagnetic devices. In theory, a sheet of negative-index material could act as a superlens to create images of exceptional detail.

Although most early experiments with these kinds of exotic materials were performed with microwaves, in 2007 a team led by Henri Lezec of the California Institute of Technology in Pasadena achieved negative refraction for visible light. In order to create an object that acted as if it were made of negatively refracting material, Lezec's team built a prism of layered metals perforated by a maze of nanoscale channels.

This was the first time that physicists had devised a way to make visible light travel in a direction opposite from the way it traditionally bends when passing from one material to another. Some physicists suggest that the phenomenon may someday lead to optical microscopes for imaging objects as small as molecules and for creating cloaking devices that render objects invisible.

In 2005, Akhlesh Lakhtakia of Pennsylvania State University and Tom Mackay of the University of Edinburgh determined that negative refraction around a rotating black hole can change the apparent location of stars as viewed from Earth—at least in theory. In other words, a traditional material like glass or water is not required to cause refraction; for example, the space around such a black hole can have refractive properties and can have a negative index of refraction that causes light to refract in a direction opposite from traditional materials.

In 2007, researchers created a "superblack" surface that was virtually free of reflections. Ordinarily, light reflects from a surface when it strikes the boundary between two materials that have different refractive indices. The greater the difference between the refractive indices of two materials, the more light is reflected. To prevent such reflections in their exotic material, a research team at Rensselaer Polytechnic

Institute in Troy, New York, created a multilayer surface composed of nanoscale filaments. The topmost surface layer had a refractive index of 1.05, which is close to air's index of 1.0. The researchers' especially low-reflection coating works for wavelengths between near ultraviolet and near infrared.

Willebrord Snell, also known as Snellius, Snell, Snel van Royen, or Willebrord van Snel van Royen (1580–1626), Dutch astronomer and mathematician famous for his Law of Refraction.

CURIOSITY FILE: Snell's biggest contribution to science, Snell's Law of Refraction, was not published until almost seventy years after he died. • Sophisticated fishermen use the term "Snell's Window" when referring to angles above a lake that are within a fish's vision. Fishermen avoid being seen by fish by staying outside this window, the dimensions of which are controlled by Snell's Law. • Raindrops create rainbows via refraction. Because various colors of light have different wavelengths, the colors refract differently when they pass through the water. A rainbow is actually a circle centered on a point directly opposite the sun from the observer; however, the observer does not see the full circle because the landscape gets in the way.

> Snell's law may be derived from Fermat's principle, which states that the light travels the path which takes the least time.... In a classic analogy by Feynman, the area of lower refractive index is replaced by a beach, the area of higher refractive index by the sea, and the fastest way for a rescuer on the beach to get to a drowning person in the sea is to run along a path that follows Snell's law.
>
> —"Snell's Law," *Wikipedia*

> In 1621, a really smart guy with a really funny name (Willebrord Snell) figured out, through careful experimentation, that the angles between the surface normal and the original and bent light are related mathematically.
>
> —Mason McCuskey, *Special Effects Game Programming with DirectX*

Willebrord Snell was born in Leiden in the Netherlands. His father was a mathematics professor. Although Snell studied law, he was fascinated by mathematics and in 1600 began teaching mathematics at the University of

Leiden. Shortly thereafter, he met such famous contemporaries as Tycho Brahe and Johannes Kepler.

In 1602, he studied law in Paris, and a few years later he produced his first major work related to science—a Latin translation of Simon Stevin's (1548–1620) *Wisconstighe Ghedachtenissen* (Mathematical Memoirs, 1605–1608). Stevin was a Flemish mathematician and engineer whose memoirs contained his treatises on mathematics, mechanics, the theory of music, bookkeeping, optics, astronomy, and geography. In 1608, Snell's translation was titled *Hypomnemata mathematica* (Mathematical Memoranda).

In 1608, Snell married, and over the course of his life had eighteen children, only three of whom survived to adulthood. In 1613, Snell succeeded his father as professor of mathematics at the University of Leiden. In 1615, he investigated a new method of finding the radius of Earth, which involved a triangulation method. His work *Eratosthenes Batavus* (The Dutch Eratosthenes), published in 1617, describes his triangulation approach that made use of his own house and two local towns, and the distances between them, to aid in his computations. The value he obtained for Earth's circumference was 38,500 kilometers, which is relatively close to the actual figure of 40,000 kilometers. He continued to improve upon his calculations with the help of his students, but his early death in 1626 meant that his additional calculations were never formally published.

In 1619, he published papers on comets. Two years later, in *Cyclometricus*, Snell reported his discovery of a new method for calculating π by using polygons. In particular, he calculated π to thirty-four decimal places by imagining a polygon with 1,073,741,824 sides, a method that had been previously used by German mathematician Ludolph Van Ceulen (1540–1610) but that had never been published. Snell improved upon the traditional methods of calculating approximations of π by using polygons, so that, for example, he could use 96-sided polygons to give the digits of π correct to seven places while the classical method yielded π only to two places. [For the mathematically inclined reader, Snell used $\pi \sim (2/3)n\sin(\pi/n) + (1/3)n\tan(\pi/n)$ to estimate π, where n is the number of sides of a polygon that circumscribes a circle.]

In 1624, Snell became an expert on navigation. He investigated a mathematical curve called the *loxodrome*, a path on the sphere that makes a constant angle with the meridians. (A *meridian* corresponds to a great circle on the surface of Earth that passes through the north and south geographic poles.) His work on loxodromes and navigation is published in *Tiphys batavus*. The loxodrome has the shape of a spherical spiral and is a path taken while traveling when a magnetic compass needle is kept pointing in a constant direction.

Today, Snell is most famous for his research on the refraction of light. Interest in refraction started centuries earlier. For example, Ptolemy (85–165 A.D.), the most influential of Greek astronomers and geographers of his time, constructed tables of angle of refraction and the corresponding angle of incidence for various media. He estimated that the ratio of the two angles was a constant for the two media forming an interface, or, in other words, $\theta_1/\theta_2 = k_{12}$. Johannes Kepler also made numerous measurements of angles of incidence and corresponding angles of refraction for various media interfaces, but he was unable to find the precise relationship between the angles he measured. In 1621, Snell, who was then a professor of mathematics at Leiden, found that Ptolemy's simple equation was inaccurate, and he soon discovered the correct law through his experiments. Unlike Ptolemy's formulation, Snell's Law employs the ratio of the *sines* of the angles instead of the angles themselves. Snell's work was circulated privately in manuscript form and was not published.

According to popular physics text books, such as Arnold Aron's *Development of Concepts of Physics*, the law apparently came to the attention of both René Descartes and Huygens, and Descartes published the relation in modern form in 1637. Many believe that Descartes independently derived the law, and in France, Snell's Law is referred to as Descartes's Law.

As mentioned in the Introduction of this book, Snell's Law was discovered by various investigators over the centuries. Perhaps the first person to understand the basic relationship expressed by Snell's Law was the Arabian mathematician Ibn Sahl in the year 984. In 1602, English astronomer and mathematician Thomas Harriot also discovered the law, but he did not publish his work. In 1621, Snell discovered the law; his unpublished notes on the subject were discovered by the Dutch scholar and manuscript collector Isaac Vossius around 1662, and Huygens discussed the law in his *Dioptrica*, published in 1703.

Snell, never having formally published his work on the law, died just a few years after the discovery. Descartes was actually the first person to publish the law explicitly in terms of sine functions in his 1637 *Discourse on Method* (which was originally published in Leiden in French together with his work *Dioptrique*). Descartes did not experimentally verify the law. Huygens and others actually accused Descartes of plagiarism, given that Descartes visited Leiden during and after Snell's work, but no evidence exists to support this assertion.

A lunar crater named Snellius with a diameter of 82 kilometers was named after Snell and approved in 1935 by the International Astronomical Union General Assembly.

FURTHER READING

Arons, Arnold, *Development of Concepts of Physics* (Reading, Mass.: Addison-Wesley, 1965).

Barry, Patrick L., "The New Black: A Nanoscale Coating Reflects Almost No Light," *Science News* 171(9): 132, March 3, 2007.

Brooks, Michael, "Illusions of a Starry, Starry Night," *New Scientist*, no. 2502: 30–33, June 4, 2005; contains information on black holes and negative refractive indices.

Hung, Edwin, *Beyond Kuhn: Scientific Explanation, Theory Structure, Incommensurability and Physical Necessity* (Aldershot, Hampshire, U.K.: Ashgate Publishing, 2005).

Leutwyler, Kristin, "New Material Reverses Snell's Law," *Scientific American*, April 9, 2001; see www.sciam.com/article.cfm?articleID=000A27C4-0C1B-1C5E-B882809EC588ED9F.

Lezec, Henri J., Jennifer A. Dionne, and Harry A. Atwater, "Negative Refraction at Visible Frequencies," *Science* (express online publication), March 12, 2007; see www.sciencemag.org/cgi/content/abstract/1139266v1.

Livio, Mario, "Ask the Experts: Why Do Rainbows Form?" *Scientific American* 295(1): 104, July 2006.

Livio, Mario, "The Quest for the Superlens," *Scientific American*, 295(1): 60–70, July 2006.

Mackay, Tom, Akhlesh Lakhtakia, and Sandi Setiawan, "Gravitation and Electromagnetic Wave Propagation with Negative Phase Velocity," *New Journal of Physics* 7(75): 1–14, 2005; see www.iop.org/EJ/abstract/1367-2630/7/1/075/.

Rashed, Roshdi "A Pioneer in Anaclastics: Ibn Sahl on Burning Mirrors and Lenses," *Isis*, 81: 464–491, 1990; discusses Ibn Sahl's possible discovery of Snell's Law.

Struik, Dirk, "Snel," in *Dictionary of Scientific Biography*, Charles Gillispie, editor-in-chief (New York: Charles Scribner's Sons, 1970).

INTERLUDE: CONVERSATION STARTERS

> It seemed that Earth was being taken and remade, not by ETs from another arm of the Milky Way . . . but by beings from another universe where all the laws of nature were radically different from those in this one. Humanity's reality, which operated on Einstein's laws, and the utterly different reality of humanity's dispossessors had collided, meshed at this Einstein intersection. All things seemed possible now in this worst of all possible new worlds.
>
> —Dean Koontz, *The Taking*

> The notion of causation does not enter the equation of Snell's law. There is no conditional dependence, no temporal asymmetry between cause and effect expressed in

this law. The incoming beam does not cause the refraction. It can be asked: "What *causes* the planets to perform Keplerian elliptical orbits"? "What *causes* light beams to follow Snell's Law"? . . . In Kepler's Laws and Snell's Law we find examples of deterministic relation, in classical physics, without any causal component.

—Friedel Weinert, *The Scientist As Philosopher: Philosophical Consequences of Great Scientific Discoveries*

Physicists had to invent words and phrases for concepts far removed from everyday experience. It was their fashion to avoid pure neologisms and instead to evoke, even if feebly, some analogous commonplace. The alternative was to name discoveries and equations after one another. This they did also. But if you didn't know it was physics they were talking, you might very well worry about them.

—Carl Sagan, *Contact*

We may in fact regard geometry as the most ancient branch of physics. Without it I would have been unable to formulate the theory of relativity.

—Albert Einstein, in his address "Geometry and Experience" to the Prussian Academy of Sciences, 1921

HOOKE'S LAW OF ELASTICITY

England, 1660. The size of a material deformation is directly proportional to the deforming force.

CROSS REFERENCE: NEWTON'S LAW OF UNIVERSAL GRAVITATION AND BOYLE'S GAS LAW.

> During the year that Hooke discovered his law, Dutch peasants (Boers) settled in South Africa, and water closets arrived in England from France. The Irish natural philosopher Robert Boyle described his research showing that the removal of air from a chamber extinguishes a flame and kills small animals, which suggested that combustion and respiration may be related processes.

Hooke's Law of Elasticity states that if an object, such as a metal rod or spring, is elongated by some distance, x, the restoring force F exerted by the object is proportional to x:

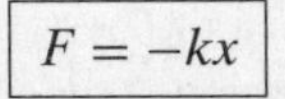

$$F = -kx$$

Here, k is a constant of proportionality that is often referred to as the spring constant when Hooke's Law is applied to springs. Hooke's Law is an approximation that applies for certain materials, such as steel, which are called Hookean materials because they obey Hooke's Law under a significant range of conditions. For other materials, such as aluminum, Hooke's Law has a more restricted use that applies only to a portion of the elastic range of the material. Rubber objects are non-Hookean because of their very complex responses to applied forces. For example, the stiffness of rubber is very sensitive to temperature and the rate at which a force is applied.

Students most often encounter Hooke's Law in their study of springs where the law relates the force F, exerted by the spring, to the distance x that the spring is stretched. The spring constant k is measured in force per length. The negative sign in $F = -kx$ indicates that the force exerted by the spring opposes the direction of displacement. For example, if we were to pull the end of a spring to the right, the spring exerts a "restoring" force to the left. The displacement of the spring refers to its displacement from equilibrium position at $x = 0$.

The spring constant provides an indication of the stiffness of the spring. A large value for k indicates that the spring is stiff, whereas a low value for k means that the spring is loose. As another example, consider a mass hanging from a spring. The initial position of the end of the spring is located at, for example, 0.300 meters. When a 0.200 kilogram mass is added to the end of the spring, the spring is stretched to a new location at 0.330 meters. Therefore, the displacement is 0.030 meters. The restorative force of the spring must balance the weight of the added mass. For this example, the weight is $m \times g = 1.96$ N, where N stands for newtons. We can then determine an approximate value for the spring constant k, given that the spring required 1.96 N to move it a distance of 0.030 meters. Thus, $k = 1.96/0.030 = 65.33$ N/m.

Hooke's Law is most accurate for small deformations of an object. The law is sometimes expressed in terms of stress (the force within a material that develops as a result of the externally applied force) and strain (the deformation produced by the stress). Stress is proportional to strain for small values of stress. Note also that the value of k depends on the material that composes the object and usually on the dimensions and shape of the object. When considered in terms of stress and strain, Hooke's Law is often formulated as stress/strain = E, where E is the modulus of elasticity, also known as Young's modulus, which, for example, may be measured in pounds per square inch.

We have been discussing movements and forces in one direction. French mathematician Augustin Louis Cauchy (1789–1857) generalized Hooke's Law to three-dimensional forces and elastic bodies, and this more complicated formulation relies on six components of stress and six components of strain. The stress-strain relationship forms a 36-component stress–strain tensor when written in matrix form.

If a metal is lightly stressed, a temporary deformation may be achieved by an elastic displacement of the atoms in the three-dimensional lattice. Removal of the stress results in a return of the metal to its original shape and dimensions.

Robert Hooke (1635–1702), English physicist and polymath famous for Hooke's Law of Elasticity and a variety of experimental and theoretical work.

CURIOSITY FILE: Hooke was one of the first proponents of biological evolution during a time when many learned people relied on the book of Genesis and were confused by the existence of fossils. • Many of Hooke's inventions have been lost, partly due to Isaac Newton's dislike for the man. In fact, Newton had Hooke's portrait removed from the Royal Society and attempted to have Hooke's Royal Society papers burned. • Hooke's design for a marine chronometer was rediscovered in 1950 at the Library of Trinity College, Cambridge. • Hooke's remains were exhumed in the eighteenth century, and their present location is shrouded in mystery. • Hooke invented an early form of hearing aid called the otocousticon. • In 2006, the Royal Society purchased a seventeenth-century manuscript by Hooke for $1.75 million. The manuscript was filled with 500 pages of notes recorded during Royal Society meetings. In the notes, Hooke castigates his rivals Newton and Robert Boyle, whom he claims stole his ideas. Hooke also wrote that Dutch microscopicst Anton van Leeuwenhoek found "a vast number of small animals in his Excrements which were most abounding when he was troubled with a Looseness and very few or none when he was well."

> Robert Hooke is one of the most neglected natural philosophers of all time. The inventor of...the iris diaphragm in cameras, the universal joint used in motor vehicles, the balance wheel in a watch, the originator of the word "cell" in biology, he was...architect, experimenter, [and astronomer]—yet is known mostly for Hooke's Law. He was Europe's last Renaissance man, and England's Leonardo.
>
> —Robert Hooke Science Centre, www.roberthooke.org.uk

> Hooke was an unattractive man, disfigured [by smallpox], orphaned at 13 years of age [by a suicidal father], robbed of credit for his greatest inspirations and ideas, with many of his creations almost certainly willfully destroyed or lost after his death in 1703.
>
> —Maurice Smith, "Robert Hooke: The Inspirational Father of Modern Science in England"

> The tendency to flit from idea to insight without pause was Hooke's innate characteristic....He pours out a continuous stream of brilliant ideas.
>
> —Richard Westfall, "Robert Hooke," in *Dictionary of Scientific Biography*

> Instruments enlarge the senses and make them more precise and reliable; Hooke speaks of "supplying of their infirmities with instruments, and...the adding of artificial organs to the natural." He included here not only the obvious examples like microscopes and telescopes, but also instruments related...to magnetism...to investigate a phenomenon not directly sensible at all.
>
> —J. A. Bennett, "Robert Hooke as Mechanic and Natural Philosopher"

Robert Hooke was born on the Isle of Wight, a British island close to the southern coast of England. His father was religious and expected his son to enter the ministry. However, because Hooke was a sickly child, with constant headaches, he was not expected to reach adulthood, and his parents decided not to bother educating him. Left to pursue his own interests, Hooke fell in love with mechanical contraptions such as toys and clocks. When he was young, he created a wooden clock and an intricate model of a fully rigged ship with working guns.

In 1648, Hooke's father, who was afflicted with jaundice, decided that he did not wish to suffer any longer, and he hanged himself. Given that Hooke was an excellent drawer, the rest of his family decided that he should move to London to become an apprentice to a portrait painter.

Hooke eventually lost interest in art because he wanted a more comprehensive education. He enrolled in Westminster School, where he devoured the contents of the six books of Euclid's *Elements* during the first week of school. He learned Latin, Greek, and some Hebrew and was an exceptional organist.

In 1653, he entered Christ College, Oxford, where he studied astronomy and mechanics. During his life, Hooke's interests ranged far and wide, covering such topics as physics, astronomy, chemistry, biology, geology, architecture, and even naval technology. He often had so many ideas in his head that he simultaneously worked on numerous projects in different fields.

In 1655, Hooke was employed by Robert Boyle to help construct an air pump that Boyle used to conduct the experiments necessary to formulate Boyle's Gas Law. Some historians of science suggest that it is possible that it was Hooke who formally stated Boyle's Law, but Hooke's precise role in the experiments is unclear.

At the same time that he was working with Boyle on gases, Hooke also worked on clocks, particularly those that could keep fairly accurate time while at sea. Realizing that the pendulum clock could hardly be used on a rocking ship, he suggested that "springs instead of gravity" be used to drive the clock mechanism. Beginning his experiments around 1658, Hooke constructed a clock with a spiral spring and an improved escapement, that is, the mechanical device that regulates movement.

In 1660, at the same time he worked on clocks, Hooke discovered what we now call Hooke's Law of Elasticity, which, among other things, describes the variation of tension with extension in an elastic clock spring. However, he made his law public only in 1678. Although Hooke's Law may not appear to be a profound discovery, it seems that no one before him stated the law explicitly.

Hooke, along with Italian astronomer Giovanni Domenico Cassini (1625–1712) and Dutch mathematical physicist Christiaan Huygens

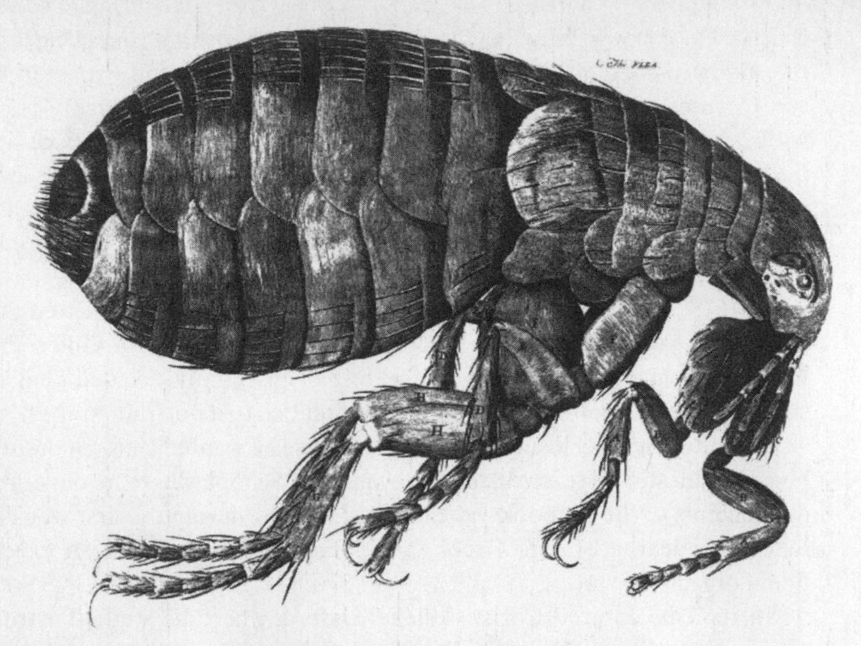

Flea, from Robert Hooke's *Micrographia*, published in 1665.

(1629–1695), was among the first astronomers to carefully observe the surface of Jupiter. In 1664, Hooke reported a small spot on the biggest Jovian belt, which he believed to be a permanent feature of Jupiter and not simply the shadow of a moon.

In 1665, Hooke became professor of geometry at Gresham College, London. The position gave him living space at the college and required him to give lectures in both English and Latin. He was required to be unmarried during his stay at the college.

In 1665 he published *Micrographia*, a book that featured breathtaking microscopic observations and biological speculation on specimens that ranged from plants to fleas. Hooke was the first to coin the word "cell" to describe the basic units of all living things. His choice was motivated by his observations of plant cells that reminded him of "cellula," which were quarters in which monks lived. When describing his microscopic observation of thin slices of cork, he wrote in *Micrographia*:

> I could exceedingly plainly perceive it to be all perforated and porous, much like a Honey-comb, but that the pores of it were not regular. . . . these pores, or cells, . . . were indeed the first microscopical pores I ever saw, and perhaps, that were ever seen, for I had not met with any Writer or Person, that had made any mention of them before this. . . .

About this magnificent work, Robert Westfall writes in the *Dictionary of Scientific Biography*:

> Robert Hooke's *Micrographia* remains one of the masterpieces of seventeenth century science, [presenting] a bouquet of observations with courses from the mineral, animal and vegetable kingdoms. Above all, the book suggested what the microscope could do for biological science.

Robert Hooke was also a Surveyor to the City of London and helped to rebuild London after the Great Fire in 1666. He was a famous architect and designed many buildings such as the Bethlem Royal Hospital and the Royal College of Physicians. The dome in St. Paul's Cathedral in London used Hooke's method of construction.

Hooke was also fascinated by fossils and geology. His fellow scientists had proposed a number of theories to explain the origin of fossils. One strange but commonly held theory suggested that fossils grew within Earth, somewhat like the incubation of an embryo in the womb. A mysterious "shaping force" could create the images of living creatures within stones. Hooke was the first person to use a microscope to study fossils, and he observed that the structures of petrified wood and fossil seashells bore a striking similarity to actual wood and shells. In *Micrographia*, he compared petrified wood to rotten wood and concluded that wood could be turned to stone by a gradual process:

> This petrify'd Wood having lain in some place where it was well soak'd with petrifying water (that is, such water as is well impregnated with stony and earthy particles) did by degrees separate abundance of stony particles from the permeating water, which stony particles, being by means of the fluid vehicle convey'd, not only into the Microscopical pores... but also into the pores or Interstitia... of that part of the Wood, which through the Microscope, appears most solid....

He also believed that many fossils represented extinct creatures: "There have been many other Species of Creatures in former Ages, of which we can find none at present; and that 'tis not unlikely also but that there may be divers new kinds now, which have not been from the beginning."

The public was fascinated by *Micrographia*, because it provided a new look at familiar objects such as a fine needle point that looked like a rough carrot under the microscope. The observations of molds and insects (including a flea, louse, a bee stinger, and an eye of a fly) provided some of the most spectacular images and stimulated the imagination of both scientists and lay people.

Hooke's contributions to the field of geology are extolled in Ellen Drake's *Restless Genius: Robert Hooke and His Earthly Thoughts*:

> The few geologists who have read Robert Hooke's *Discourse of Earthquakes* have been astonished by his almost clairvoyant postulations concerning the formation of geomorphological features, the origin and usefulness of fossils, biological evolution, and all the dynamic changes that constantly take place on this planet.... Hooke was therefore important in the history of Earth science, as he was in the development of many other fields of scientific and technological endeavor.

Hooke was intrigued by the science of respiration and the workings of the lungs. In one experiment, he was placed in a sealed vessel, from which the air was gradually pumped! I have not been able to ascertain the precise purpose of this experiment, but he damaged his ears and experienced deafness in the process. Most likely, Hooke's aim was to study, in a general fashion, the effect of low atmospheric pressure on a human being.

He also opened the chest of a dog, destroyed the motion of its lungs, and then used a bellows to provide a stream of air that passed through the lungs in order to better understand the function of the lungs in the process of respiration.

As discussed above, Hooke was an extremely prolific inventor in the area of clocks and probably invented the balance spring that coils and uncoils with a natural periodicity. He investigated the colors of membranes and of thin plates of mica. He invented or improved meteorological instruments such as the barometer (for measuring atmospheric pressure), anemometer (for measuring wind velocity), and hygrometer (for measuring humidity). His invention of the hygrometer stemmed from his observations of goat beard hairs that would bend when dry and straighten when wet. Despite his facility with invention, Hooke was sometimes unhappy with the credit he received. For example, Jim Bennett writes in "Robert Hooke as Mechanic and Natural Philosopher" that "Hooke's later lectures are punctuated by bitter outbursts on the fate of inventors—reviled and ridiculed, plagued by difficulties and conservatism, denied any benefit from their work, in the end only to see their inventions carried off by others."

We can discuss additional examples of Hooke's innovations. For example, he suggested that scientists use thermometers that assigned 0 degrees to the freezing temperature of water. He invented the air pump in a form that lasted for many years. He constructed the most powerful microscope at the time, which achieved a 30-fold magnification, as well as the first practical universal joint and the first Gregorian reflecting telescope (a telescope with a parabolic primary mirror). He was the first person in history to discover a binary star, that is, two stars orbiting about a common center of mass. Hooke also postulated the inverse-square principle of gravity but lacked the mathematical know-how to prove it. Although Hooke did not discover the Law of Universal Gravitation, he did appear to contribute to Newton's thinking on the subject.

Hooke and Newton disliked each other for many years. For example, Newton was furious when, in 1672, Hooke criticized Newton's demonstration of the use of prisms to split white light into its various colored components. In 1679, as mentioned above, Hooke mused to Newton about a possible inverse-square principle of gravitation, but when Newton published his *Principia mathematica* in 1687, he did not credit Hooke. Regarding the inverse-square principle, Newton told Hooke, "Merely because one says something might be so, it does not follow that it has been proved that it is."

In addition to this snub by Newton, 1687 was a particularly sad and frustrating year for Hooke. Hooke's niece, with whom he had a romantic relationship, died this year, and Hooke's health quickly declined.

Biographer Richard Westfall writes in the *Dictionary of Scientific Biography*,

> His frame was badly twisted. Add to his wretched appearance, wretched health. He was a dedicated hypochondriac who never permitted himself the luxury of feeling well for the length of a full day. Hooke's spiny character was nicely proportioned to the daily torment of his existence.

Today, some physicians have speculated that Hooke was inflicted with scoliosis, a crippling degenerative disease that causes an unnatural curvature of the spine. Hooke finally died on March 3, 1703, having been blind from diabetes and bedridden the last year of his life. He left behind his huge library of more than 3,000 books in Latin, French, and Italian. Although he was financially well off from the work he performed as a surveyor, he had never married. Today, the location of his grave is unknown. A lunar crater with a diameter of 36 kilometers was named after Hooke and approved in 1935 by the International Astronomical Union General Assembly. A crater on Mars has also been named after Hooke.

Stephen Inwood, author of *The Forgotten Genius: The Biography of Robert Hooke 1635–1703*, describes the final scene of Hooke's death:

> Dr. Robert Hooke, Gresham Professor of Geometry and Curator of Experiments for the Royal Society, lay dead on his bed. In death, as in life, he was not an attractive sight. His ragged clothes were twisted about his emaciated body like a winding sheet, and the lice were so thick on his corpse that "there was no coming near him." Hooke's property passed to his next of kin (probably his cousin), Elizabeth Stephens, an illiterate woman whose signature was a pirate's hook.

Despite all of Hooke's achievements, Lisa Jardine writes in *The Curious Life of Robert Hooke* that he is not remembered for much today:

> He is most likely to be remembered, though, as a boastful, cantankerous, physically misshapen know-all, who was somehow involved with the early Royal Society and was Sir Isaac Newton's sworn enemy.... Although his name crops up regularly in English seventeenth-century histories of ideas, from anatomical dissection to cartography, and from architecture to scientific instrument-making, no single major discovery or monument (apart from the law of elasticity) is any longer securely attributed to him.

FURTHER READING

Bennett, Jim, "Robert Hooke as Mechanic and Natural Philosopher," *Notes and Records of the Royal Society of London* 35(1): 33–48, July 1980.

Bennett, Jim, Michael Cooper, Michael Hunter, and Lisa Jardine, *London's Leonardo: The Life and Work of Robert Hooke* (New York: Oxford University Press, 2003).

Drake, Ellen, *Restless Genius: Robert Hooke and His Earthly Thoughts* (New York: Oxford University Press, 1996).

Hooke, Robert, *Micrographia* (digital download version); see www.gutenberg.org/etext/15491. Available on CD-ROM (Oakland, California: Octavo, 1998).

Inwood, Stephen, *The Forgotten Genius: The Biography of Robert Hooke 1635–1703* (San Francisco: MacAdam/Cage Publishing, 2005).

Jardine, Lisa, *The Curious Life of Robert Hooke: The Man Who Measured London* (New York: Harper Perennial, 2005).

"Robert Hooke"; see www.roberthooke.com.

Westfall, Richard, "Robert Hooke," in *Dictionary of Scientific Biography*, Charles Gillispie, editor-in-chief (New York: Charles Scribner's Sons, 1970).

INTERLUDE: CONVERSATION STARTERS

A flock of sheep consisted of several individual sheep and was a flock only by convention—the quality of flockness was put on it by humans—it existed only in some human's mind as a perception. Yet Hooke had found that the human body was made up of cells—therefore, just as much an aggregate as a flock of sheep. Did this mean that the body, too, was just a figment of perception?

—Neal Stephenson, *Quicksilver*

The universe we inhabit, and its operational principles, exist independently of our observation or understanding; mathematical models of the universe . . . are descriptive tools that exist only in our minds. Mathematics is at root a formal description of orderliness, and since the universe is orderly (at least on scales of space-time . . . which [we can] observe), it should come as no surprise that the real world is well modeled mathematically.

—Keith Backman, "The Danger of Mathematical Models," *Science*, October 20, 2006

In the beginning, God said the four-dimensional divergence of an antisymmetric, second rank tensor equals zero, and there was Light, and it was good.

—Message on a Berkeley University T-shirt, as told by Michio Kaku, "Parallel Universes, the Matrix, and Superintelligence," KurzweilAI.net

The core of science is not a mathematical model; it is intellectual honesty.

—Sam Harris, La Jolla meeting, "Beyond Belief 2006: Science, Religion, Reason, and Survival," November 2006

I don't believe there are any fundamental laws or any final theory. I think the only laws we'll ever find are the ones we impose on nature by the way we look at it.

—David Ambrose, *Superstition*

If we live in a simulated reality, we should expect occasional sudden glitches, small drifts in the supposed constants and laws of Nature over time, and a dawning realization that the flaws of Nature are as important as the laws of Nature for our understanding of true reality.

—John Barrow, "Living in a Simulated Universe"

As a Jew, I have no problem whatsoever believing in intelligent design.... I chose to believe that evolution was packaged with the original matter that resulted in the big bang. The "designer" can no longer intervene because he is held by the laws of the universe he willed.

—Marysia Meylan, *New York Times*

BOYLE'S GAS LAW

Ireland, 1662. The pressure of a gas varies inversely as its volume at constant temperature.

CROSS REFERENCE: ROBERT HOOKE, CHARLES'S GAS LAW, AVOGADRO'S GAS LAW, DALTON'S LAW OF PARTIAL PRESSURES, GRAHAM'S LAW OF EFFUSION, AND THE IDEAL GAS LAW.

In 1662, Charles II sold the harbor city of Dunkirk to France for £400,000, and the last silver pennies were minted in London. The British merchant John Graunt published the first book on statistics. The book contained the London Life Table, the first table showing the ages at which people are likely to die in a city. Graunt's work was a starting point for both statistics and demography.

In 1662, Robert Boyle—Irish chemist, physicist, and inventor—studied the relationship between the pressure P and the volume V of a gas in a container held at a constant temperature. Boyle observed that the product of the pressure and volume are nearly constant:

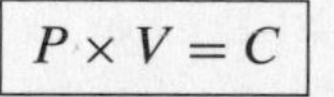

$$P \times V = C$$

This relationship between pressure and volume is called Boyle's Law in his honor. For an example application, suppose we have a gas confined in a jar with a piston at the top. The initial state of the gas has a volume of 5.0 cubic meters, and the pressure is 1.0 kilopascal. While holding the temperature and amount of gas (number of moles) constant, we add weights to the top of the piston to increase the pressure. (The concept of *mole* is more fully explained in the entry on Avogadro's Gas Law.) When the pressure reaches 4.0 kilopascals, we find that the volume has decreased to 1.25 cubic meters. The product of pressure and volume remains a constant: $5\ \text{m}^3 \times 1\ \text{kPa} = 1.25\ \text{m}^3 \times 4\ \text{kPa}$.

Note that Boyle's Law is sometimes called the Boyle-Mariotte Law, because French physicist Edme Mariotte (1620–1684) discovered the same law independently of Boyle but did not publish it until 1676. Here are Boyle's exact words on his own experiments with air that led to Boyle's Law, from the second edition of *New Experiments Physico-Mechanicall, Touching The Spring of the Air, and Its Effects*:

> The pressures and expansions [are] in reciprocal proportion.... Common air, when reduced to half its wonted extent, obtained near about twice as forcible a spring as it had before, so

this thus compressed air being further thrust into half this narrow room, obtained thereby a spring about as strong again as that it last had, and consequently four times as strong as that of the common air.

An ordinary syringe provides another example of a practical application of Boyle's Law. When a physician pushes the plunger on a syringe, he decreases the volume inside the syringe, increasing the pressure and causing the medicine to be injected. A balloon inflated at sea level will expand as it rises in the atmosphere and encounters decreased pressure. Similarly, when we inhale, our diaphragms move downward, increasing the lung volume and reducing the pressure so that air flows into the lungs. In a sense, Boyle's Law keeps us alive with each breath we take.

Boyle's Law is most accurate for an *ideal gas*, which consists of identical particles of negligible volume, with no intermolecular forces, and with atoms or molecules that collide elastically with the walls of the container. Real gases obey Boyle's Law at sufficiently low pressures, and the approximation is often sufficiently accurate for practical purposes when describing real gases.

Scuba divers learn about Boyle's Law because it helps to explain what happens during ascent and descent with respect to the lungs, mask, and buoyancy control device (BCD). For example, as a person descends, pressure increases, causing any air volume to decrease. Divers notice that their BCDs appear to deflate, and the pressure in the airspace behind the ears changes. To equalize the ear space, air must flow through the diver's Eustachian tubes to compensate for the reduction in air volume.

Other gas laws in this book describe the relationship between temperature, pressure, and volume of gases. If we consider Avogadro's Law, we can derive the Ideal Gas Law, $PV = nRT$, from these other gas laws. Here, P is the pressure, V is the volume, n is the number of moles of gas, R is the ideal gas constant (usually in units of L-atm/mol-K or Pa-m^3/mol-K), and T is the temperature in degrees kelvin.

Other important gas laws discussed in this book include Charles's Gas Law, Dalton's Law of Partial Pressures, and Graham's Law of Effusion. No real gas obeys these gas laws exactly, because, for example, these laws assume that gas particles are much smaller than the distance between particles, and therefore the volume of a gas is assumed to be mostly empty space and the volume of the gas molecules themselves to be negligible. These laws also assume that there is no force of attraction between gas molecules or between the molecules and the walls of the container. To account for some of the complexities of actual gases, the formula $PV = nRT$ can be rewritten as a more realistic Van der Waals Equation: $(P + an^2/V^2)(V - nb) = nRT$, where a is a constant used to correct for intermolecular attractive forces that may exist, and b is a constant to correct for volume of individual gas molecules. For example, for helium, $a = 0.034$ L^2-atm/mol^2 and $b = 0.024$ L/mol. For ammonia (NH_3), $a = 4.17$ L^2-atm/mol^2 and $b = 0.037$ L/mol. Deviations from ideal gases are greatest when the intermolecular attractive forces of gas molecules are greatest and/or when the mass (and subsequently volume) of gas molecules is large.

Robert Boyle (1626–1691), Irish natural philosopher, famous for his work on the properties of gases and for his support of the "corpuscular" view of matter that was a precursor to the modern theory of chemical elements.

CURIOSITY FILE: Boyle conducted research in alchemy—his goal was not only to transmute base metals into gold but also to attract angels. For Boyle, alchemy was both pure science and a defense against growing numbers of atheists. • Boyle proved that sound transmission was impossible in a vacuum.

> Boyle is charitable to ingenious men that are in want, and foreign chemists have had large proof of his bounty, for he will not spare for cost to get any rare secret. At his own cost and charges he got translated and printed the New Testament in Arabic, to send into the Mahometan countries.
>
> —John Aubrey, *Brief Lives*

> Boyle's interest seems to have been fueled more by his constant desire to acquire knowledge of God.
>
> —J. J. MacIntosh, "Robert Boyle," *Stanford Encyclopedia of Philosophy*

Robert Boyle was born at Lismore Castle in Munster, Ireland—the fourteenth child of the wealthy Richard Boyle, First Earl of Cork. Boyle wrote of his father, who was probably the richest man in Great Britain, "He, by God's blessing on his prosperous industry, from very inconsiderable beginnings, built so plentiful and so eminent a fortune, that his prosperity has found many admirers, but few parallels." Boyle's father had constructed mills, founded towns, and establishing ironworks and other industries.

As a child, Boyle learned to speak Latin and French, and at age 8 he was sent to Eton College. After spending a few years at the college, he traveled abroad with a French tutor. In Italy, Boyle had the honor of meeting the aging astronomer Galileo. Boyle cherished this meeting, and it provided an impetus for Boyle to discover more about workings of the world.

Boyle was tutored privately in practical mathematics and other areas of a liberal education. He also became interested in medicine and chemistry.

In 1654, Boyle joined a small group of eminent English scientists, mathematicians, and philosophers who had been meeting weekly in London and in Oxford since 1645. In 1662, the group became the Royal Society, which exists today as the oldest continuous scientific society in the world. The Society's motto, *Nullius in Verba*, means "Nothing in Words," which suggests that science must be experimentally based.

Boyle had many interests. As just one example, in 1654, he displayed his fascination with the organ systems of fish when he wrote to a friend:

> I am exercising myself in making anatomical dissections of living animals: wherein I have satisfied myself of the circulation of the blood . . . and have seen (especially in the dissections of fishes) more of the variety and contrivances of nature, and the majesty and wisdom of her author, than all the books I ever read in my life could give me convincing notions of.

However, he was a little squeamish about dissections, and his physiological studies were hampered by the "tenderness of his nature," which kept him from doing many anatomical dissections, especially of living animals, though he knew this kind of work would be "most instructing."

Boyle wrote on many subjects, including theology, hydrostatics, philosophy, and other areas of science. Although his first love was chemistry, his first published scientific book, *New Experiments Physico-Mechanicall, Touching the Spring of the Air and Its Effects* (1660), was on pneumatics (the use of pressurized gases to do work). The text of *New Experiments* was the result of three years of experimenting using an air pump with the assistance of Robert Hooke, the English experimental philosopher. Hooke designed the apparatus, and Boyle used it to make several discoveries—for example, that a flame required air and that sound did not travel in a vacuum. In particular, he showed that the sound of a watch in a bell jar grew fainter as the air was pumped out. The second edition of *New Experiments*, published in 1662, contained the pressure–volume inverse relationship which today we call Boyle's Law.

He also performed various other experiments:

- He proved that many fruits and vegetables contain air and carbon dioxide.
- He discovered new chemical reactions and substances. For example, he produced hydrogen from steel filings and a strong mineral acid.

- He found that certain plant extracts can be used to distinguish acids from bases. For example, he observed that all acids turned the blue syrup of violets red, and that all alkalies turned syrup of violets green.
- He studied the force that freezing water can produce as it expands.

Boyle emphasized the application of mathematics to the study of chemistry, a field that he believed exhibited complexity merely as a result of simple mathematical laws applied to fundamental particles. Today, Boyle is most famous for his law that states that if the volume of a gas is decreased, the pressure increases proportionally. Realizing that his results could be explained if all gases were made of tiny particles, Boyle tried to construct a universal *corpuscular theory* of chemistry. In his 1661 *The Sceptical Chymist*, Boyle denounced the Aristotelian theory of the four elements (earth, air, fire, and water) and developed the concept of primary particles that came together to produce corpuscles.

Although Boyle believed that the universe could be understood using mechanical principles, he also believed that this mechanical model was not counter to a belief in God or that it somehow downgraded God to a mere mechanic. According to Boyle, a God who could create a mechanical universe that obeyed laws was to be revered more than a God who created a universe without scientific laws. Boyle had also come to believe that angels were created "before the visible World... was half completed," but, in contrast, God created new human souls daily and worked a "physical miracle" to attach them to their respective bodies.

Boyle never married, and from the age of 41 lived in his sister Katherine's house, where he frequently had visitors. Before he died, he specified that his money should be used to found the Boyle lectures that were intended to refute atheism and religions that competed with Christianity.

Firm in his belief in Christianity, Boyle wanted the words of the Bible to spread throughout the world; thus, he ensured that the Bible was translated into a variety of languages, such as Turkish and various Native American languages. He wrote, "To convert Infidels to the Christian Religion is a work of great Charity and kindness to men."

Boyle believed that angels existed, that they were generally smarter than humans, and that it was possible that God's primary goal in making the universe was to provide a universe for the angels. This angel-centric creation suggested that it was possible that the universe might be forever too complex for humans to understand. He wrote in 1680, "We presume too much of our own abilities, if we imagine that the omniscient God can have no other Ends in the framing & managing of Things Corporeal, than such as we Men can discover."

During the course of his life, he continually sought to improve the well-being of humanity. For example, he invented ways to improve agriculture and medicine and was interested in the possibilities of producing fresh water from salt water and preserving food by vacuum packing.

The following is a sampling of some of his works. I selected titles from a long list in order to give a sampling of the diversity of his interests:

- 1660, *New Experiments Physico-Mechanicall: Touching the Spring of the Air and Its Effects*
- 1661, *The Sceptical Chymist*
- 1663, *Experiments and Considerations upon Colours, with Observations on a Diamond That Shines in the Dark*
- 1666, *Hydrostatical Paradoxes*
- 1670, *Cosmical Qualities of Things*
- 1664, *Excellence of Theology Compared with Natural Philosophy*
- 1675, *Some Considerations about the Reconcileableness of Reason and Religion, with a Discourse about the Possibility of the Resurrection*

A lunar crater with a diameter of 57 kilometers was named after Boyle and approved in 1970 by the International Astronomical Union General Assembly. Gilbert Burnet, Bishop of Salisbury, gave a sermon at Boyle's funeral that emphasized Boyle's love for both religion and science. According to Burnet, Boyle, like several other scientists of his era,

> directed all their enquiries into Nature to the Honour of its great Maker: And have joined two things, that how much soever they may seem related, yet have been found so seldom together, that the World has been tempted to think them inconsistent; A constant looking into Nature, and yet more constant study of Religion, and a Directing and improving of the one by the other.

Michael Hunter in *Robert Boyle Reconsidered* sums up Boyle's life:

> By any standards, Robert Boyle is one of the commanding figures of seventeenth-century thought. His writings are remarkable for their range, their significance, and their sheer quantity: during his life he published over forty books.... In his blending of a commitment to scientific work with deep piety, Boyle presented almost an ideal type of "the Christian virtuoso,"... a great intellectual innovator who was at the same time a paragon of godliness and probity.

FURTHER READING

Anstey, Peter, *The Philosophy of Robert Boyle* (London: Routledge, 2000).

Boyle, Robert, *New Experiments Physico-Mechanicall, Touching the Spring of the Air, and Its Effects*, second edition (Oxford, 1662).

Hall, Marie, "Boyle," in *Dictionary of Scientific Biography*, Charles Gillispie, editor-in-chief (New York: Charles Scribner's Sons, 1970).

Hunter, Michael, *Robert Boyle Reconsidered* (Cambridge, U.K.: Cambridge University Press, 2003).

MacIntosh, J. J., "Robert Boyle," *Stanford Encyclopedia of Philosophy*; see plato.stanford.edu/entries/boyle/.

Principe, Lawrence, *The Aspiring Adept: Robert Boyle and His Alchemical Quest* (Princeton, N.J.: Princeton University Press, 2000).

Wojcik, Jan, *Robert Boyle and the Limits of Reason* (Cambridge, U.K.: Cambridge University Press, 2002); this book is the source for Gilbert Burnet's sermon preached at the funeral of Boyle.

INTERLUDE: CONVERSATION STARTERS

It would be entirely wrong to suggest that science is something that knows everything already. Science proceeds by having hunches, by making guesses, by having hypotheses, sometimes inspired by poetic thoughts, by aesthetic thoughts even, and then science goes about trying to demonstrate it experimentally or observationally. And that's the beauty of science, that it has this imaginative stage but then it goes on to the proving stage, the demonstrating stage.

—Richard Dawkins, in John Brockman's *What We Believe but Cannot Prove*

If we go back to our checker game, the fundamental laws are rules by which the checkers move. Mathematics may be applied in the complex situation to figure out what in given circumstances is a good move to make. But very little mathematics is needed for the simple fundamental character of the basic laws. They can be simply stated in English for checkers.

—Richard Feynman, *The Character of Physical Law*

Biology occupies a position among the sciences at once marginal and central. Marginal because—the living world constituting but a tiny and very "special" part of the universe—it does not seem likely that the study of living beings will ever uncover general laws applicable outside the biosphere. But if the ultimate aim of the whole of science is indeed, as I believe, to clarify man's relationship to the universe, then biology must be accorded a central position....

—Jacques Monod, *Chance and Necessity*

God, we are told, is not a puppet-master in regard either to human actions or to the processes of the world. If we are to exist in an environment where we can live lives of productive work and consistent understanding...the world has to have a regular order and pattern of its own. Effects follow causes in a way that we can chart, and so can make some attempt at coping with. So there is something odd about expecting that God will constantly step in if things are getting dangerous.

—Rowan Williams, "Of Course This Makes Us Doubt God's Existence," *Sunday Telegraph*, January 2, 2005

NEWTON'S LAWS OF MOTION, GRAVITATION, AND COOLING

England, 1687 (a publication date for the laws of motion and gravitation), 1701 (for the law of cooling). Newton's Laws of Motion concern relations between forces acting on objects and the motion of these objects. His Law of Universal Gravitation states that objects attract one another with a force that varies as the product of the masses of the objects and inversely as the square of the distance between the objects. His Law of Cooling states that the rate of heat loss of a body is proportional to the difference in temperatures between the body and its surroundings.

CROSS REFERENCE: ROBERT HOOKE, JOHANN BERNOULLI, KEPLER'S LAWS OF PLANETARY MOTION, NEWTON'S LAW OF VISCOSITY, AND EINSTEIN'S GENERAL AND SPECIAL THEORIES OF RELATIVITY.

> In 1687, the men who were led by French explorer Robert Cavelier de La Salle murdered La Salle while they were desperately searching for the mouth of the Mississippi River. Five years earlier, La Salle had claimed the entire Mississippi basin for France. According to the *Catholic Encyclopedia*, "La Salle's schemes of empire and of trade were far too vast for his own generation to accomplish."

NEWTON'S LAWS OF MOTION, 1687

Newton's Laws of Motion revolutionized our basic concepts of physics and how objects move in the universe. Dudley Williams and John Spangler write in *Physics for Science and Engineering*:

> These principles form the basis not only of classical dynamics but of classical physics in general. Although they involve certain definitions and can in a sense be regarded as axioms, Newton asserted that they are based on quantitative observation and experiment; certainly, they cannot be derived from other more basic relationships. The test of their validity involves predictions. . . . The validity of such predictions was verified in every case for more than two centuries.

Aristotle's view that a body could be kept in motion only by applying a force was accepted for more than a thousand years, until Newton demolished this thinking with his First Law of Motion. Interestingly, Aristotle's ideas did provide a reasonable explanation for his qualitative observations, but Galileo understood that rigorous measurements of positions as a function of time required a better theory and a more accurate way of looking at the world. Of Newton's Laws of Motion, Ernest Abers and Charles F. Kennel write in *Matter in Motion*:

> It is a mark of Newton's genius that of all the possible statements about motion, he recognized that three and only three completely define a logically consistent framework within which all problems of motion can be analyzed quantitatively. These are Newton's three laws.

Newton's First Law of Motion (Law of Inertia)

The First Law states that bodies do not alter their motions unless forces are applied to them. A body at rest stays at rest. A moving body continues to travel with the same speed and direction unless acted upon by a net force. In other words, an object such as a bowling ball traveling in uniform motion (i.e., traveling in a straight line at a constant speed) will remain traveling in uniform motion unless acted upon by a net force.

The term "net force" is important because an object is often acted upon by numerous forces, but it will remain at a constant velocity whenever the forces are balanced. For example, a coffee cup sitting on a table has constant zero velocity because the downward force caused by the cup's weight is exactly counterbalanced by the upward force that the table applies to the cup. The net force is zero. However, the cup will obviously move if I suddenly unbalance the forces acting on it by giving it a push. Consider a ball that is subject to a constant force of gravity. If I give the ball an initial push on an infinitely long frictionless road, the ball will never stop rolling. Of course, in reality, friction always exists between the ball, the surface of the road, and the air molecules in the path of the ball.

According to the First Law, if an object has constant velocity, we conclude there is no net external force. A state of rest is a special case of constant velocity motion where the speed is 0. If a body's velocity is changing, we conclude that a net force is acting on the body.

Conceptually, the First Law of Motion would have been considered quite novel before Newton's era. As mentioned above, before the time of Galileo, researchers believed that bodies, such as balls, would move only as long as a force was applied to the body, and the body would stop moving when the force was removed. (Of course, this made perfect sense to them, given that if one stops pushing a ball, it does indeed stop moving. The ancients did not think of friction as a force.) A body was considered in its "natural state" when not moving. Similarly, it was believed that the application of a continual force was required in order to keep the planets from coming to rest. However, if we view gravity as acting via a gravitational force, this force only serves to change the direction of motion of a planet or motion, but it is not required to keep a planet traveling at a constant speed—assuming no frictional effects are caused by particles in interplanetary space.

This law of motion should probably not be attributed solely to Newton, even though it is commonly referred to as Newton's First Law. Arnold Arons notes in *Development of Concepts of Physics*:

> In Newton's day... [some erroneous ideas concerning] the physics of motion continued to be taught from scholastic textbooks; pedantry is slow to change in any era. But by the latter part of the seventeenth century, Galileo's conception of inertia, refined and corrected, was accepted and taken for granted by most active, productive physical scientists.... [However,] Newton set the law of inertia at the head of the laws of motion and gave it the tone of a proclamation of emancipation from scholastic theory.

In common usage, *inertia* usually refers to an object's amount of resistance to change in velocity. If it is not apparent to you that a moving object does not "naturally" stop moving without an applied force, you can imagine an experiment in which the face of a penny is sliding along a smooth horizontal table. Obviously, the penny eventually slows down and stops. Next, we add a thin film of oil to the surface, and the penny decreases its speed at a slower rate and travels farther. If we use an even better lubricant, the penny travels farther still. We can extrapolate this experiment to a point in which all friction is gone in order to realize that the penny would continue sliding along such an imaginary surface forever. In fact, an external force would be needed to change the velocity of the sliding penny, but no more force is needed for it to continue with constant velocity.

Newton's Second Law of Motion

According to Newton's Second Law of Motion, when a net force acts upon an object, the rate at which the momentum changes is proportional to the force applied. Today, we express this as

$$\boxed{\mathbf{F} = \mathrm{d}\mathbf{p}/\mathrm{d}t,}$$

where $\mathbf{p}$ is the momentum, which is equal to the mass times the velocity of the object; $\mathbf{F}$ is the applied force, and $d\mathbf{p}/dt$ is the rate of change in momentum. (Bold letters indicate vector quantities that have both a magnitude and direction.) Thus, force is simply defined in terms of momentum. Two forces are equal if they cause the same rate of change in the momentum of the object. Notice how the use of $d\mathbf{p}/dt$ does not assume that the mass of the object is constant. In fact, the mass can change when we study problems such as those involving a raindrop growing as it falls or a rocket expending fuel as it travels.

The law also ensures that the direction of change in momentum caused by a force is in the same direction as the force. For an object with constant mass m, this law may be expressed succinctly as

$$\boxed{\mathbf{F} = m\mathbf{a},}$$

where $\mathbf{F}$ is the net force and $\mathbf{a}$ is the acceleration (the rate of change of the object's velocity). The units of force are typically given in units of newtons (N). One N is the magnitude of force acting on a body with a mass of one kilogram when the particle has an acceleration of one meter per second per second (m/s^2).

For example, if a catapult exerts a constant force of 200 N on a stone of mass 0.10 kilograms, then the acceleration of the stone is 2,000 m/s^2 just as it leaves the catapult. For a particular value of force, the smaller the mass, the larger the acceleration.

If many forces act on a body, one can still use this formula to calculate the size and direction of the resultant acceleration by first determining the vector sum of all the forces $\mathbf{F}$ and using this vector in $\mathbf{F} = m\mathbf{a}$. If an object falls to Earth with zero air resistance, the only force acting on the object is its weight, which would produce a downward acceleration equal to the acceleration of gravity (~9.8 m/s^2 near Earth's surface).

Newton's Third Law of Motion

Abers and Kennel write in *Matter in Motion*, "Newton's first two laws had been proposed in various forms by Galileo, Hooke, and Huygens. Newton's third law is completely original, and makes the laws of mechanics logically complete." According to Newton's Third Law of Motion, for every action, there is always an equal and opposite reaction. In other words, all forces occur in pairs of forces that are equal in magnitude and opposite in direction.

This law is perhaps most apparent when considering objects that touch: The downward force of a spoon on the floor is equal to the upward force of the floor on the spoon. The law also holds for objects that gravitationally attract one another. For example, a bird in flight actually pulls up on Earth with the same force that Earth pulls down on the bird. If a person falls to the ground, the person's force on Earth is the same as Earth's force on the person. However, due to the much larger mass of Earth, Newton's Second Law predicts that the acceleration of Earth will be much smaller than the person's acceleration. In outer space, not only does a comet accelerate toward the Sun, but the Sun accelerates toward the comet.

Notice how these examples of Newton's Third Law employ forces of the same kind. As another example, if a grassy lawn exerts a frictional force on the accelerating tires of the toy wagon, Newton's Third Law predicts that a frictional force exists corresponding to the tires pushing backward on the lawn. Newton himself gave an example in his own words: "If you press a stone with your finger, the finger is also pressed by the stone. If a horse draws a stone tied by a rope, the horse will be equally drawn back towards the stone."

The law is sometimes mathematically written as

$$\boxed{\mathbf{F}_{BA} = -\mathbf{F}_{BA}.}$$

If body A exerts a force $\mathbf{F}_{BA}$ on body B, an equal but opposite force $\mathbf{F}_{BA}$ is exerted by body B on A.

One consequence of the Third Law is that the sum of the momenta of two objects that exert force upon each other remains constant through time. This statement assumes that no other forces are interacting with the objects, and it applies reasonably well, for example, to the study of two billiard balls just before and after collision. In fact, Newton studied the momenta $\mathbf{p}_1$ and $\mathbf{p}_2$ of bodies 1 and 2 before and after collisions, and he understood that the momentum was conserved such that the sum of $\mathbf{p}_1 + \mathbf{p}_2$ was constant.

Newton's Third Law even implies that when a basketball player throws a basketball at the floor, Earth should move! However, the basketball player does not observe Earth move for several reasons. For simplicity, assume that the basketball weighs 1 kilogram and moves at 100 kilometers an hour when it hits the basketball court floor that is coated with glue so that the basketball sticks upon impact. The momentum of the ball just before it hits the floor is $\mathbf{p} = m\mathbf{v} = 100$ kg · km/s. After the collision, Earth acquires this momentum. Let's calculate the change in Earth's speed. Let M_E be the mass of Earth, and v_E its change in speed. Then we have $100 = M_E v_E$ or $v_E = 100/M_E$. Given that the mass of Earth is very roughly $M_E = 5.9742 \times 10^{24}$ kilograms, the change in the speed of Earth is roughly $v_E = 1/(5.9742 \times 10^{22})$ km/h—a very small change indeed! In fact, this speed is less than two billionths of a billionth of a centimeter per hour or a speed equivalent to about a proton-width movement a year. Because the mass of Earth is so large, the tiny change in speed is immeasurable. In reality, Earth is not moved at all by the basketball. The interaction between the ball and Earth is dominated by countless other effects, including the conversion of some of the energy of interaction to heating the floor at the point of contact—and the prancing of animals, the crashing of raindrops, and movement of ocean waves.

Today, if you ask your friends if the planets in our Solar System revolve around the Sun, most will say yes. However, because of Newton's Third law (for every action there is an equal and opposite reaction), we know that a planet does not does not orbit around a stationary Sun as Johannes Kepler believed. Instead, Newton suggested that both the planet and the Sun orbit around the common center of mass located between the planet and the Sun, and this requires scientists to modify Kepler's Third Law to make it somewhat more accurate. However, Newton's correction is small because the Sun is much more massive than any of the planets.

For a large planet such as Jupiter, the common center point is actually located outside of the surface of the Sun. Of course, when viewed from an even more complete perspective, the situation becomes more complex. Like all stars, the Sun itself moves through space. It's fast—20 km/s (45,000 miles/hour) with respect to nearby stars. Think of the Sun as a Ferrari that drags along nine planets in its interstellar race. When considering the entire galaxy, the Sun also moves in a nearly circular orbit around the galactic center—with a speed of 220 km/s.

Many physicists whom I consulted said that Newton's Third Law was the most novel and original of Newton's three Laws of Motion. Arons writes in *Development of Concepts of Physics*,

> All the other concepts we have encountered had a prior history of development and discussion, but historians can find no precedent for Law III in the writings of other investigators, nor is there any explicit indication of it in any of Newton's own writings prior to the *Principia* in 1687.

These three laws are the foundations of classical dynamics and were mostly unquestioned until the early 1900s. Today, we know that when objects have velocities approaching the speed of light, Newtonian dynamics can fail and must be considered within the framework of Einstein's Special Theory of Relativity. In particular, for high-speed objects, momentum is not simply the classical $\mathbf{p} = m\mathbf{v}$ but a slightly more complicated expression, namely, $\mathbf{p} = m\mathbf{v}/(1 - \mathbf{v}^2/c^2)^{1/2}$, where c is the speed of light in a vacuum. Similarly, the postulates of quantum theory become important when considering the realm of the ultrasmall, such as size scales involving atoms and subatomic particles.

Technically speaking, the laws of motion are valid only in *inertial reference frames*, which are moving at constant velocity. Although the surface of Earth is not an inertial reference frame—because Earth is rotating on its axis and revolving around the Sun—for many real-world experiments, the surface of Earth can be treated as inertial.

Newton's genius was that he provided a complete system and framework for our understanding of how the universe works for everyday objects traveling at speeds we encounter in our daily lives. We still use these laws for practical problems involving bullets, baseballs, and rockets. While it is true that researchers such as Galileo made significant discoveries in classical mechanics, it was Newton who formulated a complete system.

NEWTON'S LAW OF UNIVERSAL GRAVITATION, 1687

Every material object attracts any other material object with a force that varies directly as the product of the masses of the objects and inversely as the square of the distance that separates the objects. Newton's Law of Universal Gravitation is usually written as follows for two point masses, that is, idealized bodies whose size is very small compared to separation distances:

$$\boxed{F = G\frac{m_1 m_2}{r^2},}$$

where F is the magnitude of the gravitational force between the two masses, G is the gravitational constant, m_1 is the mass of one of the point masses, m_2 is the mass of the other point mass, and r is the distance between the two masses. The value of G is usually given as 6.67×10^{-11} $\mathrm{N \cdot m^2/kg^2}$. Although Newton knew how, in theory, one could attempt to measure G accurately, he lacked the precise instrumentation; however, he did devise a proof for the *constancy* of G. Henry Cavendish, more than 100 years after the publication of Newton's *Principia*, was able to determine a good approximation using delicate torsion balances.

Gravitational interactions do not exist only between Earth and nearby objects or between the Sun and other planets. Gravitational interactions exist between all objects with an intensity that is proportional to the product of their masses. Thus, as you read this book, you are gravitationally attracted to it, and the book to you. Paul Tipler in *Physics* writes:

> It is gravity that binds us to the Earth and that binds the Earth and the other planets to the Solar System. The gravitational force plays an important role in the evolution of stars and in the behavior of galaxies. In a sense, it is gravity that holds the universe together.

Newton used this law to explain the motions of planets, which Kepler described in a rigorous fashion but did not explain using the forces of gravity. Newton justified his law, in part, by demonstrating that an inverse-square force could cause elliptical orbits, thus deriving Kepler's Laws of Planetary Motion. Newton, however, never did understand the "cause" of gravity and how it could propagate through the vacuum of space. Not until the theories of Albert Einstein did we view gravity as a consequence of curved space. (Of course, some scientists may similarly suggest that Einstein did not understand the ultimate "cause" of curved space and that most scientific explanations simply attempt to organized and predict observed patterns.) In 1693, Newton admitted his lack of understanding in a letter he wrote to the minister and theologian Richard Bentley (1662–1742):

> I have not yet been able to discover the cause of these properties of gravity from phenomena, and I frame no hypotheses.... It is enough that gravity does really exist and acts according to the laws I have explained, and that it abundantly serves to account for all the motions of celestial bodies. That one body may act upon another at a distance through a vacuum without the mediation of anything else, by and through which their action and force may be conveyed from one another, is to me so great an absurdity that, I believe, no man who has in philosophic matters a competent faculty of thinking could ever fall into it.

Newton contemplated the possibility that some kind of magnetic-like force held the planets to the Sun. French philosopher and scientist René Descartes (1596–1650), one of his contemporaries, thought that giant tornado-like forces made the planets spin around the Sun—the planets were caught in the Sun's whirlwind like little boats going round and round in a maelstrom.

Today, astrologers sometimes suggest that the planetary arrangements can influence our lives due to their gravitational effects. However, using Newton's Law of Universal Gravitation, we can show that the gravitational pull from the planets on our bodies is very weak because of their great distances. How could the arrangement of planets at the time of your birth affect your temperament and interests, given that the doctor in the delivery room had a much greater gravitational effect on you than Mars or

Venus? If gravity were the force behind astrology, then the Moon would dominate all the planets combined. Yet, astrologers do not give the Moon a dominant role in their prognostications.

Returning our attention to Newton and the third book of his *Principia*, we find that Newton used his gravitational law to explain the orbits of planets and moons, the paths of comets, and the motion of the Moon as perturbed by the gravity of the Sun. His gravitational calculations showed him that many other possible paths for bodies in outer space existed in addition to elliptical orbits. As discussed in the entry for Kepler's Laws of Planetary Motion, other paths in the shape of conic sections are possible, depending on the speed of the object in a gravitational field. Some comets travel on parabolic paths, looping around our Sun and never returning. If the velocity of Earth were suddenly increased by a factor of roughly 1.4, its elliptical orbit would become parabolic, and we would shoot off out of the Solar System. How long could we survive if some cosmic hand gave us this dreadful push?

Throughout history, Newton's Law of Universal Gravitation allowed scientists to make valuable predictions. For example, the law was used to help discover the planet Neptune. For many years, scientists had observed strange variations in the motion of Uranus. The French astronomer Urbain Leverrier (1811–1877) and British astronomer John Couch Adams (1819–1892) independently used Newton's Law of Universal Gravitation to predict the existence of a more distant planet that was affecting the orbit of Uranus. Neptune was discovered in 1864, traveling in an orbit close to its predicted position. Arons writes in *Development in Concepts of Physics* about the predictive and explanatory value provided by Newton's Law of Universal Gravitation:

> The *theory* of gravitational forces, whose main hypothesis is the attraction of all particles of matter for one another, yields the derived *law* of universal gravitation, which in turn explains, as we have seen, Kepler's empirical laws and a wealth of other phenomena. Since one purpose of any theory is this type of explanation and summary, Newton's theory strikes us as eminently satisfactory.

Note that Newton's Law of Universal Gravitation applies to point masses or objects that can be considered as point masses. For more complicated shapes, the law is an approximation—and the distance r that is used in the equation is usually not obvious. In examples that involve a uniform spherical shell, the shell gives rise to gravitational forces outside the sphere that are identical with the force exerted by a point mass at the center of the sphere. More generally, an object with a spherically symmetric distribution of mass exerts the same gravitational attraction on other objects as if all of the mass of the object were concentrated at a point at its center.

If we want to accurately use this law for objects with various shapes, we can regard each extended body as being made of numerous point masses, and use integral calculus to compute the gravitational forces between such oddly shaped objects. Interestingly, Newton's Law of Universal Gravitation is independent of the presence of other bodies in the space between two objects, and there is no such thing as a "gravity screen" that shields the gravitational attraction of one object for another.

Although we now know that Einstein's General Theory of Relativity—which represents gravitation as curved space—does a better job at describing motions in high gravitational fields (e.g., the revolution of Mercury about the Sun), Newtonian mechanics still describes the world of ordinary experience.

Einstein also suggested that gravitational effects move at the speed of light. Thus, if the Sun were suddenly plucked from the Solar System, Earth would not leave its orbit about the Sun until about eight minutes later, the time required for light to travel from the Sun to Earth. Many physicists today believe that gravitation must be quantized and take the form of particles called gravitons, just as light takes the form of photons, which are tiny quantum packets of electromagnetism.

Let's review Einstein's theory of gravity in slightly greater detail. In 1915, ten years after Einstein proclaimed his *Special* Theory of Relativity (which suggested that distance and time are not absolute and that measurements of the ticking rate of a clock depend on your motion with respect to the clock), Einstein gave us his *General* Theory of Relativity, which explained gravity from a new perspective. In particular, Einstein made the startling suggestion that gravity is not really a force like other forces, but results from the curvature of space-time caused by masses in space-time. The result is that certain objects, such as the planet Mercury, have slightly different trajectories in Newton's and Einstein's universe. In fact, before Einstein's General Theory of Relativity, the 1910 edition of the *Encyclopaedia Britannica* actually stated that Newton's Law of Universal Gravitation has a relationship involving $r^{2.0000001612}$ instead of r^2 in order to better predict the motions of the planet Mercury.

To best understand Einstein's theory of gravity, consider that wherever a mass exists in space, it warps space. Imagine a bowling ball sinking into a rubber sheet. This is a convenient way to visualize what stars do to the fabric of the universe. If you were to place a marble into the depression formed by the stretched rubber sheet, and give the marble a sideways push, it would orbit the bowling ball for a while, like a planet orbiting the Sun. If you rolled a marble on the sheet far from the bowling ball—where there is no obvious depression in the rubber sheet—the marble would not be affected. The warping of the rubber sheet by the bowling ball is a metaphor for a star warping space. Physicists describe this with the motto, "Matter makes space bend. Space tells matter how to move." Even light rays are bent by the curvature of space. As a result, the apparent positions of stars that we see in the night sky can be far from their actual positions.

Time is also distorted in regions of large masses, and, according to Einstein, the very existence of time depends on the presence of space. Einstein's General Theory of Relativity can be used to understand how gravity warps and slows time, and it explains why time moves slightly slower for you in the basement of your house than on the top floor where gravity is slightly weaker. Not only does General Relativity permit time travel, but it actually seems to encourage time travel in strange ways that I discuss in many of my books, such as *Time: A Traveler's Guide*. All of these theories are not speculations, but rather they have been verified by experiments and observations.

I should point out that today a small number of physicists are exploring a controversial alternative to Newton's Law of Universal Gravitation called modified Newtonian dynamics, or MOND, in order to explain certain mysteries of gravity at galactic scales. For many decades, astronomers have realized that galaxies are spinning so quickly that they should tear themselves apart, and these astronomers have postulated the existence of *dark matter* in order to provide the additional gravity needed to hold galaxies together. Supporters of MOND suggest an alternate explanation—that galaxies may stay intact not because of unseen matter but because gravity itself does not weaken as suggested by the $1/r^2$ term in Newton's Law of Universal Gravitation. However, most astronomers feel that in order to posit such a revolutionary change in our understanding of gravity, firmer evidence is required than we have today.

Today, many physicists suggest that universes exist that are parallel to ours, like layers in an onion, and that we might detect them by gravity leaks from one layer to an adjacent layer. For example, light from distant stars might be distorted by the gravity of invisible objects residing in parallel universes only millimeters away. Since 1997, scientists at the University of Colorado at Boulder have conducted experiments to search for these possible nearby universes. If they can observe deviations in the inverse-square principle of Newton's Law of Universal Gravitation, they feel that this might be evidence for matter in parallel universes or in hidden dimensions. The whole idea of multiple universes is not as far-fetched as it may sound. According to a recent poll of seventy-two leading physicists conducted by the American researcher David Raub, 58% of physicists (including Stephen Hawking) believe in some form of multiple-universes theory.

NEWTON'S LAW OF COOLING, 1701

According to Newton's Law of Cooling, the rate of heat loss of a body is proportional to the difference in temperature between the body and its surroundings. Today, we often write this law as

$$\boxed{T(t) = T_{env} + [T(0) - T_{env}]e^{-kt},}$$

where T is the temperature, t is time, T_{env} is the temperature of the environment, $T(0)$ is the initial temperature of the object, and k is a positive constant. For example, imagine that you have a pot of boiling tomato soup and place it in a sink of cold water kept at a constant 4°C by running water through the sink. You stir the soup as it cools. The rate of change of the temperature of the soup is governed by $T(0) = 100$ and $T_{env} = 4$, so $T(t) = 4 + 96e^{-kt}$. If we had an additional measurement, such as the temperature of the soup after 5 minutes, we could calculate the value of k and then have a complete equation that specifies the temperature of the soup at any time t.

My more macabre teachers once explained to me that we could use Newton's Law of Cooling to find the time of death for a corpse discovered in a motel room if we know the temperature of the room, which is assumed to be constant. For example, you are called to a crime scene at the edge of town, and you find a woman's body lying on the carpet of some rundown motel room. Her name is Monica. She is dead. Her corpse is at a temperature of 80°F. The temperature of the room is a cool 60°F. An hour later, the temperature of the corpse drops to 75°F. This is all the information you need to find the approximate time of Monica's death if you assume she had a normal body temperature of about 98.6°F while alive.

In the case of Monica's corpse, the time of death will only be approximate. Newton's Law of Cooling assumes a uniform temperature of the cooling body. However, in reality, a human body is not uniformly warm, and the skin is certainly cooler than the internal organs. Nevertheless, Newton's Law of Cooling gives a good approximation of the time of death for an individual.

Isaac Newton (1642–1727), British mathematician and physicist famous for his laws of motion, gravitation, and cooling, his theories on light and color, and his development of calculus.

CURIOSITY FILE: When Newton's body was exhumed, scientists discovered that it contained large amounts of mercury, probably resulting from his research in alchemy. • Newton would have kept many of his discoveries to himself had he not been coaxed by colleagues to publish. • As he became older, Newton had many portraits painted and seemed concerned with leaving the image of his face for posterity. • Newton believed that metals could be considered "living opposites" of trees, growing underground rather than above ground. • Many thousands of print and Internet sources suggest that Diamond, Newton's dog, jumped onto a table and knocked over a candle, setting Newton's papers on fire and destroying many years worth of work. However, biographer Milo Keynes suggests that Newton never owned a dog and that the story is probably apocryphal. • In the year 628, the Indian astronomer Brahmagupta suggested that gravity was a force of attraction. He used the Sanskrit term "gruhtvaakarshan" for gravity.

> Had Newton not been steeped in alchemical and other magical learning, he would never have proposed forces of attraction and repulsion between bodies as the major feature of his physical system.
>
> —John Henry, "Newton, Matter, and Magic," in John Fauvel et al.'s *Let Newton Be!*

> Newton was the greatest genius that ever existed and the most fortunate, for we cannot find more than once a system of the world to establish.
>
> —Joseph Louis Lagrange, *Oeuvres de Lagrange*, 1867

> God gave . . . the Prophecies of the Old Testament, not to gratify men's curiosities by enabling them to foreknow things, but that after they were fulfilled, they might be interpreted by the event, and his own Providence, not the Interpreters, be then manifested thereby to the world. For the event of things predicted many ages before will then be a convincing argument that the world is governed by Providence.
>
> —Isaac Newton, *Observations upon the Prophecies of Daniel and the Apocalypse of St. John*, 1733

> This law [of gravitation] has been called "the greatest generalization achieved by the human mind". . . . I am interested not so much in the human mind as in the marvel of a nature which can obey such an elegant and simple law as this law of gravitation. Therefore our main concentration will not be on how clever we are to have

> found it all out, but on how clever nature is to pay attention to it.
>
> —Richard Feynman, *The Character of Physical Law*

> What water is to a fish or air is to a bird, mathematics was to Newton, the element through which he moved without effort.
>
> —Richard Westfall, "Newton's Scientific Personality"

Isaac Newton was an English mathematician, physicist, and astronomer who invented calculus, proved that white light was a mixture of colors, explained the rainbow, built the first reflecting telescope, discovered the binomial theorem, introduced polar coordinates, and showed the force causing apples to fall is the same as the force that drives planetary motions and produces tides. He was also author of treatises on Biblical subjects such as Biblical prophecies. In fact, he devoted more time to the study of the Bible, theology, and alchemy than to science—and he wrote more on religion than he did on natural science.

Newton spent a major portion of his life analyzing ancient Biblical texts, and his scientific discoveries are even more striking considering the relative amount of time he devoted to them. Economist and philosopher John Maynard Keynes (1883–1946), who studied Newton's writings on alchemy and authored a biographical essay on Newton, wrote "Newton was not the first of the age of reason: he was the last of the magicians."

Perhaps less well known is the fact that Newton was a creationist who wanted to be remembered as much for his theological writings as for his scientific and mathematical texts. Newton believed in a Christian unity as opposed to a trinity. In particular, he believed that Jesus was created by God as a mortal human and not as God incarnate, and he noted, "It is the temper of the hot and superstitious part of mankind in matters of religion ever to be fond of mysteries, and for that reason to like best what they understand least."

Some of Newton's nontraditional Christian beliefs are echoed in the faiths of America's founding fathers. None of the first five U.S. presidents was a conventional Christian. For example, John Adams, a Unitarian, did not accept the notion of the trinity or the divinity of Christ. In 1804, Thomas Jefferson used a razor to remove all passages of the King James Version of the New Testament that had supernatural content—such as the virgin birth, resurrection, or turning water into wine. About one-tenth of the bible remained, which he pasted together and published as *The Philosophy of Jesus of Nazareth*. Apparently, Jefferson admired Jesus as a teacher and prophet but was not always interested in the cloak of divinity.

Newton respected the Bible and accepted its account of Creation while at the same time looking for various Bible codes and hidden messages. He also wrote works that criticized various Biblical interpretations, such as *An Historical Account of Two Notable Corruptions of Scripture*. He determined that date of the crucifixion of Jesus Christ to be April 3, 33 A.D., a date accepted by some today. Newton developed calculus as a means of describing motion, and perhaps for understanding the nature of God through a clearer understanding of nature and reality.

Before addressing the events of his life, it is intriguing to achieve a wider understanding of his diverse interests. An examination of the 1,752 books in Newton's personal library after his death has verified his strong interest in occult and religious subjects. The most complete breakdown I have seen comes from John Harrison's *The Library of Isaac Newton*, which focuses on certain categories of his books and is summarized in table 4. Notice that only about 12% of the books in the library deal with the topics for which he is most famous today. He was indeed a biblical fundamentalist, believing in the reality of angels, demons, and Satan. He believed in a literal interpretation of Genesis and that Earth was only a few thousand years old. Newton spent much of his life trying to prove that the Old Testament is accurate history. His book on the Bible suggested that Christianity took a wrong turn in the fourth century A.D., when the first Council of Nicaea promoted erroneous doctrines on the nature of Jesus.

TABLE 4 Number of Books in Isaac Newton's Personal Library, by Topic

Topic	Number of Books
Theology	477 (27.2%)
Alchemy	169 (9.6%)
Mathematics	126 (7.2%)
Physics	52 (3.0%)
Astronomy	33 (1.9%)

Source: John Harrison, *The Library of Isaac Newton* (Cambridge, U.K.: Cambridge University Press, 1978).

One wonders how many more problems in physics Newton would have solved if he had spent less time on his Biblical studies. Newton appears not to have had the slightest interest in sex, never married, and, according to his contemporaries, almost never laughed (although he sometimes smiled). Newton suffered a massive mental breakdown, and some have conjectured that throughout his life he had manic depression (bipolar disorder), with alternating moods of sadness and happiness. Some recent historians of science have suggested that Newton had Asperger syndrome, a high-functioning form of autism. Milo Keynes writes in "The Personality of Isaac Newton":

> Isaac Newton was a humorless, solitary, anxious, insecure, and private man with obsessional traits. He was poor at human relationships, such as the expression of gratitude, and held unorthodox and heretical religious beliefs. He was clearly puritanical, with feelings of guilt, and had little capacity for enjoyment—his only strong liking appears to have been one for roast beef.... He never used the word "love"....

According to biographer Anthony Storr, "Newton's preoccupation with place seeking... may be traced to his ambition: his fear of being embroiled with women to his almost total suppression of sexuality."

Bill Bryson in *A Short History of Nearly Everything* focuses on some of Newton's quirks:

> Newton was a decidedly odd figure... famously distracted (upon swinging his feet out of bed in the morning he would reportedly sometimes sit for hours, immobilized by the sudden rush of thoughts to his head), and capable of the most riveting strangeness.... Once he inserted a bodkin [a long leather sewing needle] into his eye socket just to see what would happen.

Isaac Newton was born in Woolsthorpe-by-Colsterworth, England, on Christmas Day in 1642, according to the calendar in use when he was born. Today, we would give the date as January 4, 1643, according to the corrected Gregorian calendar date now in use. Newton never knew his father, also named Isaac Newton, who died three months before his son's birth. Newton's father was illiterate and could not even sign his own name. Richard Westfall writes of the year of Newton's birth in *Never at Rest: A Biography of Isaac Newton*:

> Since Galileo, on whose discoveries much of Newton's own career in science would squarely rest, had died that year, a significance

> attaches itself to 1642. . . . Born in 1564, Galileo had lived nearly to eighty. Newton would live nearly to eighty-five. Between them, they virtually spanned the entire scientific revolution, the central core of which their combined work constituted.

Newton was so small at birth that the villagers expected him to die. In the first few years of his life, his neck was weak and required a neck brace to give his head additional support.

Newton's mother remarried when he was two and sent him away to be raised by his grandmother. Newton always hated his mother and stepfather and as a teenager threatened to burn them alive in their house.

While away from his mother's home, Newton tended to be rather shy and quiet. Michael Guillen notes in *Five Equations That Changed the World*, "Whenever Newton did socialize, it was with girls; they were tickled by the doll furniture and other toys he made for them using his customized kit of miniature saws, hatchets, and hammers."

When he went to the local Grantham's Free Grammar School, his teachers initially called him both idle and inattentive, and he had the next-to-lowest rank in the entire school! However, later he decided to change his standing in his class and made his way to first place. A teacher said of 18-year-old Newton, "His genius now begins to mount upwards apace and shine out with more strength."

In 1661, Newton entered Trinity College, Cambridge. Although his mother was wealthy at this point in her life, she refused to pay for Newton's tuition, which forced Newton to earn his keep by emptying chamber pots and grooming the hair of older, richer students. Like other lawgivers in this book, his initial intent was to seek a law degree, but his interest in science soon pushed aside everything else. Although he was shy among people, he was not shy about performing outrageous experiments on himself—such as the time when he wedged probes between his eye and his eye socket "as near to the backside of his eyes as possible" so that he could better understand the workings of the human visual system.

Newton's interest in mathematics germinated in 1663 when he purchased an astrology book and discovered that he could not understand the mathematics in the book. He then voraciously tackled several geometry and algebra books, and received his Bachelor's Degree from Cambridge in 1665. His mathematical prowess was not fully apparent at this point in life—all it would take was a plague to release his inner genius.

In 1665, the Great Plague, also known as the Black Death, struck the people of London. The plague germs were carried by fleas, which were carried by rats. At first, authorities ignored the deaths in London, but as summer approached, more people died, and people panicked. The rich left the city for their estates in the country, and the merchants followed soon after. By June, the roads were flooded with people fleeing London. By July, more than a thousand people died in London each week. Dogs and cats were suspected to be carriers, so the Lord Mayor had an estimated 40,000 dogs and 200,000 cats killed. Alas, this caused an increase in the rat population, so the germs spread more rapidly. An estimated 100,000 people died in London and its vicinities.

Without the Great Plague, Newton may never have discovered his great laws. Cambridge closed the university in the summer of 1665, and during this time Newton went home to Woolsthorpe. Here, in relative isolation, he cogitated upon what he had learned at Cambridge. Newton returned to Cambridge in March of 1666 when the school reopened after the plague subsided over the winter. The plague reappeared, so Cambridge was closed again in June of 1666 until April of 1667, when Newton again returned to Trinity College.

Today, Newton's biographers sometimes refer to the 1666 date as his famous "annus mirabilis," or year of miracles. During that year, according to Newton's later writings, he discovered the chromatic composition of light and discovered the inverse-square principle of his Law of Universal Gravitation. He established the foundations of differential and integral calculus, several years before its independent discovery by German mathematician Gottfried Wilhelm Leibniz (1646–1716). However, upon critical analysis, it is likely that 1666 may be a little early to anchor some of his great ideas, and it was in Newton's own self-interest to place his discoveries as early as possible. It is more likely that some of his ideas on mechanics did not fully gel until around 1685–1687, when he was actually composing his book the *Principia*.

Nevertheless, Newton did indeed devote much of his time during the plague to the intense study of mathematics and physics, and the seeds of many of his greatest ideas started to take form. In a period of less than two years, while Newton was still younger than 25, he began his important advances in mathematics, optics, and astronomy.

Following his death, Newton's draft letter to the biographer and journalist Pierre Des Maizeaux (1673–1745) was found, and it describes Newton's recollections of his accomplishments during the plague years. His letter first describes the year 1665, during which he says he invented calculus and discovered methods for approximating mathematical series. In 1666, he says he developed his theories of colors. He explains further about his discovery of his Law of Universal Gravitation:

> [In 1666] I began to think of gravity extending to the orb of the Moon, and (having found out how to estimate the force with which a globe revolving within a sphere presses the surface of the sphere) from Kepler's rule [Kepler's Third Law]..., I deduced that the

> forces which keep the Planets in their Orbs must be reciprocally as the squares of their distances from the centers about which they revolve: and thereby compared the force requisite to keep the Moon in her Orb with the force of gravity at the surface of the earth, and found them answer pretty nearly. All this was in the two plague years of 1665 and 1666. For in those days, I was in the prime of my age of invention, and minded Mathematics and Philosophy more than at any time since.

In other words, in 1666, Newton began to consider that Earth's gravity influenced the Moon, counterbalancing its centripetal force. From these thoughts on centripetal force and Kepler's Third Law of Planetary Motion, Newton deduced the inverse-square principle.

In the field of mathematics, Newton is generally credited with the discovery of the binomial theorem, an important formula giving the expansion of powers of sums. Newton discovered what we now refer to as Newton's identities (which describe the roots of a polynomial) and Newton's method (a procedure for finding approximations to the roots of a function). In 1666, Newton also determined π to sixteen places using twenty-two terms of this series:

$$\pi = \frac{3\sqrt{3}}{4} + 24\left(\frac{1}{12} - \frac{1}{5 \cdot 2^5} - \frac{1}{28 \cdot 2^7} - \frac{1}{72 \cdot 2^9} - \ldots\right)$$

Regarding this queer formula, in 1666 he wrote, "I am ashamed to tell you to how many figures I carried these computations, having no other business at the time."

In 1687, Newton finally published the *Philosophiae naturalis principia mathematica* (Mathematical Principles of Natural Philosophy), now called the *Principia* for brevity. During his writing of the *Principia*, Newton is said to have spent eighteen months in which he often neglected to eat or sleep and in which he sometimes remained motionless for hours lost in thought, never leaving his room. Although Newton had thought about the behavior of moving bodies and gravitation in 1666, his complete mathematical theory was set down in these eighteen months of writing. According to Abers and Kennel in *Matter in Motion*, from Newton's pen suddenly burst forth

> definitions, laws, theorems, the law of universal gravitation, the explanation of Kepler's three laws, the theory of motion in a vacuum, the theory of motions in fluids (a forerunner of modern aerodynamics), the theory of waves in fluids, the study of the tides and of comets, a theory describing the small deviations of the earth's shape from a perfect sphere, an exact calculation of the precession of the

> equinoxes.... Single-handedly, Newton demonstrated that science need not be qualitative, as it was for Aristotle, or mathematically precise only about ideal situations, as it was for Galileo, but that it could describe God's real universe with great precision.

Book I of the *Principia* discusses the motions of objects in a vacuum, subject to various forces. In this book, Newton states his three laws of motion. Book II discusses the motion of fluids and the motions of objects in resistive media, discussing, for example, how air resistance slows the fall of a ball as it plummets to Earth. Newton discovered that the faster the ball falls from a great height, the larger the force of air resistance. At some point during the descent of the ball, the force of air resistance balances the force of gravity and the ball stops accelerating. We are lucky to have this effect, which prevents raindrops from hitting our skulls at hundreds of miles per hour even though they have fallen for several miles.

Book III of the *Principia* discuses the force of gravity and how a mutual attraction is exerted between any two masses, regardless of their sizes. In Book III, Newton never explained how gravity propagated between objects. According to Abers and Kennel,

> Newton would guess the laws of physics, and then check his guess by proving that the laws explain everything to which they are to apply. Newton's method does not penetrate to "ultimate" reality: it merely strives to put the observed phenomena of nature into a logically consistent order.

The *Principia* concludes with a prose hymn to God, because Newton envisioned that the *Principia* was a testament both to science and to God, and as much a contribution to science as to theology.

In his later letters to Minister Bentley, Newton explained *why* he believed that an intelligent being created the cosmos. Newton marveled at the fact that most of the objects that orbit the Sun are contained with an "ecliptic" plane. Most planets are contained in the same plane as Earth's orbital plane, offset by just a few degrees. He reasoned that natural processes could not create such behavior. This, he argued, was evidence of design by a benevolent and artistic creator. Today, scientists believe that the formation process of the Solar System naturally produced a disk of material out of which formed the Sun and the planets.

Newton also suggested that God was responsible for giving the planets their initial velocities in orbit, without which the planets would have fallen into the Sun. What else could have given the planets just the right push to ensure that they assumed concentric orbits? When he looked at the Solar System and the stars in the sky, he simply saw too much order to neglect the evidence of the helping hand of God. In his book *Opticks* (1706), he wrote:

> Whence is it that the Sun and Planets gravitate toward one another, without dense Matter between them? . . . What hinders the fixed stars from falling upon one another? . . . Does it not appear from Phenomena that there is a Being, incorporeal, living, intelligent, omnipresent, who in infinite space . . . sees the things themselves intimately and thoroughly perceives them, and comprehends them wholly?

At one point in his life Newton thought of the universe as "God's Sensorium" in which the objects in the Universe—their motions and their transformations—were the thoughts of God, literally the activities of His mind. God's thoughts were grand, and the speed at which God thought determined the pace of the universe's evolution.

Newton's important contribution to the field of fluid dynamics involved the development of calculus and his fundamental laws of physics. His name is also attached to a linear relationship, involving stresses in certain kinds of liquids. A "Newtonian fluid" flows like water. In particular, a Newtonian fluid's shear stress is linearly proportional to the velocity gradient in the direction perpendicular to the plane of shear. In equation form, we have $\tau = \mu(d\mathbf{v}/dx)$, where τ is the shear stress exerted by the fluid, μ is the fluid viscosity and is considered a constant of proportionality, and $d\mathbf{v}/dx$ is the velocity gradient. This is sometimes referred to as Newton's Law of Viscosity (1687). The viscosity of Newtonian fluids depends on temperature and pressure but not on the forces acting on the fluid. On the other hand, an example of a non-Newtonian fluid is the popular Silly Putty toy. It bounces. It breaks like a solid when a rapid force is applied to it. It can flow like a liquid. It appears to melt into a puddle over a long period of time.

Quicksand is another example of a non-Newtonian fluid. If you should someday find yourself in quicksand and move slowly, the quicksand will act like a liquid and you will be able to escape more easily than if you move around very quickly, because fast motions can cause the quicksand to act more like a solid from which it is harder to escape.

Let us return to Newton's years immediately after the Great Plague and after he had returned to Cambridge—a period of time during which his mathematical abilities started to become well known. In 1669, at the age of 26, he became Lucasian Professor of Mathematics. (The current Lucasian Professor of Mathematics is the famous physicist Stephen Hawking, who was appointed in 1980.) Newton's interest in optics continued, and he became convinced that white light was not the single entity that

Aristotle believed it to be but rather was a mixture of many different rays corresponding to different colors. The English physicist Robert Hooke (see "Hooke's Law of Elasticity," above) criticized Newton's work, which filled Newton with a rage that seemed out of proportion to the comments Hooke had made. As a result, Newton withheld publication of his *Opticks* until after Hooke's death in 1703—so that Newton could have the last word on the subject of light and could avoid all arguments with Hooke. In 1704, Newton's *Opticks* was finally published. In this work, Newton discusses his investigations of colors and the diffraction of light.

I should note that Book III of the *Principia* almost did not get published. Hooke had claimed (perhaps with some justification) that the content of his letters written to Newton during 1679 and 1680 should justify Hooke being acknowledged as playing a role in Newton's discoveries relating to gravity. Newton was so angry with Hooke that he threatened to suppress Book III. Newton finally published the work, but systematically deleted nearly every mention of Hooke's name from the book. Hooke had considered the idea of an inverse-square principle of gravity, but he did so on intuitive grounds and without derivation as Newton had done.

In 1672, Newton was elected a fellow of the Royal Society. In the years during which he was writing the *Principia*, his lecture skills at Cambridge were evidently not impressive—often so few students went to hear Newton's lectures that he read to the walls. According to William Bixby's *The Universe of Galileo and Newton*:

> To some students, he was the personification of an absent-minded professor. He was careless in his manner of dress, often appearing before his students in a state of disarray. He was unconcerned with externals—so much so that even if nobody attended one of his classes, he would deliver his lecture with as much satisfaction as if the hall had been crowded with listeners.

Newton had more than one nervous breakdown, and after suffering one in 1693, he retired from research. Some have speculated that his breakdown was caused by chemical poisoning as a result of his alchemy experiments (also known as *chrysopoeia*, the attempt to turn other metals into gold) or, discussed above, that he simply suffered from clinical depression or bipolar disorder. In 1696, Newton left Cambridge to take a government position in London, becoming Warden of the Royal Mint. It is interesting to note that Newton would spend more time at the mint than he did as a productive scientist, and we can wonder how much more the world would have gained had his depression not plagued him, and thus he might have been able to spend more of his time at scientific research. At the mint, he proposed measures to prevent counterfeiting of coins. Unscrupulous individuals sometimes trimmed the edges off coins in order to use the clippings to create new coins. To prevent this trimming, Newton had the mint create engraved edges, or serrations, on coins so that merchants could easily confirm that their coins were intact. Today, many of our own coins have ridged edges.

Legend has it that Newton sometimes put on a disguise and went into bars and brothels where he would quietly listen to other people's conversations in order to trap counterfeiters. One of the most famous counterfeiters begged Newton for his life before he was hung, drawn, and quartered.

What are we to make of Newton's total contribution to mathematics and his mathematical abilities as a young man and later in his life? According to I. B. Cohen in the *Dictionary of Scientific Biography*,

> Newton appears to have had no contact with higher mathematics until . . . age 23. . . . [Yet] any summary of Newton's contributions to mathematics must take account not only of his fundamental work in the calculus and other aspects of analysis—including infinite series (and most notably the general binomial expansions)—but also his activity in algebra and number theory, classical and analytical geometry, finite differences, the classification of curves, methods of computation and approximation, and even probability.

Newton's mathematical prowess was still great, even in his later years. One story suggests that in 1696, Swiss mathematician Johann Bernoulli (1667–1748) (see "Bernoulli's Law of Fluid Dynamics" in part II) posed a set of difficult mathematical problems intended for mathematicians to ponder in the coming century. Bernoulli's challenge started,

> I, Johann Bernoulli, address the most brilliant mathematicians in the world. Nothing is more attractive to intelligent people than an honest, challenging problem, whose possible solution will bestow fame and remain as a lasting monument. . . . If someone communicates to me the solution of the proposed problem, I shall publicly declare him worthy of praise.

Newton received the list and, one evening after work as Master of the Mint, chose a problem and solved it before going to bed. He gave the solution to his secretary at the Royal Society and told him to send it to Bernoulli anonymously. When Bernoulli saw the solution, he easily guessed that Newton was the author, and said, "The lion is known by his claw."

In 1701, Newton's Law of Cooling was published anonymously in a paper titled "Scala graduum caloris." The law states that the rate of heat loss of an object is proportional to the difference in temperatures between the object and its surroundings. Newton's paper on cooling, published in the *Philosophical Transactions of the Royal Society*, also focused on his attempts to establish a temperature scale, using a thermometer containing linseed oil.

In 1703 Newton was elected president of the Royal Society, and he was knighted in 1705. His remaining years were often filled with rage as a result of his argument with Leibniz as to who discovered calculus. Some have speculated that Newton's angry personality had its seeds in his childhood. John Fauvel and colleagues in *Let Newton Be!* note, "This separation from his mother, between the ages of three and ten, [was] crucial in helping to form the suspicious, neurotic, tortured person of the adult Isaac Newton."

Newton and Leibniz are generally credited with the co-invention of calculus, but various earlier mathematicians explored the concept of rates and limits, starting with the ancient Egyptians who developed rules for calculating the volume of pyramids and approximating the areas of circles.

During their lives, both Newton and Leibniz puzzled over problems of tangents, rates of change, minima, maxima, and infinitesimals (unimaginably tiny quantities that are almost but not quite zero). Both men understood that differentiation (finding tangents to curves) and integration (finding areas under curves) are inverse processes. Newton's discovery (1665–1666) started with his interest in infinite sums; however, he was slow to publish his findings. For example, his early work on calculus, *De methodis serierum et fluxionum*, was circulated in a manuscript to several peers in 1671, but it did not officially appear in print until 1736 as an English translation.

Leibniz had published his discovery of differential calculus in 1684 and of integral calculus in 1686. He had said, "It is unworthy of excellent men, to lose hours like slaves in the labor of calculation.... My new calculus... offers truth by a kind of analysis and without any effort of imagination." Newton was outraged. Debates raged for many years on how to divide the credit for the discovery of calculus, and as a result, progress in calculus was delayed.

I conclude this entry on Newton with some additional enlightening comments on his personality, his extreme sensitivity to criticism, and his scientific achievements. In 1908, W. W. Rouse Ball wrote in the fourth edition of *A Short Account of the History of Mathematics*:

> As to his manners, he dressed slovenly [and had] extreme absence of mind when engaged in any investigation.... On the few occasions when he sacrificed his time to entertain his friends, if he left them to get more wine or for any similar reason, he would as often as not be found after the lapse of some time working out a problem, oblivious alike of his expectant guests and of his errand.

He took no exercise, indulged in no amusements, and worked incessantly, often spending eighteen or nineteen hours out of the twenty-four in writing.... With the exception of his papers on optics, every one of his works was published only under pressure from his friends and against his own wishes. There are several instances of his communicating papers and results on condition that his name should not be published.

John Maynard Keynes wrote in "Newton, the Man":

> I believe that the clue to his mind is to be found in his unusual powers of continuous concentrated introspection.... His peculiar gift was the power of holding continuously in his mind a purely mental problem until he had seen straight through it. I fancy his pre-eminence is due to his muscles of intuition being the strongest and most enduring with which a man has ever been gifted.... I believe that Newton could hold a problem in his mind for hours and days and weeks until it surrendered to him its secret.

Abers and Kennel suggest in *Matter in Motion*:

> Newton was the one man who was in equal measure a creative mathematician and a creative physicist. He was one of the few physicists equally adept at theory and at experiment. His invention of the reflecting telescope, much less celestial mechanics, would have ensured him a prominent place in the history of astronomy. Newton's contemporaries remarked upon his extraordinary intuition. He seemed to know things that even he could not prove.

Toward the end of his life, Newton was troubled with a painful kidney stone and perhaps urinary incontinence. He had gout and inflammation of the lungs. On February 28, 1727, he went to London to preside at a meeting of the Royal Society, but his health was in sharp decline and his kidney stone was still torturing him. On March 20, 1727, he died in London at the age of 84, in great pain from the stone. He was buried in Westminster Abbey, the first scientist to be accorded this honor. A lunar crater with a diameter of 78 kilometers was named after Newton and approved in 1935 by the International Astronomical Union General Assembly. Among his most important published works are the following:

- 1671, *De methodis serierum et fluxionum* (On the Methods of Series and Fluxions). ("Fluxion" was Newton's term for derivative in the field of calculus.) As mentioned, this work did not officially appear in print until 1736.

- 1672, Newton's edition of *Geographia generalis* (General Geography) by the German geographer Varenius.
- 1672–1676, letters on optics.
- 1684, *De motu corporum in gyrum* (On the motion of bodies in an orbit).
- 1687, *Philosophiae naturalis principia mathematica* (published in Latin in 1687; revised in 1713 and 1726; and translated into English in 1729).
- 1704, *Opticks* (a revised edition appeared in Latin in 1706). Appendices contained discussions of cubic curves, infinite series, and the method of fluxions.
- 1701–1725, *Reports as Master of the Mint.*
- 1707, *Arithmetica universalis.*

The following is a sampling of his works published after his death:

- 1728, *The Chronology of Ancient Kingdoms Amended*
- 1728, *The System of the World*
- 1728, *Optical Lectures*
- 1728, *Universal Arithmetic* (English version of *Arithmetica universalis*)
- 1733, *Observations upon the Prophecies of Daniel and the Apocalypse of St. John*
- 1754, *An Historical Account of Two Notable Corruptions of Scripture*

Philosopher David Hume (1711–1776) in *History of England* remarked, "In Newton, this island may boast of having produced the greatest and rarest genius that ever rose for the ornament and instruction of the species." French mathematician Guillaume L'Hôpital (1661–1704) wondered about Isaac Newton, writing, "Does he eat, drink and sleep like other men? I cannot believe otherwise than that he is a *genius*, or a *celestial intelligence* entirely disengaged from matter." Patricia Fara in *Newton: The Making of Genius* sums up Newton's unusual breadth of interests and his contentment to always stay close to home:

> Even the briefest survey of Newton's life unsettles his image as the idealized prototype of a modern scientist.... A renowned expert on Jason's fleece, Pythagorean harmonies, and Solomon's temple, his advice was also sought on the manufacture of coins and remedies for headaches.... Newton had no laboratory team to supervise... and never traveled outside Eastern England....

FURTHER READING

Abers, Ernest, and Charles F. Kennel, *Matter in Motion* (Boston, Massachusetts: Allyn & Bacon, 1977).

Arons, Arnold, *Development of Concepts of Physics* (Reading, Massachusetts: Addison-Wesley, 1965).

Ball, W. W. Rouse, *A Short Account of the History of Mathematics,* fourth edition (New York: Dover Publications, 1960).

Bixby, William, *The Universe of Galileo and Newton* (New York: Harper & Row, 1964).

Bryson, Bill, *A Short History of Nearly Everything* (New York: Random House, 2003).

Cohen, I. B., "Newton," in *Dictionary of Scientific Biography*, Charles Gillispie, editor-in-chief (New York: Charles Scribner's Sons, 1970).

Fara, Patricia, *Newton: The Making of Genius* (New York: Columbia University Press, 2004).

Fauvel, John, Raymond Flood, Michael Shortland, and Robin Wilson, editors, *Let Newton Be!* (New York: Oxford University Press, 1989).

Guillen, Michael, *Five Equations That Changed the World* (New York: Hyperion, 1995).

Harrison, John, *The Library of Isaac Newton* (Cambridge, U.K.: Cambridge University Press, 1978).

Jennings, Byron, "On the Nature of Science," July 27, 2006; see xxx.lanl.gov/abs/physics/0607241.

Keynes, John Maynard, "Essays in Biography: Newton, the Man," in *The Collected Writings of John Maynard Keynes*, volume 10 (New York: Macmillan/St. Martin's Press, 1972), 363–364.

Keynes, Milo, "The Personality of Isaac Newton," *Notes and Records of the Royal Society of London*, 49(1): 1–56, January 1995.

Pickover, Clifford, *Time: A Traveler's Guide* (New York: Oxford University Press, 1998).

Storr, Anthony, "Isaac Newton," *British Medical Journal*, 291: 1779–1784, 1985.

Tipler, Paul, *Physics* (New York: Worth Publishers, 1976).

Westfall, Richard, *Never at Rest: A Biography of Isaac Newton* (New York: Cambridge University Press, 1983).

Westfall, Richard, "Newton's Scientific Personality," *Journal of the History of Ideas*, 48(4): 551–570, October/December 1987.

Williams, Dudley, and John Spangler, *Physics for Science and Engineering* (New York: Van Nostrand, 1981).

INTERLUDE: CONVERSATION STARTERS

> Newton was not the first of the age of reason. He was the last of the magicians, the last of the Babylonians and Sumerians, the last great mind which looked out on the world with the same eyes as those who began to build our intellectual inheritance rather less than ten thousand years ago. . . .

[Newton saw] the whole universe and all that is in it as a riddle, as a secret which could be read by applying pure thought to certain evidence, certain mystic clues which God had laid about the world to allow a sort of philosopher's treasure hunt.... He regarded the universe as a cryptogram set by the Almighty....

—John Maynard Keynes, "Newton, the Man," *The Collected Writings of John Maynard Keynes*

Newton's law of universal gravitation was based on a variety of observations: the paths of planets in their motion about the sun, the acceleration of objects near the earth.... Physical laws are usually expressed as mathematical equations... that can be used to make predictions.... It is usually easiest to learn the physics and the necessary mathematics at about the same time since the immediate application of mathematics to a physical situation helps you understand both the physics and the mathematics.

—Paul Tipler, *Physics*

Some laws are not laws at all but simply definitions. For example, $\mathbf{F} = \mathrm{d}\mathbf{p}/\mathrm{d}t$ (Newton's second "law" of mechanics) is not a law at all but is a mathematical definition of force (introduced first by Newton himself).

—"Physical Law," *Wikipedia*

Historically there was a change in the nature of scientific models with Newton. Ptolemy, Copernicus and Newton all developed models that correctly predicted planetary motion. While it is sometimes claimed that the Ptolemaic and Copernican models were only descriptive in contrast to Newton's which was explanatory, it is more precise to say that Newton introduced a higher level of abstraction using ideas farther removed from the observations....

—Byron Jennings, "On the Nature of Science"

Descartes and Newton both thought of the laws of nature [as being imposed by God].... According to this traditional view, material things are bound to act according to the laws of nature, because such things are themselves powerless. Only spiritual beings, such as God, human minds, or angels, were thought to be able to act on their own account....

In the eighteenth century, secularized versions of the divine command theory were developed. Instead of thinking of God as the source of all power and order, some natural philosophers of the period began to speak of the "forces of nature" as the source of nature's activity.

—Brian Ellis, *The Philosophy of Nature: A Guide to the New Essentialism*

Newton's aim was to unravel nothing less than God's secret messages.... Above all Newton was intent on finding out when the world would come to an end. Then, he believed Christ would return and set up a 1,000-year Kingdom of God on Earth and he—Isaac Newton, that is—would rule the world as one among the saints.... [Newton] had calculated the year of the apocalypse: 2060.

—George G. Szpiro, *The Secret Life of Numbers*

Newton was probably responsible for the concept that there are seven primary colours in the spectrum—he had a strong interest in musical harmonies and, since there are seven distinct notes in the musical scale, he divided up the spectrum into spectral bands with widths corresponding to the ratios of the small whole numbers found in the just scale.

—Malcolm Longair, "Light and Colour," in Trevor Lamb and Janine Bourriau's *Colour: Art & Science*

That's still all we have: an understanding of the effect [of gravity], with almost no grasp of the cause. Is gravity carried by an elementary particle?... Right now, mathematics is the best investigative tool for getting gravity to square with subatomic forces like electromagnetism....

—John Hockenberry, "What Causes Gravity?" *WIRED* magazine, February, 2007

Newton was the greatest creative genius that physics has ever seen. None of the other candidates for the superlative (Einstein, Maxwell, Boltzmann, Gibbs, and Feynman) has matched Newton's combined achievements as theoretician, experimentalist, *and* mathematician.... If you were to become a time traveler and meet Newton on a trip back to the seventeenth century, you might find him something like the performer who first exasperates everyone in sight and then goes on stage and sings like an angel....

—William H. Cropper, *Great Physicists*

The law of gravity and gravity itself *did not exist* before Isaac Newton. . . . And *what that means* is that that law of gravity exists *nowhere* except in people's heads. It's a ghost!

—Robert Pirsig, *Zen and the Art of Motorcycle Maintenance*

1700–1800

Where should we look to discover the principles that underpin reality? In Einstein's view, while facts reside in the world, principles reside in the mind.... Einstein insisted that great theories are those that explain the most facts from the least number of principles.... The simpler the theory, the less it will look anything like the world we see. The ultimate view of science, he believed, is to find one all-encompassing, self-evident principle ... from which the whole of reality can be deduced.

—Amanda Gefter, "Power of the Mind," *New Scientist*, 188(2529): 54, December 10, 2005

According to the General Theory of Relativity, the laws of motion can be expressed in any inertial or accelerated frame. Thus the choice between a heliocentric model and an earth-centric one is not a matter of right or wrong but one of convention and convenience. What is a model assumption and what is convention is not always clear.

—Byron Jennings, "On the Nature of Science"

Hofstadter's Law states: It always takes longer than you think even when you take Hofstadter's Law into account.

—Douglas R Hofstadter, *Metamagical Themas: Questing for the Essence of Mind and Pattern*

Some men still believe in the mathematical design of nature. They may grant that many of the earlier mathematical theories of physical phenomena were imperfect, but they point to the continuing improvements that not only embrace more phenomena but offer far more accurate agreement with observations. Thus Newtonian mechanics replaces Aristotelian mechanics, and the theory of relativity improved on Newtonian mechanics. Does not this history imply that there is design and that man is approaching closer and closer to the truth?

—Morris Kline, *Mathematics: The Loss of Certainty*, 1980

BERNOULLI'S LAW OF FLUID DYNAMICS

Switzerland (where Bernoulli spent much of his life), 1738. The total energy of fluid pressure, gravitational potential energy, and kinetic energy of a moving fluid remains constant. For liquid flowing in a pipe, an increase in velocity occurs simultaneously with decrease in pressure.

CROSS REFERENCE: BERNOULLI-EULER LAW.

> In 1738, German clockmaker Franz Ketterer invented the cuckoo clock, and Joseph Guillotin (the French physician who invented the guillotine) was born. When the French government persisted in naming the execution machine after Guillotin, his relatives decided to change their family name.

Imagine a fluid flowing steadily through a pipe that carries the liquid from the top to the bottom of a hillside. The pressure of the liquid will change along the pipe. Daniel Bernoulli (1700–1782) discovered the law that relates pressure, flow speed, and height for a fluid flowing in a pipe. Today, we write Bernoulli's Law as

$$\boxed{\frac{v^2}{2} + gz + \frac{p}{\rho} = C,}$$

where v is the fluid velocity, g the acceleration due to gravity, z the elevation (height) of a point in the fluid, p the pressure, ρ the fluid density, and C is a constant. Scientists prior to Bernoulli had understood that a moving body exchanges its kinetic energy for potential energy when the body gains height. Bernoulli realized that, in a similar way, a moving fluid exchanges its kinetic energy for pressure.

As with most laws in this book, Bernoulli's Law holds for an idealized situation. For example, the formula assumes a steady (nonturbulent) fluid flow in a closed pipe. The fluid must be incompressible. Because most fluids are only slightly compressible, Bernoulli's Law is often a useful approximation. The fluid should not be viscous, which means that the fluid should not have internal friction. Although no real fluid meets all these criteria, Bernoulli's relationship is generally very accurate for free-flowing regions of fluids that are away from the walls of pipes or containers, and especially useful for gases and light liquids. The equation can be generalized to a steady *compressible* flow (in which changes in density play a role) by adding the internal energy per unit mass to the left-hand side of the formula.

Bernoulli's Law often makes reference to a subset of the information included in the above equation, namely, that the decrease in pressure occurs simultaneously with an increase in velocity. The idea that an increase in the speed of a fluid results in a decrease in the pressure is at the core of many everyday phenomena. Bernoulli's Law predicts correctly that a shower curtain is pulled inward when the water first comes out of the shower head because the increase in water and air velocity inside the shower causes a pressure drop. The pressure difference between the outside and inside of the curtain causes a net force on the shower curtain that sucks the curtain inward.

Bernoulli's formula has numerous practical applications in the fields of aerodynamics, where it is considered when studying flow over airfoils—such as wings, propeller blades, rudders (whose shapes control stability or propulsion)—and flow in supersonic nozzles. Bernoulli's Law is used when designing a *Venturi throat*—a constricted region in the air passage of a carburetor that causes a reduction in pressure, which in turn causes fuel vapor to be drawn out of the carburetor bowl. The term "venturi" is also applied to a short tube with a constricted region that facilitates the measurement of fluid pressures and velocities as a fluid flows through the tube. The fluid increases speed in the smaller diameter region, reducing its pressure and producing a partial vacuum via Bernoulli's Law. This Venturi effect is named after the Italian physicist Giovanni Battista Venturi (1746–1822). In carburetors, the Venturi effect sucks gasoline into an engine's intake air stream. Additionally, Bernoulli's Law plays a role in the functioning of Pitot tubes used for aircraft speedometers.

I have frequently seen the Venturi effect in action when squeezing a flexible hose through which water flows. If the flow is sufficiently strong, the constriction I put in the hose remains in the hose, even when I remove my hand, because the partial vacuum produced in the constriction is sufficient to keep the hose collapsed.

The fact that pressure falls with increasing velocity is exploited by an airplane wing, which is designed to create an area of fast flowing air on its upper surface. The pressure near this area is lower; thus, the wing tends to be pulled upward.

Daniel Bernoulli (1700–1782), Dutch-born Swiss mathematician, physicist, and medical doctor famous for his wide variety of work in mathematics, hydrodynamics, vibrating systems, probability, and statistics.

CURIOSITY FILE: Both Daniel Bernoulli and his father, Johann Bernoulli, had to secretly study mathematics against their respective fathers' strict orders to forget about mathematics and to pursue more prosperous careers. • Bernoulli wrote on whatever subjects struck his fancy—one of his papers discussed formulas for computing the relationship between the number of oarsmen on a ship and the resultant ship velocity. • In 1738, Bernoulli published "Exposition of a New Theory on the Measurement of Risk," dealing with the economic theory of risk aversion and overall happiness gained from a good or service.

> There is no philosophy which is not founded upon knowledge of the phenomena, but to get any profit from this knowledge it is absolutely necessary to be a mathematician.
>
> —Daniel Bernoulli, Letter to John Bernoulli III, January 7, 1763

> Just as a great river is fed by small streams, some even barely noticeable . . . , so science and technology proceeds from small individual contributions until it becomes an ever-increasing flow of knowledge and techniques. This big river of fluid mechanics is closely associated with Daniel Bernoulli, the author of the first textbook in this field.
>
> —G. A. Tokaty, *History and Philosophy of Fluid Mechanics*

> Daniel Bernoulli was a member of a truly remarkable family which produced no fewer than eight mathematicians of ability within three generations, three of whom—James I (1654–1705), John I (1667–1748), and Daniel—were luminaries of the first magnitude.
>
> —S. L. Zabell, in John Eatwell et al.'s *Utility and Probability*

Daniel Bernoulli is one of the most versatile scientists presented in this book and comes from a family of extraordinary Swiss mathematicians. Not only did he study fluid flow, as discussed in the preceding section, but he also investigated a variety of topics in mathematics, biology, physics, and astronomy.

Before discussing Daniel, let me discuss his famous father, Johann (1667–1748). Mathematical genius seemed to be the very essence of the Bernoulli brain. Swiss mathematician Johann conducted pioneering work with his brother Jacob Bernoulli (1654–1705) in calculus and many areas of applied mathematics. Johann's father tried hard to push Johann into a business career, but Johann preferred mathematics and the academic life. Johann and Jacob's relationship gradually deteriorated. In particular, Jacob was jealous of Johann's friendship with Leibniz and worried that his younger brother was a better mathematician than he. Jacob did all he could to slow down the rise of his brother. For example, in 1695, Jacob convinced the members of the University of Basel's academic senate to reject Johann's application for professorship. When Johann learned of his brother's betrayal, he was furious.

Both the meanness and mathematics that were so apparent in Johann's generation progressed into the next. Daniel Bernoulli was born in Groningen, the Netherlands, while his father taught mathematics at the University of Groningen, and the Bernoulli family returned to their native city of Basel, Switzerland, when Daniel was five years old. The poor relationship between Daniel and his dad started from an early age. Just as Johann's father had tried to force Johann into a merchant career, Johann did the same to his son. Johann virtually forbade his son to become educated in mathematics, which Johann said did not make sufficient money. Moreover, Johann mapped out Daniel's entire future and even selected the woman that Daniel must eventually marry. However, Daniel was stubborn. According to Michael Guillen, author of *Five Equations That Changed the Word*:

> Daniel Bernoulli decided to stop altogether his pretending to go along with his father's astrology-like notions of what God expected of him—that included the business of becoming a merchant, marrying some preselected girlfriend, *and*, the nonmathematical charade he had been carrying on for several years now. Consequently, the young man broke the bad news to his father and begged for permission to pursue his love of mathematics.

Johann finally told Daniel that he did not have to become a merchant, but instead Daniel now must become a medical doctor—and no mathematics would be allowed. Time passed, and finally Johann relented and agreed to personally teach Daniel mathematics, as long as Daniel continued his medical education.

While growing up in Basel, Switzerland, Daniel had managed to study philosophy and logic, obtaining his master's degree in 1716. In 1721, he obtained his doctorate in medicine. His dissertation was on the mechanics of breathing.

In the meantime, he did pursue mathematics, and his various mathematics papers, published in 1724, caused the St. Petersburg Academy, along with Empress Catherine I of Russia, to invite him for a visit. Daniel's

St. Petersburg years (1725–1733) were his most creative, leading to famous papers on hydrodynamics, oscillations, and probability.

Johann Bernoulli sent the Swiss mathematician Leonhard Euler (1707–1783) to St. Petersburg in order to work with Daniel, which led to fruitful and creative discussions. Euler would turn out to be one of the greatest and most prolific mathematicians who ever lived.

While in Russia, Daniel Bernoulli invented ways for measuring the pressure flowing through pipes by punching a hole in the wall of a pipe and attaching a small glass tube to the hole. As the water flowed through the pipe, it would also enter the glass tube, and the height that it rose in the upward-pointing tube was a measure of the pressure of the flowing water. Bernoulli recognized that this approach might be used to measure blood pressure, and he told his friend Christian Goldbach, "I made a new discovery that will be very useful in the design of the water supply, but mainly, it will open a new era in physiology." Physicians throughout Europe began to measure blood pressure by sticking pointed-end glass tubes directly into patients' arteries. Ouch!

Bernoulli's fascinating paper on probability, which was finally published in 1738, described a paradox now known as the "St. Petersburg Paradox." The puzzle involves coin flips and money that a gambler is to receive depending on the outcome of the flips. Philosophers and mathematicians have wondered: What is the fair price for joining this game? How much would you be ready to pay for joining this game?

Here's one way to view the St. Petersburg scenario: Flip a penny until it lands tails. The total number of flips, n, determines the prize, which equals $\$2^n$. Thus, if the penny lands tails the first time, the prize is $\$2^1 = \$2$, and the game ends. If the penny comes up heads the first time, it is flipped again. If it comes up tails the second time, the prize is $\$2^2 = \$4$, and the game ends. And so on. A detailed discussion on the paradox of this game is beyond the scope of this book, but a rational gambler would enter a game if and only if the price of entry was less than the expected value of the financial payoff. According to some analyses of the St. Petersburg game, any finite price of entry is smaller than the expected value of the game, and a rational gambler might desire to play the game no matter how large we set a finite entry price to play the game!

Peter L. Bernstein in *Against the Gods: The Remarkable Story of Risk* comments on the mystery and profundity of Bernoulli's St. Petersburg Paradox:

> His paper is one of the most profound documents ever written, not just on the subject of risk but on human behavior as well. Bernoulli's emphasis on the complex relationships between measurement and gut touches on almost every aspect of life.

Daniel's primary work, *Hydrodynamica*, was completed around 1734 but not published until 1738. In this book, he discussed fluid pressure and velocity and presented his famous law stating that fluid pressure increases as its velocity decreases. The word "hydrodynamics"—which refers to the branch of modern science that deals with the motion of fluids—derives from the title of Daniel's work. His father, jealous of Daniel's work, published his own *Hydraulica*, which may have been predated to 1732 to make it appear that his work was written before his son's *Hydrodynamica*. Johann had even persuaded Euler to write in the preface of *Hydraulica*, "I was thoroughly astounded by the very fluent application of Your principles to the solution of the most intricate Problems, because of which... Your very distinguished Name will forever be revered among future generations." According to Guillen,

> Daniel Bernoulli could never prove it, but he would always suspect his father of plagiarism and his alleged friend Euler of duplicity. "Of my entire *Hydrodynamica*, of which indeed I in truth need not credit one iota to my father," Bernoulli lamented, "I am robbed all of a sudden, and therefore in one hour I lose the fruits of work of ten years."

Daniel shared the 1735 prize from the Paris Academy of Sciences for his work on planetary orbits with his father, who promptly threw Daniel out of the house for obtaining a prize he felt should be his alone. According to John J. O'Connor and Edmund F. Robertson's entry on Bernoulli in *The MacTutor History of Mathematics Archive*:

> Daniel's father was furious to think that his son had been rated as his equal, and this resulted in a breakdown in relationships between the two. The outcome was that Daniel found himself back in Basel but banned from his father's house. Whether this caused Daniel to become less interested in mathematics or whether it was the fact that his academic position was a non-mathematical one, certainly Daniel never regained the vigor for mathematical research that he showed in St. Petersburg.

In 1750, Daniel was appointed chair of physics at Basel, where he taught until his death in 1782. In many ways, Daniel turned out to be the "Carl Sagan" of his era, as he enjoyed clarifying science for a general public and did not confine himself to a single field of knowledge. Ten of his essays entered in competitions of the Paris Academy won awards. Essay topics included marine navigation and technology, magnetism, astronomy,

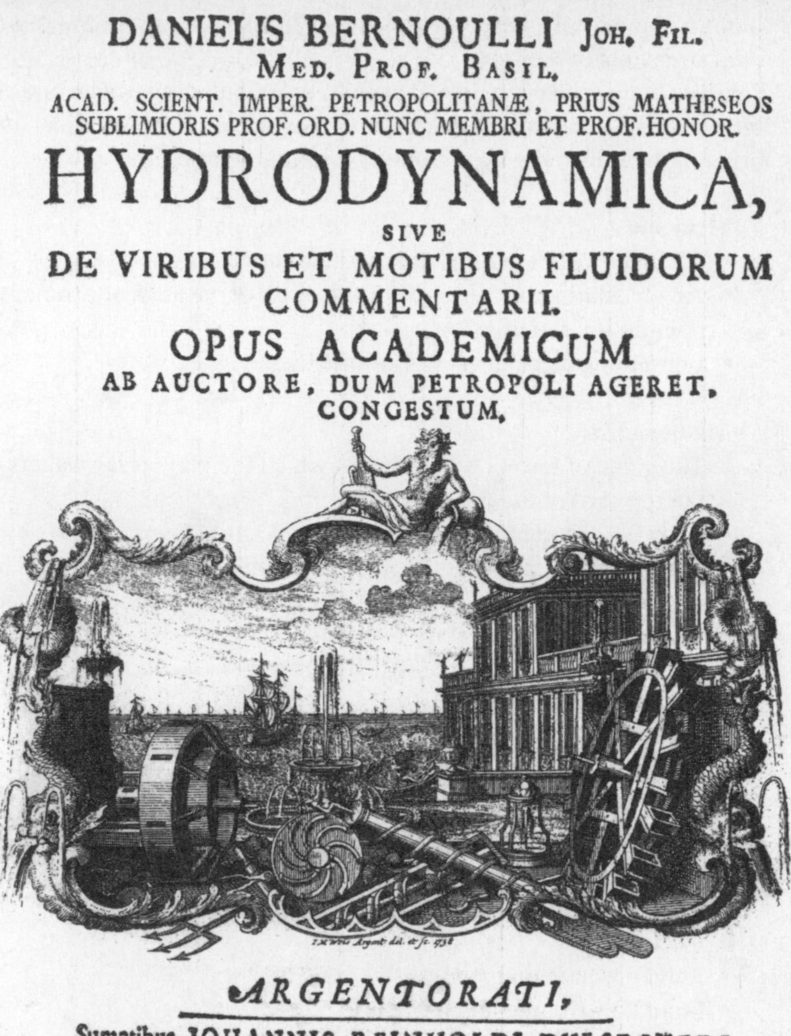

DANIELIS BERNOULLI Joh. Fil.
Med. Prof. Basil.
ACAD. SCIENT. IMPER. PETROPOLITANÆ, PRIUS MATHESEOS SUBLIMIORIS PROF. ORD. NUNC MEMBRI ET PROF. HONOR.

HYDRODYNAMICA,

SIVE

DE VIRIBUS ET MOTIBUS FLUIDORUM COMMENTARII.

OPUS ACADEMICUM

AB AUCTORE, DUM PETROPOLI AGERET, CONGESTUM.

ARGENTORATI,

Sumptibus JOHANNIS REINHOLDI DULSECKERI,
Anno M D CC XXXVIII.

Typis Joh. Henr. Deckeri, Typographi Basiliensis.

Frontispiece for Chapter 1 of Daniel Bernoulli's *Hydrodynamica*, published in 1738.

planetary orbits, and the optimal shapes for sand-filled hourglasses and boat anchors.

His hourglass was special in that it was designed to keep good time even when a ship was rocking in heavy seas. To achieve this, Bernoulli's award-winning invention involved mounting an hourglass atop an iron slab floating in mercury. Even when a ship was in a storm, the density of the mercury would help keep the timepiece from becoming too agitated by the movements of the ship.

Sadly, the precise content of many of his popular science lectures is not known today. His papers covered so many areas of science that I give a flavor of their diversity by presenting a sampling of topics:

Medicine

- Mechanics of breathing and muscular contraction
- An elucidation of the shape of the optic nerve at its attachment point to the retina
- Computing mechanical work done by the heart

Mathematics

- The game of faro (a card game in which the players lay wagers on the top card of the dealer's pack)
- Riccati's differential equations [named after Count Jacopo Francesco Riccati (1676–1754), these mathematical equations have the form $y' = q_0(x) + q_1(x)y + q_2(x)y^2$ and are not easily solved using standard elementary techniques]
- Lunulae (a class of crescent shapes)
- Divergent sine and cosine series
- Infinite continued fractions
- Probability and statistics applied in the areas of economics, disease spread, and population statistics

Mechanics

- Theory of rotating bodies
- Friction
- Fluid pressure on pipe walls
- Fluid flow from container holes
- Oscillations of fluids in tubes immersed in water tanks
- Investigations of pumps, windmill sails, and the Archimedean screw
- Atmospheric pressures
- Refraction of light
- Air flow from small openings
- Theories of ocean tides
- The actions of sails and oars
- The mechanics of flexible bodies
- Velaria, lintearia, and catenaria (geometrical curves sometimes exhibited by natural processes)
- Oscillations of ropes loaded with weights
- Vibrations of threads of uneven thicknesses, plates in water and organ pipes, and musical instrument strings

Another less famous law that bears Bernoulli's name is the Bernoulli-Euler Law, which is useful in studying, for example, a horizontal beam that supports a vertical load that causes the beam to bend. The law states that an elastic beam of thickness t, bent to a radius of curvature R ($R > t$), has a bending moment M given by $M = EI/R$, where E is Young's modulus for the material of the beam, and I is the second moment of area of the cross-section of the beam about an axis that is normal to the plane of bending. Young's modulus is a measure of the stiffness of a given material. Bernoulli suggested the law in 1742 and Leonhard Euler derived it in 1744.

In his 1738 textbook on hydrodynamics, Bernoulli suggested his famous Bernoulli's Law of Fluid Dynamics, in which he states that "proportionality" existed between pressure and velocity. He wrote, "It is clearly very amazing that this very simple rule, which nature affects, could remain unknown up to this time." It was not until 1755 that Euler derived the fuller expression that relates pressure, velocity, density, and height.

I think of both Bernoulli and Euler whenever I squeeze a bulb that sits atop a perfume bottle. Squeezing the bulb over the liquid perfume creates a low pressure area due to the higher speed of the air, which then sucks the perfume up to the opening of the bottle. Similarly, Bernoulli's Law helps us understand why house windows often explode outward in a hurricane. The high speed of the air outside the window pane results in lower pressure outside than inside, thus pulling the glass outward. If you believe that a hurricane is approaching, it may wise to open a few windows to equalize the pressure.

Graham Cleverley, my colleague and a historian of science, in a personal communication to me, comments on Bernoulli's Law as it relates to airplane wings:

> Conventional airplanes fly more economically because their design takes Bernoulli's principle into account. Without the principle . . . , sailing ships wouldn't be able to sail against the wind. Europeans would not have discovered America until the nineteenth century. And when we got there and played baseball, we wouldn't be able to throw curveballs or sliders. . . . You wouldn't be able to bend it like Beckham. Or serve like Sampras. And table tennis would be very dull.

Ray Kurzweil writes in John Brockman's *What We Believe but Cannot Prove*:

> It is the nature of engineering to take a natural, often subtle effect and control it, with a view toward greatly leveraging and magnifying it. . . . Consider, for example, how we have focused and amplified the

subtle properties of Bernoulli's principle (that air rushing over a curved surface has a slightly lower pressure than it does over a flat surface) to create the whole world of aviation.

Perhaps it may be an exaggeration to say that Bernoulli's Law "creates the whole world of aviation." We do not need Bernoulli's principle to fly or to have aircraft. After all, we could all be flying around in Zeppelins, which do not use the principle. However, standard commercial passenger aircraft—either jet or propeller-driven—make use of Bernoulli's principle to improve stability, control, and efficiency. John Anderson writes in *A History of Aerodynamics*:

> The fundamental advances in aerodynamics in the eighteenth century began with the work of Daniel Bernoulli. Newtonian mechanics had unlocked, but not opened, the door to modern hydrodynamics. Bernoulli was the first to open that door, though just a crack. Leonhard Euler and others who would follow would fling the door wide open.

FURTHER READING

Anderson, John, *A History of Aerodynamics: And Its Impact on Flying Machines* (Cambridge, U.K.: Cambridge University Press, 1999).

Bernstein, Peter L., *Against the Gods: The Remarkable Story of Risk* (Hoboken, N.J.: John Wiley & Sons, 1998).

Brockman, John, *What We Believe but Cannot Prove* (New York: Harper Perennial, 2006).

Guillen, Michael, *Five Equations That Changed the World* (New York: Hyperion, 1995).

Kobes, Randy, and Gabor Kunstatter, "Bernoulli's Principle," Physics Department, University of Winnipeg; see theory.uwinnipeg.ca/mod_tech/node68.html.

Martin, Robert, "The St. Petersburg Paradox," in *Stanford Encyclopedia of Philosophy*; see plato.stanford.edu/entries/paradox-stpetersburg/.

O'Connor, John J., and Edmund F. Robertson, "Daniel Bernoulli, 1700–1782," in *MacTutor History of Mathematics Archive*, School of Mathematics and Statistics, University of St. Andrews, Scotland; see www-groups.dcs.st-and.ac.uk/~history/Mathematicians/Bernoulli_Daniel.html.

Quinney, D. A., "Daniel Bernoulli and the Making of the Fluid Equation," *plus* magazine, January 1997; see pass.maths.org.uk/issue1/bern/.

Straub, Hans, "Daniel Bernoulli," in *Dictionary of Scientific Biography*, Charles Gillispie, editor-in-chief (New York: Charles Scribner's Sons, 1970).

INTERLUDE: CONVERSATION STARTERS

Until Einstein's time, scientists typically would observe things, record them, then find a piece of mathematics that explained the results. Einstein exactly reverses that process. He starts off with a beautiful piece of mathematics that's based on some very deep insights into the way the universe works and then, from that, makes predictions about what ought to happen in the world. It's a stunning reversal to the usual ordering in which science is done.... [Einstein demonstrated] the power of human creativity in the sciences....

—Sylvester James Gates, quoted in Peter Tyson's "The Legacy of $E = mc^2$"

Mathematics, rightly viewed, possesses not only truth, but supreme beauty—a beauty cold and austere, like that of sculpture.

—Bertrand Russell, *Mysticism and Logic*, 1918

Mathematicians are only dealing with the structure of reasoning, and they do not really care what they are talking about. They do not even need to *know* what they are talking about.... But the physicist has meaning to all his phrases.... In physics, you have to have an understanding of the connection of words with the real world.

—Richard Feynman, *The Character of Physical Law*

Enlightenment "natural theology," presuming the Creator to have had our best interests at heart when He instituted nature's laws and then retired, made no allowance for either Satanic influence of divine playback for wickedness. God's indifference... was more complete than any deist had dared to conceive.

—Frederick Crews, "Follies of the Wise," *Skeptical Inquirer*, March/April, 2007

There is no reason why the most fundamental aspects of the laws of nature should be within the grasp of human minds, which evolved for quite different purposes, nor why those laws should have testable consequences at the moderate energies and temperatures that necessarily characterize life-supporting planetary environments.... As we probe deeper into the intertwined logical structures that underwrite the nature of reality, we can expect to find more deep results which limit what can be known. Ultimately, we may even find that their totality characterizes the universe more precisely than the catalogue of those things that we can know.

—John Barrow, *Boundaries and Barriers: On the Limits of Scientific Knowledge*

Ghosts are unscientific. They contain no matter and have no energy and therefore according to the laws of science, do not exist except in people's minds. Of course, the laws of science contain no matter and have no energy either and therefore do not exist except in people's minds.... It's best to refuse to believe in either ghosts or the laws of science.

—Robert Pirsig, *Zen and the Art of Motorcycle Maintenance*

LAMBERT'S LAW OF EMISSION

 *Switzerland/Germany, 1760*. The intensity emitted in any direction from a region of a diffuse surface is proportional to the cosine of the angle between the direction of radiation and the normal to the surface.

CROSS REFERENCE: BEER'S LAW OF ABSORPTION, THE LAMBERT-BEER LAW, AND THE BOUGUER-BEER LAW.

> In 1760, during the French and Indian War, Cherokee natives allied with French forces and attacked a North Carolina militia stationed at Fort Dobbs. The Russians occupied and burned Berlin. German cabinetmaker Kaspar Faber made preparations for the first commercial production of pencils.

Lambert's Law of Emission, also known as Lambert's Cosine Law or the Cosine Law of Emission, states that the intensity (flux per unit solid angle) emitted in any direction from a region of a perfect diffuse radiating surface is proportional to the cosine of the angle between the direction of radiation and the normal to the surface. (An ideal diffuse surface is usually a rough surface, like chalk, such that the small variations in the surface cause an incoming light ray to be reflected in all directions equally.) Thus, the region, or element, of the surface that obeys Lambert's Cosine Law will appear equally bright when observed from any direction.

Here's another way of stating the law: The total radiant power observed from a perfect radiating surface is proportional to the cosine of the angle θ between the observer's line of sight and a line drawn perpendicular to the surface. The radiating surface appears equally bright regardless of the viewing angle because, purely from geometrical considerations, the apparent size of a portion of the surface is proportional to the cosine of the angle.

Lambert's Cosine Law may be stated as

$$\boxed{I_e \propto \cos\theta,}$$

where I_e is the intensity of emitted light, θ is the angle between the observed emitted intensity and the normal to the surface, and $\propto$ means "is proportional to."

An untreated piece of lumber from the lumber yard exhibits nearly Lambertian reflectance (i.e., it obeys Lambert's Cosine Law), but the same piece of wood with a glossy coat of varnish is probably not a Lambertian reflector because the observer will see specular highlights (bright spots of light that appear on shiny objects when illuminated) when viewing the wood from specific angles.

Textbooks frequently give additional examples of Lambertian reflectors that include "sand-blasted opal glass" and "scraped plaster of Paris." Such surfaces are said to have a "matte" finish. When we have a matte surface, luminance is sometimes expressed in terms of total luminous flux, in units of lumens, emitted by a unit area of surface. A surface emitting one lumen per square centimeter has a luminance of one lambert, named in honor of Johann Lambert.

An example of Lambert's law in action can be found in our observations of visible light from the Sun. Because the Sun is nearly a Lambertian radiator, its brightness is almost the same everywhere on an image of the solar disk.

Johann Heinrich Lambert (1728–1777), Swiss-German mathematician and physicist famous for his work on the mathematical constant π and for his laws of reflection and absorption of light.

CURIOSITY FILE: Philosopher Immanuel Kant called Lambert "the greatest genius of Germany." • Lambert developed theorems regarding conic sections that made it possible to simplify calculation of the orbits of comets. • Lambert invented the first practical hygrometer and photometer. • He introduced the word "albedo" when studying the brightness of planets. • Lambert's accomplishments are particularly impressive, considering he was almost entirely self-taught.

> The first object of my endeavors was the means to become perfect and happy. I understood that the will could not be improved before the mind had been enlightened.
>
> —Johann H. Lambert, quoted in *Dictionary of Scientific Biography*

> In geometry, Lambert goes beyond the previously assumed concept of space, by establishing the properties of incidence. Lambert's physical erudition indicates yet another clear way in which it would be possible to eliminate the traditional myth of three-dimensional geometry through the parallels with the physical dependence of functions. A number of questions that were formulated

> by Lambert in his metatheory in the second half of the 18th century have not ceased to remain of interest today.
>
> —J. Folta, "Remarks on the Axiomatic Development of Mathematics in the Second Half of the Eighteenth Century"

Johann Lambert is another example of an extreme polymath, having published more than 150 works on topics ranging from geometry and probability to optics, cartography, philosophy, meteorology, astronomy, and perspective in art. He is also considered to be the greatest eighteenth-century logician. When Lambert was asked by the King of Prussia, Frederick the Great, in which science he was most proficient, Lambert modestly replied, "All."

Lambert was born in Mülhausen, which is now Mulhouse, Alsace, France. Encyclopedias list Lambert variously as a German scientist, a Swiss-German scientist, or a German-French scientist. (In my country count in the introduction for this book, I counted him as German.) The town of Mülhausen was a member of an association of ten free towns in Alsace that were allied to the Swiss Confederation, which was a free republic until it was absorbed into France in 1798.

Lambert was one of five sons and two daughters. His father was a tailor. During Lambert's preteen years, he had a diverse education with studies that included Latin and French. At age 12, he left school to help his father in the tailor shop but studied science on his own whenever time permitted. At age 17, Lambert worked as a secretary to the editor of a conservative newspaper, and he studied science, philosophy, and mathematics after work. His keen interest in these subjects is obvious in a letter that he wrote while still in his teens:

> I bought some books in order to learn the first principles of philosophy. The first object of my endeavors was the means to become perfect and happy.... I studied Christian Wolff's "On the power of the human mind," Nicolas Malebranche's "On the investigation of truth," and John Locke's "Essay concerning human understanding." The mathematical sciences, in particular algebra and mechanics, provided me with clear and profound examples to confirm the rules I had learned. Thereby, I was able to penetrate into other sciences more easily and more profoundly, and to explain them to others, too.

While in the town of Chur, which was at the time part of the Swiss Confederation, he was elected to the Literary Society of Chur and to the

Swiss Scientific Society based in Basel. He made regular meteorological observations as part of his society responsibilities. In 1755, he published his first paper, which was on his theories of caloric heat. In 1758, he published a book dealing with the passage of light through various substances. The book was followed by his 1760 book *Photometria*—one of his most famous works because it contained the emission law that often bears his name. Christoph Scriba in the *Dictionary of Scientific Biography* writes:

> Lambert carried out his experiments with few and primitive instruments, but his conclusions resulted in laws that bear his name. The exponential decrease of the light in a beam passing through an absorbing medium of uniform transparency is often called "Lambert's law of absorption," although Bouguer discovered it earlier. "Lambert's cosine law" states that the brightness of a diffusely radiating plane surface is proportional to the cosine of the angle formed by the line of sight and the normal to the surface.

As alluded to by Scriba, Lambert's name is also associated with an absorption law of light. Under "Beer's Law of Absorption" in part III, I discuss how, in 1729, French mathematician Pierre Bouguer (1698–1758) formulated an absorption law for light—namely, that the fraction of light absorbed by a particular material is directly proportional to the thickness of the material. In Bouguer's 1729 paper *Essai d'optique sur la gradation de la lumière* ("Optical Experiment on the Gradation of Light"), he defined the quantity of light lost by passing through a given extent of the atmosphere, and perhaps Bouguer should be considered as the first known discoverer of "Beer's Law." Lambert—a scientist more prominent than Bouguer—rediscovered and published Bouguer's law. When additional careful experiments were made, scientists noticed that the amount of light absorbed by solutions also probably depended on additional factors. In 1852, August Beer (1825–1863) announced a more complete law of absorption that is known variously as Beer's Law, the Lambert-Beer Law, and the Bouguer-Beer Law. See the entry on "Beer's Law" for further discussion.

In 1761, Lambert published his cosmological theories in *Cosmologische Briefe über die Einrichtung des Weltbaues* (Cosmological Letters on the Arrangement of the World Structure). Here, he proposed that we live in a finite universe composed of galaxies of stars. Lambert came to believe that all planets, comets, and moons in the universe were likely to contain life. In his *Cosmologische Briefe*, Lambert asserted,

> The Creator is much too efficient not to imprint life, forces and activity on each speck of dust. . . . [I]f one is to form a correct notion of the world, one should set as a basis God's intention in its true

extent to make the whole world inhabited.... All possible varieties which are permitted by general laws ought to be realized....

According to Lambert, an omnipotent God would populate all parts of the universe with diverse beings, and humanoids were likely to be everywhere. In order to protect such life forms, God would rarely allow collisions between bodies such as planets and comets. Both Immanuel Kant (1724–1804) and Lambert conceived of a fractal, or hierarchical, universe of stars clustered into larger systems, which today we call galaxies. These clusters are clustered into superclusters, and so on, at different size scales.

Lambert's philosophical works focus on the nature of human knowledge and thought, mathematical logic, and methods for scientific proof. In the field of mathematics, Lambert is most famous for being the first to prove that π is irrational, that is, cannot be written as the ratio of two integers.

Lambert developed a means of organizing colors by using a triangular pyramid representation. The triangular base is black at its center with vertices colored cinnabar (red), yellow, and azurite (blue). As one gazes upward along the pyramid, the colors increase in brightness until reaching the white tip at the top. Lambert suggested that his system could help textile merchants decide which colors they had available. He also hoped that printers would get ideas for aesthetic combinations of colors by studying his color pyramid.

Around the year 1772, Lambert developed a map projection that is now called the Lambert conformal conic projection. The shapes of the countries on a globe are well preserved when represented in this flat map. Cartographers still use this projection today and consider it one of the more useful projections for regions of Earth near the middle latitudes with an east–west orientation.

Lambert died of tuberculosis in Berlin at age 49, having never married. A lunar crater with a diameter of 30 kilometers was named after Lambert and approved in 1935 by the International Astronomical Union General Assembly. A Martian crater is also named in his honor.

Throughout his life, Lambert had been awed by the power of science and scientists—and he was concerned about potential dangers from outer space. Sara Schechner, in *Comets, Popular Culture, and the Birth of Modern Cosmology*, comments on Lambert and other scientists of his era with similar interests:

In affirming a role for comets in the beginning and end of the world, natural philosophers derived a new sense of power. If comets were indeed divine tools for reforming the world, it became conceivably possible for astronomers to predict when certain scriptural prophecies might be fulfilled.

Lambert had written on these thoughts in 1761 in *Cosmologische Briefe*:

> I was not far from looking at astronomers as authorized prophets and . . . from seeing in the invention of the telescope and in the rapid growth of astronomy the herald of an impending disaster. How could, I thought, a genie suggest to Copernicus the structure of the world, to Kepler its laws, and to Newton that terrible attraction and the doctrine about the course and impacts of comets, so that everything might be available for prediction of the calamity, and the inhabitants of the Earth might see to it that instead of having all come to an end, a seed for propagation might remain alive on the changed Earth.

FURTHER READING

Folta, J., "Remarks on the Axiomatic Development of Mathematics in the Second Half of the Eighteenth Century," *DVT-Dejiny Ved a Techniky*, 6: 189–205, 1973.

Lambert, Johann, *Cosmologische Briefe über die Einrichtung des Weltbaues* (Augsburg, 1761).

Schechner, Sara, *Comets, Popular Culture, and the Birth of Modern Cosmology* (Princeton, N.J.: Princeton University Press, 1999).

Scriba, Christoph, "Johann Lambert," in *Dictionary of Scientific Biography*, Charles Gillispie, editor-in-chief (New York: Charles Scribner's Sons, 1970).

INTERLUDE: CONVERSATION STARTERS

> Nature is showing us only the tail of the lion, but I have no doubt that the lion belongs to it even though, because of its large size, it cannot totally reveal itself all at once. We can see it only the way a louse that is sitting on it would.
>
> —Albert Einstein to Heinrich Zangger, March 10, 1914

> Our attempts at modeling physical reality normally consist of two parts: (1) A set of local laws that are obeyed by the various physical quantities. These are usually formulated in terms of differential equations. (2) Sets of boundary conditions that tell us the state of some regions of the universe at a certain time. . . . Many people would claim that the role of science is confined to the first of these and

that theoretical physics will have achieved its goal when we have obtained a complete set of local physical laws.

—Stephen Hawking, *Black Holes and Baby Universes*

Science, like life, feeds on its own decay. New facts burst old rules; then newly divined conceptions bind old and new together into a reconciling law.

—William James, *The Will to Believe and Other Essays in Popular Philosophy*

We materialists don't deny the force of ideas; we merely say that the minds precipitating them are wholly situated within brains that…seem to have emerged without any need for miracles.

—Frederick Crews, "Follies of the Wise," *Skeptical Inquirer*, March/April, 2007

BODE'S LAW OF PLANETARY DISTANCES

Germany, 1766. The mean distances of planets from the Sun can be predicted using a simple numerical relationship.

CROSS REFERENCE: ASTRONOMER JOHANN DANIEL TITIUS, THE TITIUS-BODE LAW, AND SWISS-GERMAN PHYSICIST JOHANN HEINRICH LAMBERT.

> In 1766, two English surveyors, Charles Mason and Jeremiah Dixon, drew the Mason-Dixon Line between Pennsylvania and Maryland, which would later be used to mark the boundary between free and slave regions of the United States. In this same year, the British Parliament repealed the unpopular Stamp Act that had taxed all documents in the American colonies by requiring the documents to carry a tax stamp. English chemist John Dalton (see "Dalton's Law of Partial Pressures" in part III) was born.

Bode's Law, also known as the Titius-Bode Law, expresses a relationship that describes the mean distances of the planets from the Sun. Consider the simple sequence 0, 3, 6, 12, 24, . . . , in which each successive number is twice the previous number. Next, add 4 to each number and divide by 10 to form the sequence 0.4, 0.7, 1.0, 1.6, 2.8, 5.2, 10.0, 19.6, 38.8, 77.2, . . . Remarkably (and perhaps strangely!), Bode's Law states that this sequence gives the mean distances D of the known and yet-to-be-discovered planets from the Sun, expressed in astronomical units (AU). An *AU* is the mean distance between Earth and the Sun, which is approximately 92,960,000 miles (149,604,970 kilometers). For example, Mercury is approximately one-third of an AU from the Sun, and Pluto is about 39 AU from the Sun.

Bode's Law can be expressed by

$$\boxed{D = (N + 4)/10,}$$

where $N = 0, 3, 6, 12, 24, 48 \ldots$ We also sometimes see the law expressed as

$$\boxed{D = A + BC^n,}$$

where $A = 0.4$, $B = 0.3$, $C = 2$, and $n = 0, 1, 2, 3 \ldots$

This relationship was discovered in 1766 by the German astronomer Johann Daniel Titius (1729–1796) of Wittenberg and published by Bode six years later. At the time, the law gave a remarkably good estimate for the mean distances of the planets that were then known—Mercury (0.39), Venus (0.72), Earth (1.0), Mars (1.52), Jupiter (5.2), and Saturn (9.55). Uranus, discovered in 1781, has a mean orbital distance of 19.2, which also

TABLE 5 Accuracy of Bode's Law

Object	Bode's Calculation	Prediction	Actual Distance
Mercury	$0.4 + 0 \cdot 0.3$	0.4	0.39
Venus	$0.4 + 1 \cdot 0.3$	0.7	0.72
Earth	$0.4 + 2 \cdot 0.3$	1.0	1.00
Mars	$0.4 + 4 \cdot 0.3$	1.6	1.52
Ceres	$0.4 + 8 \cdot 0.3$	2.8	2.77
Jupiter	$0.4 + 16 \cdot 0.3$	5.2	5.20
Saturn	$0.4 + 32 \cdot 0.3$	10.0	9.54
Uranus	$0.4 + 64 \cdot 0.3$	19.6	19.19
Neptune	$0.4 + 128 \cdot 0.3$	38.8	30.07

Distances are expressed in astronomical units.

agrees with the law, and its distance was taken as strong evidence of the correctness and reliability of the law. The large asteroid Ceres, discovered 1801, has mean orbital distance 2.77, which seemed to fill the apparent gap between Mars and Jupiter. Some astronomers were so impressed by the apparent success of Bode's Law that they proposed the name Ophion for a large planet predicted to lie beyond Uranus at a distance of 38.8 AU. Table 5 gives an indication of the accuracy of Bode's Law.

Alas, Neptune, discovered in 1846, has a mean orbital distance of 30.07, and Pluto, discovered in 1930, has a mean orbital distance of 39.5; these two planets provide large discrepancies from the predicted values of 38.8 and 77.2, respectively. (Note, however, that 38.8 is close to the actual value of 39.5 for Pluto, as if Bode's Law skips the location of Neptune.) In 2006, the International Astronomical Union gave Ceres and Pluto the designation of "dwarf planets."

Astronomers have wondered why such a simple sequence should seem to explain so much about the Solar System. Some astronomers have even conjectured that the deviation of Neptune and Pluto from their predicted positions means that they are no longer in their original orbits in the Solar System. Today scientists have major reservations about Bode's Law, and the law is clearly not as universally applicable as other laws in this book. The relation may be purely empirical and perhaps represents a coincidence.

Some astronomers hypothesize that a phenomenon of "orbital resonances," which is caused by orbiting bodies that gravitationally interact with other orbiting bodies, can create regions around the Sun that are free of long-term stable orbits and thus, to some degree, can account for the spacing of planets. Orbital resonances can occur when two orbiting bodies have periods of revolution that are in a simple integer ratio such that the bodies exert a regular gravitational influence on each other.

Johann Elert Bode (1747–1826), German astronomer famous for his statement of Bode's Law expressing the proportionate distances of several planets from the Sun.

CURIOSITY FILE: Bode was responsible for naming the planet Uranus after the ancient Greek god that came every night to mate with the Earth Mother, Gaia. The god Uranus had imprisoned Gaia's youngest children deep within Earth, fearful of their power. In order to prevent Uranus from fathering any more children, Gaia had their son Cronus ambush Uranus and castrate him, tossing the torn testicles into the ocean.

> Professor Bode could not explain *why* the rule worked . . . but anybody who could add and multiply had no doubt that it did work.
>
> —Willy Ley, *Watchers of the Skies*

> Bode's Law is neither Bode's nor a law.
>
> —Mark Littmann, *Planets Beyond*

> From Mars outward there follows a space . . . in which, up to now, no planet has been seen. Can we believe that the Creator of the world has left this space empty? Certainly not!
>
> —Johann Bode, *Instruction for the Knowledge of the Starry Heavens*

Johann Elert Bode (pronounced bō'duh) was born in Hamburg, Germany. His father was a merchant. Little has been published about Bode's boyhood. We do know that in 1768, Bode published his popular book, *Anleitung zur Kenntnis des gestirnten Himmels* (Instruction for the Knowledge of the Starry Heavens), in which he popularized the empirical law on planetary distances, originally discovered by German astronomer Johann Titius.

Titius had merely presented the law in a footnote to his 1766 German translation of naturalist Charles Bonnet's *Contemplation de la Nature*. Bode discovered the footnote and inserted it into the new edition of his own astronomy book, without reference to Titius. Mark Littmann writes in *Planets Beyond: Discovering the Outer Solar System*:

> This exercise by Titius, buried in another author's book, would probably have attracted no attention had not Bode happened across it. Bode was a young, energetic, self-taught astronomer who, at the age of 21, had published a very popular introduction to the heavens. In 1772, he had just been hired by the Berlin Academy of Sciences to work on its annual astronomical almanac [which was] selling poorly. Bode quickly transformed it from a money loser to a high-profit item by correcting the publication's inaccuracies and by [supplementing it with] general science news from around the world.

Bode was also employed by the Berlin Academy of Sciences to perform laborious mathematical calculations, and he was a director of the Berlin Observatory. In 1801, while still at the observatory, he published the *Uranographia*, a gorgeous celestial atlas showing the positions of stars and other astronomical objects—and also included artistic depictions of the constellations. Many of the diagrams are a wonder to behold, with their drawings of animals and mythic heroes superimposed on the stars. Bode even introduced new constellations in his *Uranographia*, such as a set of stars forming the constellation *Machina Electrica* along with a drawing of an electrostatic generator (not quite as majestic an image as the superhuman gods of yore!). However, Bode's new constellations never achieved lasting acceptance.

Uranographia contained eighteen maps and two planispheres (polar projections with an adjustable overlay to show the stars visible at a particular time and place). The book showed almost 17,000 stars and 2,500 nebulae and represented all astronomical objects defined by the cartographers during the previous centuries. *Uranographia* marked the end of a long history of artistic representations of the constellations, because soon the star atlas would no longer be a single work aimed at both the amateur and professional astronomer. The ideal of one large book that contains all visible objects in the sky, superimposed with artful constellation drawings, slowly lost its popularity, and subsequent atlases showed fewer ornate figures.

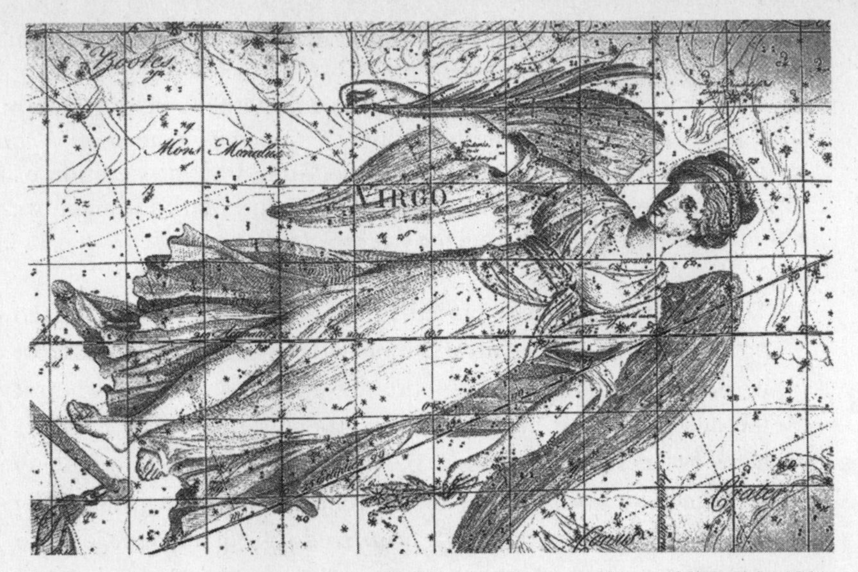

The constellation Virgo, from Johann Bode's *Uranographia*, published in 1801.

Together with Johann Heinrich Lambert (see "Lambert's Law of Emission," above), Bode founded the *Astronomisches Jahrbuch* (Astronomical Yearbook) and then the *Berliner Astronomisches Jahrbuch*, which he continued to publish until his death in 1826.

During his life, Bode discovered several nebulae and star clusters. Alas, he also added a large number of nonexistent astronomical objects without verification—more than twenty of his discoveries never actually existed. He also observed a number of actual comets and calculated cometary orbits.

Bode was fascinated by Uranus, the new planet discovered by German-born British astronomer William Herschel (1738–1822) in 1781. Although Herschel always referred to this planet as "Georgium Sidus" (George's Star) to honor King George III of England, Bode proposed the name "Uranus" after the Greek God—a name that was gradually adopted. French astronomers actually began calling the planet Herschel before Bode proposed the name Uranus, which did not come into common usage until around 1850.

Bode collected the available astronomical observations of Uranus and published many of them in his beloved *Astronomisches Jahrbuch*. As he compiled the information, Bode realized that Uranus had actually been first observed by German astronomer Tobias Mayer (1723–1762) in 1756 and even earlier in 1690 by English astronomer John Flamsteed (1646–1719), when it was cataloged as "star" 34 Tauri.

In 1786, Bode was elected as a member of the Berlin Academy. In 1825, he retired from the post of Director of the Berlin Observatory, and he died a year later while working on the *Jahrbuch* for 1830. A lunar crater with a diameter of 18 kilometers was named after Bode and approved in 1935 by the International Astronomical Union General Assembly. The asteroid Bodea, discovered in 1923, was also been named after Bode. The galaxy M81 that he discovered is popularly known as "Bode's Nebula" or "Bode's Galaxy."

Author and astronomer David Darling notes that Bode believed that all significant objccts in space—the Sun, stars, planets, moons, and comets—are inhabited by intelligent beings. Bode remarked that habitability was "the most important goal of creation" and that alien life forms throughout the universe "are ready to recognize the author of their existence and to praise his goodness."

To address the seeming inhospitality of comets, Bode said, "Who can conceive what special arrangements of the wise Creator in regard to the climate, zones, dwelling places... may not be expected for all those on a cometary body?" Bode believed that space was probably infinite in extent, but that the cosmos was finite with God beyond the cosmos.

FURTHER READING

Darling, David, "Bode, Johann Elert (1747–1826)," in *Encyclopedia of Astrobiology, Astronomy and Spaceflight*; see www.daviddarling.info/encyclopedia/B/Bode.html.

Frommert, Hartmut, and Christine Kronberg, "Johann Elert Bode (January 19, 1747–November 23, 1826)," SEDS (Students for the Exploration and Development of Space); see www.seds.org/messier/Xtra/Bios/bode.html.

Littmann, Mark, *Planets Beyond: Discovering the Outer Solar System* (New York: Courier Dover Publications, 2004).

Sticker, Bernhard, "Johann Bode," in *Dictionary of Scientific Biography*, Charles Gillispie, editor-in-chief (New York: Charles Scribner's Sons, 1970).

INTERLUDE: CONVERSATION STARTERS

> For me, a hypothesis is a statement whose *truth* is temporarily assumed, but whose *meaning* must be beyond all doubt.
>
> —Albert Einstein to Edward Study, September 25, 1918

> I have often made the hypothesis that ultimately physics will not require a mathematical statement, that in the end the machinery will be revealed, and the laws will turn out

be simple, like the checker board with all its apparent complexities.

—Richard Feynman, *The Character of Physical Law*

Scientific principles and laws do not lie on the surface of nature. They are hidden, and must be wrested from nature by an active and elaborate technique of inquiry.

—John Dewey, *Reconstruction in Philosophy*

If you're willing to answer yes to a God outside of nature, then there's nothing inconsistent with God on rare occasions choosing to invade the natural world in a way that appears miraculous. If God made the natural laws, why could he not violate them when it was a particularly significant moment for him to do so?

—Francis Collins, "God vs. Science" (interview), *Time*, November 13, 2006

The empirical basis of objective science has nothing "absolute" about it. Science does not rest upon solid bedrock. The bold structure of its theories rises, as it were above a swamp. It is like a building erected on plies. The piles are driven down from above into the swamp, but not down to any natural or "given" base; and if we stop driving the piles deeper, it is not because we have reached firm ground. We simply stop when we are satisfied that the piles are firm enough to carry the structure, at least for the time being.

—Karl Popper, *The Logic of Scientific Discovery*

If one accepts the premise that all knowledge comes to us through our senses, Hume says, then one must logically conclude that both "Nature" and "Nature's laws" are creations of our own imagination.

—Robert Pirsig, *Zen and the Art of Motorcycle Maintenance*

The supreme task of the physicist is to arrive at those universal elementary laws from which the cosmos can be built up by pure deduction. There is no logical path to these laws; only intuition, resting on sympathetic understanding of experience, can reach them. In this methodological uncertainty, one might suppose that there were any number of possible systems of theoretical physics all equally well justified; and this opinion is no doubt correct, theoretically. But the development of physics has shown that at any given moment, out of all conceivable constructions, a single one has always proved itself decidedly superior to all the rest.

—Albert Einstein, "Principles of Research"

COULOMB'S LAW OF ELECTROSTATICS

France, 1785. The force of attraction or repulsion between two electric charges is proportional to the magnitude of the charges and inversely proportional to the square of their separation distance.

CROSS REFERENCE: NEWTON'S LAW OF UNIVERSAL GRAVITATION, EINSTEIN'S SPECIAL THEORY OF RELATIVITY, AND COULOMB'S LAW OF FRICTION.

> In 1785, Louis XVI of France signed a law stating that a handkerchief must be square. Frenchman Jean-Pierre Blanchard and American John Jeffries were the first to cross the English Channel in a gas balloon. The dollar became the standard currency for the United States.

Coulomb's Law states that the magnitude of the force F between two point charges in free space is given by

$$\boxed{F = \frac{1}{4\pi\varepsilon_0}\frac{q_1 q_2}{r^2},}$$

where q_1 and q_2 are the magnitudes of the charges in coulombs, r is the distance between the charges in meters, ε_0 is the permittivity of free space (8.85×10^{-9} farads/meter), and F is given in units of newtons. A coulomb, denoted by the symbol C, is defined as the amount of charge that flows past a point in a wire in one second when the current in the wire is one ampere. In other words, 1 C = 1 A·s. If the charges have the same sign, the force is repulsive. If the charges have opposite signs, the force is attractive.

By examining the equation, we can see that the magnitude of the force is directly proportional to the magnitude of the charges of each object and inversely proportional to the square of the distance between them. The force exerted by one point charge on the other is along the imaginary line between the charges.

Charge values may be considered to be additive in instances when electrons and protons combine to form composite particles or collections of particles. Except for the case of quarks, which are considered to have fractional charges, all charges observed in nature are integer multiples of the charge on the electron (Q_e) or proton (Q_p), which have these amounts of charge:

$$Q_e = -(1.60217733 \pm 0.00000049) \times 10^{-19} \text{ coulombs}$$
$$Q_p = +(1.60217733 \pm 0.00000049) \times 10^{-19} \text{ coulombs}$$

Nuclear physicist Ernest Rutherford (1871–1937) conducted experiments with scattered alpha particles that showed that Coulomb's Law is accurate even for charged particles having nuclear dimensions and even for r values as low as 10^{-12} centimeters. (*Alpha particles* are helium nuclei, and they consist of two protons and two neutrons bound together.) In fact, today, experiments have demonstrated that Coulomb's Law is valid over a remarkable range of separation distances, from as small as 10^{-16} meters (a tenth of the diameter of an atomic nucleus) to as large as 10^{6} meters. Coulomb's Law is accurate only when the charged particles are stationary because movement produces magnetic fields that alter the forces on the charges.

Note that a coulomb is an extremely large electric charge compared to the charge of a single electron or proton. To get a feel for the magnitude, consider two objects, each with a net charge of +1 coulomb. If you were to place these objects a meter apart, the repulsive force would be about nine billion newtons, which corresponds to one million tons! Because the coulomb is such a huge charge, scientists sometimes use smaller measurement units like the microcoulomb (10^{-6} C), picocoulomb (10^{-12} C), or even simply the charge of the electron, e (1.602×10^{-19} C).

Coulomb's Law and Newton's Law of Universal Gravitation are examples of what physicists sometimes refer to as "action at a distance" laws—in the sense that when the laws were formulated, no known mediator of the interaction existed. Newton's law describes the gravitational attraction of masses m_1 and m_2 separated by a distance r and may be written $F_g = Gm_1m_2/r^2$, where F_g is the magnitude of the force due to gravity.

Even a casual examination of the mathematical formulations of Newton's law and Coulomb's Law reveals that the two formulas bear striking similarities. Both the electrostatic force and gravitational forces are directly proportional to the product of the interacting entities (mass or charge), and both the forces are inversely proportional to the square of the distance of separation.

For both Newton's Law of Universal Gravitation and Coulomb's Law, one might think that the respective forces involved are *instantaneously* affected by a change in locations of the relevant objects. However, this is not the case. For example, for Coulomb's Law, if one of the two charges is moved, then the force acting on the second charge does not immediately change. We know from Einstein's Special Theory of Relativity that signals do not propagate faster than the speed of light; thus, if one charge is moved, then a time delay must exist for the second particle to "become aware" of this movement. Moreover, if the first charge were suddenly plucked from the experiment, the second charge would be sensitive to this removal only some time later.

The same kind of delay applies to masses with a gravitational attraction, such as the case for Earth circling the Sun. If the Sun were suddenly removed, Earth would keep orbiting about the missing Sun for several minutes because the gravitational influence cannot travel faster than light speed. During the time needed for this influence to propagate, one body continues to experience an electrical or gravitational influence from the other body as if the missing object still existed.

Despite some similarities, a noticeable difference exists between the Newton's Law of Universal Gravitation and Coulomb's Law—the Coulomb force can be attractive or repulsive while the gravitational force is only attractive. Also, the magnitude of the Coulomb force depends upon the medium separating the charges, while the gravitational force is independent of the medium. For example, our first term in Coulomb's Law may be written more generally using ε instead of ε_0:

$$k = \frac{1}{4\pi\varepsilon},$$

where the permittivity ε is an electrical property of the medium that surrounds the two charges. The symbol ε_0 denotes the permittivity when the medium is a vacuum. The value of k, sometimes known as Coulomb's constant, is approximately equal to 9×10^9 N·m^2/C^2 when $\varepsilon = \varepsilon_0$. Electrically conducting media have permittivity values greater than ε_0. Because a vacuum has no charge carriers, the permittivity is lower for a vacuum than for any other medium. The permittivity value of dry air is so close to that of a vacuum that scientists usually treat experiments in the air as if performed in a vacuum.

The permittivity of a material is usually given relative to that of free space. If the relative permittivity is denoted by ε_r, permittivity is then calculated by multiplying ε_0 by ε_r. Approximate room-temperature relative permittivity values are given in table 6, and the values may vary according to temperature and the precise composition of the material under study. For example, a range of permittivity values exists for different kinds of paper.

Coulomb's Law is accurate only for point charges, that is, charges that are localized to an infinitely small region of space. However, all real-world experiments are performed with charges on objects that have finite dimensions. Coulomb's Law may be used in experiments with such objects if the dimensions of the charged objects are much smaller than the distance between their centers. Note that in modern times, the law has been generalized to integral and differential forms that may be used for nonpoint charges, and often these generalizations are also referred to as Coulomb's Law.

TABLE 6 Relative Permittivity Values

Material	Approximate Relative Permittivity Values, ε_r, at 300°K
Vacuum	1 (by definition)
Air	1.0005
Polyethylene	2.2
Lucite (trade name for a clear plastic)	2.8
Cocaine	3.1
Paper	3.3
Mica, muscovite	5.4
Rubber, Neoprene	6.6
Bone, cancellous (spongy)	26
Methyl alcohol	32
Brain, gray matter	56
Water (20°C)	80
Lead titanate	200

Source: Glenn Elert, "Dielectrics," in *The Physics Hypertextbook*; see hypertextbook.com/physics/electricity/dielectrics/.

Although the coulomb repulsive force should be quite strong for positively charged protons within a nucleus, the protons do not fly apart because they are held together by another fundamental force, the strong nuclear force, which is stronger than the coulomb force.

I conclude this section with a short problem that shows a practical calculation involving Coulomb's Law. Imagine two small balls, each with a mass of 0.20 grams. They are each attached to a separate 50-cm-long thread that is tied to the same point in the ceiling. Because the two balls have the same charge, they dangle from the ceiling and do not touch each other. In turns out that in this particular experiment, each thread makes a 37-degree angle with respect to a perpendicular to the ceiling. To help visualize this problem, draw a triangle ▲. The topmost point represents the attachment point of the threads, and the left and right vertices represent the positions of the balls. If we assume that the charges on each ball are the same, we can determine how large the charge is.

In order to solve this problem, we can use simple trigonometry, while realizing that the weight of an object is equal to its mass m times the acceleration due to gravity, g (which is 9.8 m/s^2). First, consider the ball on the left. Three forces act on the ball: its weight downward (mg), the tension T on the string, and the repulsion force F due to the charge on the ball at the right. Because the balls are not moving, forces in the x and y directions are in balance. Thus, for the force in the x direction, we have $F_x - 0.6T = 0$. Considering the forces in the y direction, we have $0.8T - (0.2)(10^{-3}\text{ kg})(9.8\text{ m/s}^2) = 0$, which yields $T = 2.45 \times 10^{-3}$ N. Then, we can calculate the force $F_x = 1.47 \times 10^{-3}$ N. This is the force of repulsion between the two balls. We may substitute this into the formula for Coulomb's Law to solve for the charge on the ball:

$$1.47 \times 10^{-3} = (9 \times 10^9)\frac{q^2}{(0.60)^2}$$

(The distance r between the two spheres is 0.60 m, which can be determined by trigonometry, given the 50-cm length of string and 37° angle.) Solving for q, we find that q is approximately equal to 2.4×10^{-7} coulombs or 0.24 μC, where μC is the symbol for microcoulombs.

In a system of many point-charges, charges exert forces on one another, and the resultant force exerted on any one charge is the vector sum of the individual forces exerted on that charge by all the other charges in the system.

Charles-Augustin de Coulomb (1736–1806), French physicist famous for the law that describes the force between two electrical charges.

CURIOSITY FILE: Coulomb's engineering skills played a large role in the fortification of Martinique, a Caribbean island. • yC is the official unit for yoctocoulomb, which is 10^{-24} coulombs. • Coulomb won a prize offered by the Académie des Sciences for the best method of constructing a compass for a ship.

Coulomb's contributions to the science of friction were exceptionally great. Without exaggeration, one can say that he created this science.

—I. V. Kragelsky and V. S. Schedrov, *Development of the Science of Friction*

Coulomb can be considered one of the great engineers in eighteenth-century Europe.

—C. Stewart Gillmor, "Charles Coulomb," in *Dictionary of Scientific Biography*

Who could forget "Chuck" Coulomb's 1773 address to the Academy of Science in Paris when he discussed pioneering soil mechanics theory?

—Terence Meany and Matthew Tirschwell, *The Complete Idiot's Guide to Electrical Repair*

Charles-Augustin de Coulomb is one of the preeminent physicists and engineers of all time who contributed to the fields of electricity, magnetism, applied mechanics, friction, and torsion. Coulomb was born to a well-to-do family in Angoulême in southwest France. His family later moved to Paris, where he entered the Collège Mazarin. Here he received a good general education in the humanities as well as in mathematics, astronomy, and chemistry.

At some point, his father lost all his money in financial speculations. This hardship, along with Coulomb's disagreement over career plans with his mother, caused a split in the family, and Coulomb and his father moved to Montpellier while his mother stayed in Paris. According to some sources, Coulomb's mother wanted him to be a medical doctor, but her son insisted on studying a more quantitative subject such as engineering or mathematics. The disagreements became heated, and his mother virtually disowned him.

In 1760, Coulomb entered the École du Génie at Mézières and later graduated as an engineer with the rank of lieutenant in the Corps du Génie (Corps of Engineers). Over the next two decades, he was posted to a variety of locations where he was involved in structural engineering, fortification design, and soil mechanics—for example, he spent several years in the West Indies as a military engineer—before returning to France, where he would begin to write his important papers on applied mechanics.

Coulomb created a torsion balance around 1777 in order to measure electrostatic forces. The torsion balance contains two metal balls attached to an insulating rod. The rod is suspended at its middle by a nonconducting filament or fiber. To measure the electrostatic force, one of the two balls is charged. A third ball with similar charge is placed near the charged ball of the balance, causing the ball on the balance to be repelled. This repulsion causes the fiber to twist by a certain amount. If we measure how much force is required to twist the wire by the same angle of rotation, we can estimate the degree of force caused by the charged sphere. In other words, the fiber acts as a very sensitive spring that supplies a force proportional to the angle of twist. Coulomb showed that the force varied as $1/r^2$ for repulsion between like charges, and attraction between unlike charges, separated by an initial distance r. It appears that he never actually demonstrated that the force between charges is proportional to the product of the charge values—he simply asserted this to be true. C. Stewart Gillmor, writing in *Dictionary of Scientific Biography*, indicates the degree to which Coulomb's balance affected science for many generations:

> Coulomb's simple, elegant solution to the problem of torsion in cylinders [with graduated scales] and his use of the torsion balance in physical applications were important to numerous physicists in succeeding years.... Coulomb developed a theory of torsion in thin silk and hair threads. Here he was the first to show how the torsion suspension could provide physicists with a method of accurately measuring extremely small forces.

In particular, Coulomb showed with his experiment that the exponent of r (the charge separation distance) was 2 within a few percent uncertainty. Today, we know that the exponent is 2 within about 2 parts in 10^9.

In 1779, Coulomb began his research into friction, which eventually led to his important publication *Théorie des machines simples, en ayant égard au frottement de leurs parties et à la raideur des cordages* ("Theory of Simple Machines, with Regard to the Friction Between their Parts and the Rigidity of the Linkages"). This work was followed twenty years later by a memoir on viscosity. Coulomb's Law of Friction states that for two surfaces in relative motion, the kinetic friction is almost independent of the relative speed of the surfaces.

Coulomb's research in friction was stimulated by a prize offered by the Academy of Science in Paris for "the solution of friction of sliding and rolling surfaces, the resistance to the bending in cords, and the application of these solutions to simple machines used in the navy." According to Peter J. Blau, writing in *Friction Science and Technology*,

> Coulomb's researches and conclusions about the nature of friction dominated thinking in the field for over a century and a half, and many of his concepts remain in use. In fact, the term "Coulombic friction" is still found in publications that interpret the results of recent experiments....

Between 1785 and 1791, Coulomb wrote seven important papers on electricity and magnetism, which he submitted to the Académie des Sciences. Topics included his continued use of the torsion balance to understand attraction, repulsion, distribution of electricity on the surface of charged objects, and his demonstration of the inverse square law for magnetic poles. His 1785 work, which described the use of his different kinds of torsion balances, was published in his *Recherches théoriques et expérimentales sur la force de torsion et sur l'élasticité des fils de metal* ("Theoretical and Experimental Studies on the Twisting Strain and Elasticity of Metal Wires"). Here, he showed that the torsion balance could be used to accurately measure extremely small forces.

In 1802, Coulomb married Louise Françoise LeProust Desormeaux, the mother of his two sons who were born before marriage. Louise was in her twenties. Toward the end of his life, Coulomb particularly enjoyed being in the country and teaching science to his youngest son, Charles. During his last days, Coulomb contracted a fever that finally killed him. His funeral services were held at Abbaye de St.-Germain-des-Prés.

Of Coulomb's scientific prowess, Ioan James writes in *Remarkable Physicists: From Galileo to Yukawa*:

> He has been described as the complete physicist, rivaled in the eighteenth century only by Henry Cavendish, combining experimental skill, accuracy of measurement, and great originality with mathematical powers adequate to all his demands.

A lunar crater with a diameter of 89 kilometers was named after Coulomb and approved in 1970 by the International Astronomical Union General Assembly.

Throughout his career, Coulomb conducted a variety of research and made contributions to our understanding of

- rupture of masonry piers and beams
- the sheer of brittle materials
- the physics of vaulted arches
- friction of machinery and fluid resistance
- design of windmills
- elasticity of metal and silk fibers and of soil mechanics
- magnetic compass design
- efficiency of human and animals workers (ergonomics)

Several people discovered aspects of Coulomb's Law before Coulomb. As far back as 1750, the British Reverend John Michell (1724–1793) published studies that showed that attraction and repulsion between the poles of magnets varied inversely as the square of the distance between them.

Michell's torsion balance was subsequently used by Henry Cavendish (1731–1810) to help measure the density of Earth. Coulomb had invented his torsion balance around 1784, and although Michell's work probably preceded Coulomb's, their discoveries were made independently.

John Robinson (c. 1725–?), a British doctor, measured the electrostatic forces of attraction and repulsion in 1769, and his experiments suggested that electrical repulsion had an $1/r^{2.06}$ dependency, and electrical attraction had an $1/r^c$ dependence in which $c < 2$. From these results, he suggested that $1/r^2$ was probably correct. Also, the English chemist Joseph Priestley (1733–1804) had suggested the $1/r^2$ law of electric force. Priestly wrote in *The History and Present State of Electricity:*

> May we not infer from this experiment [with charged hollow conductors] that the attraction of electricity is subject to the same laws with that of gravitation and is therefore according to the squares of the distances; since it is easily demonstrated that were the Earth in the form of a shell, a body in the inside of it would not be attracted to one side more than another?

Although Priestly offered no convincing proof for Coulomb's Law, his speculations were essentially correct. Priestly also independently invented the torsion balance and used it to show that the force between two magnetic poles varies as the inverse square of the distance between the poles.

Today, we call the $1/r^2$ law Coulomb's Law in honor of Coulomb's independent results gained through the evidence provided by his torsional measuring system. In other words, Coulomb provided convincing quantitative results for what was, up to 1785, often a good guess.

The Coulomb force is relevant at atomic-size scales, and in fact, it is instructive to compare the gravitational forces and Coulomb forces for a hydrogen atom. As an approximation, if we think of the electron as a point particle orbiting a point-particle proton, with the electron separated from the proton by a distance of about 5.3×10^{-11} meters on average, the Coulomb force can be calculated by

$$\frac{kq^2}{r^2} = \frac{(9 \times 10^9)(1.6 \times 10^{-19})^2}{(5.3 \times 10^{-11})^2} = 8.2 \times 10^{-8}\text{N}.$$

The magnitude of the gravitational force F_g between the proton and electron can be approximately determined using the mass of the electron m_e and proton m_p:

$$F_g = \frac{Gm_e m_p}{r^2} = \frac{(6.67 \times 10^{-11})(9.1 \times 10^{-31})(1.67 \times 10^{-27})}{(5.3 \times 10^{-11})^2}$$
$$= 3.6 \times 10^{-47}\text{N}$$

Notice that the Coulomb force is significantly greater than the gravitational force between the two subatomic particles.

In closing, note that the Eiffel Tower in Paris features the names of 72 great French scientists and other thinkers—including Coulomb. Gustave Eiffel's original engraving of these names was painted over in the early 1900s but restored in 1987. The letters in the names are approximately 60 centimeters tall. In the following list, I have highlighted in boldface those people listed on the tower who also have main entries in this book (only last names appear on the tower):

1. **Ampère** (André-Marie Ampère, mathematician and physicist)
2. Arago (Dominique François Jean Arago, astronomer and physicist)
3. Barral (Jean-Augustin Barral, agronomist, chemist, physicist)
4. Becquerel (Antoine Henri Becquerel, physicist)
5. Bélanger (Jean-Baptiste-Charles-Joseph Bélanger, mathematician)
6. Belgrand (Eugene Belgrand, engineer)
7. Berthier (Pierre Berthier, mineralogist)
8. Bichat (Marie François Xavier Bichat, anatomist and physiologist)
9. Borda (Jean-Charles de Borda, mathematician)
10. Breguet (Abraham Louis Breguet, mechanic and inventor)
11. Bresse (Jacques Antoine Charles Bresse, civil engineer and hydraulic engineer)
12. Broca (Paul Pierre Broca, physician and anthropologist)
13. Cail (Jean-François Cail, industrialist)
14. Carnot (Nicolas Léonard Sadi Carnot, mathematician)
15. Cauchy (Augustin Louis Cauchy, mathematician)
16. Chaptal (Jean-Antoine Chaptal, agronomist and chemist)
17. Chasles (Michel Chasles, geometer)
18. Chevreul (Michel Eugène Chevreul, chemist)
19. Clapeyron (Émile Clapeyron, engineer)
20. Combes (Émile Combes, engineer and metallurgist)
21. Coriolis (Gaspard-Gustave Coriolis, engineer and scientist)
22. **Coulomb** (Charles-Augustin de Coulomb, physicist)
23. Cuvier (Baron Georges Leopold Chretien Frédéric Dagobert Cuvier, naturalist)
24. Daguerre (Louis Daguerre, artist and chemist)
25. De Dion (Albert de Dion, engineer)
26. De Prony (Gaspard de Prony, engineer)
27. Delambre (Jean Baptiste Joseph Delambre, astronomer)
28. Delaunay (Charles-Eugène Delaunay, astronomer)

29. **Dulong** (Pierre Louis Dulong, physicist and chemist)
30. Dumas (Jean Baptiste André Dumas, chemist)
31. Ebelmen (Jean-Jacques Ebelmen, chemist)
32. Fizeau (Hippolyte Fizeau, physicist)
33. Flachat (Jeugène Flachat, engineer)
34. Foucault (Léon Foucault, physicist)
35. **Fourier** (Jean Baptiste Joseph Fourier, mathematician)
36. Fresnel (Augustin-Jean Fresnel, physicist)
37. **Gay-Lussac** (Joseph Louis Gay-Lussac, chemist)
38. Giffard (Henri Giffard engineer)
39. Goüin (Ernest Goüin, engineer and industrialist)
40. Haüy (René-Just Haüy, mineralogist)
41. Jamin (Jules Célestin Jamin, physicist)
42. Jousselin (Alexandre Louis Jousselin, engineer)
43. Lagrange (Joseph Louis Lagrange, mathematician)
44. Lalande (Joseph Jérôme Lefrançais de Lalande, astronomer)
45. Lamé (Gabriel Lamé, geometer)
46. Laplace (Pierre-Simon Laplace, mathematician and astronomer)
47. Lavoisier (Antoine Lavoisier, chemist)
48. Le Chatelier (Henri Louis le Chatelier, chemist)
49. Le Verrier (Urbain Le Verrier, astronomer)
50. Legendre (Adrien-Marie Legendre, geometer)
51. Malus (Etienne-Louis Malus, physicist)
52. Monge (Gaspard Monge, geometer)
53. Morin (Jean-Baptiste Morin, mathematician and physicist)
54. Navier (Claude-Louis Marie Henri Navier, mathematician)
55. Petiet (Jules Petiet, engineer)
56. Pelouze (Théophile-Jules Pelouze, chemist)
57. Perdonnet (Albert Auguste Perdonnet, engineer)
58. Perrier (François Perrier, geographer and mathematician)
59. Poinsot (Louis Poinsot, mathematician)
60. Poisson (Simeon Poisson, mathematician and physicist)
61. Polonceau (Antoine-Rémi Polonceau, engineer)
62. Poncelet (Jean-Victor Poncelet, geometer)
63. Regnault (Henri Victor Regnault, chemist and physicist)
64. Sauvage (Jean-Pierre Sauvage, mechanic)
65. Schneider (Jacques Schneider, industrialist)
66. Seguin (Marc Seguin, mechanic)
67. Sturm (Jacques Charles François Sturm, mathematician)
68. Thénard (Louis Jacques Thénard, chemist)
69. Tresca (Henri Tresca, engineer and mechanic)
70. Triger (Jacques Triger, engineer)

71. Vicat (Louis Vicat, engineer)
72. Wurtz (Charles-Adolphe Wurtz, chemist)

FURTHER READING

Blau, Peter J., *Friction Science and Technology* (New York: Marcel Dekker, 1995).

Elert, Glenn, "Dielectrics," in *The Physics Hypertextbook*; see hypertextbook.com/physics/electricity/dielectrics/.

Gillmor, C. Stewart, "Charles Coulomb," in *Dictionary of Scientific Biography*, Charles Gillispie, editor-in-chief (New York: Charles Scribner's Sons, 1970).

James, Ioan, *Remarkable Physicists: From Galileo to Yukawa* (New York: Cambridge University Press, 2004).

Kovacs, J., "Coulomb's Law," Project PHYSNET, Michigan State University; see physnet.org/modules/pdfmodules/m114.pdf.

Priestley, Joseph, *The History and Present State of Electricity* (London: J. Doddsley, J. Johnson, B. Davenport, & T. Cadell, 1767).

Shamos, Morris, *Great Experiments in Physics: Firsthand Accounts from Galileo to Einstein* (New York: Dover, 1987).

Wikipedia, "The 72 Names on the Eiffel Tower"; see en.wikipedia.org/wiki/The_72_names_on_the_Eiffel_Tower.

INTERLUDE: CONVERSATION STARTERS

> Individual facts are selected and grouped together such that their lawful connection becomes clearly apparent. By grouping these laws together, one can achieve other more general laws. . . . However . . . the big advances in scientific knowledge originated [by an inductive method] only to a small degree. For, if a research were to approach things without a preconceived opinion, how would he be able to pick the facts from the tremendous richness of the most complicated experiences that are simple enough to reveal their connections through laws?
>
> —Albert Einstein, "Induction and Deduction in Physics, *Berliner Tageblatt*

> We showed that if general relativity is correct, any reasonable model of the universe must start with a singularity . . . I now think that although there is a singularity, the laws of physics can still determine how the Universe began.
>
> —Stephen Hawking, *Black Holes and Baby Universes*

The future religion of humanity will be based on scientific laws.

—Greg Whitefield, quoted in Post B. Basnet's, "Nepal Becoming Mecca for Buddhist Studies," *Kathmandu Post*

Scientific laws are merely the algorithms of this program [of a world simulation]. . . . At the quantum level, we are looking at machine language, beneath that perhaps the machine itself, and there are no algorithms at that level, just the changes in state of the machine that allow the algorithms to work. This is why quantum particles seem to behave so erratically and are hard to pin down—they are not officially "in" the simulation; they are what's making the simulation happen.

—James Platt, personal communication, March 1, 2007

But there it was, the whole history of science, a clear story of continuously new and changing explanations of old facts. The time spans of permanence seemed completely random he could see no order in them. Some scientific truths seemed to last for centuries, others for less than a year. Scientific truth was not dogma, good for eternity, but a temporal quantitative entity that could be studied like anything else.

—Robert Pirsig, *Zen and the Art of Motorcycle Maintenance*

Science works because the universe is ordered in an intelligible way. The most refined manifestation of this order is found in the laws of physics, the fundamental mathematical rules that govern all natural phenomena. One of the biggest questions of existence is the origin of those laws: where do they come from, and why do they have the form that they do? . . . The laws of physics possess a weird and surprising property: collectively they give the universe the ability to generate life and conscious beings, such as ourselves, who can ponder the big questions.

—Paul Davies, "Laying Down the Laws," *New Scientist*

CHARLES'S GAS LAW

France, 1787. At constant pressure, the volume occupied by a fixed amount of gas is directly proportional to its absolute temperature.

CROSS REFERENCE: JOSEPH LOUIS GAY-LUSSAC, LEONARDO DA VINCI, GUILLAUME AMONTONS, AND AMONTONS'S LAW OF FRICTION.

> In 1787, American inventor John Fitch launched a steamboat on the Delaware River. William Herschel discovered two moons of Uranus, which his sons later named Titania and Oberon. The Constitutional Convention in Philadelphia adopted the U.S. Constitution.

Charles's Gas Law, also known as Gay-Lussac's Law, states that the volume occupied by a fixed amount of gas varies directly with the absolute temperature (i.e., the temperature in degrees kelvin). The law can be expressed as

$$\boxed{V = kT,}$$

where V is the volume at a constant pressure, T is the temperature, and k is a constant. French chemist and physicist Joseph Louis Gay-Lussac (1778–1850) first published the law in 1802, where he referenced unpublished work from around 1787 by French chemist and physicist Jacques Charles.

Physicists have discovered that a gas increases approximately by 1/273 (0.003663) of its volume at 0°C for each °C rise of temperature. Very slight deviations from 1/273 have been observed, but because they are so slight, the constant 1/273 is generally used as an approximate expansion coefficient for gases. For example, French physicist Henri-Victor Regnault (1810–1878) found values of 0.0036613 for hydrogen and 0.0037099 for carbon dioxide. Note that although the coefficients of expansion of different gases are nearly the same, different solids or liquids have very different coefficients.

As the temperature of the gas increases, the gas molecules move more quickly and hit the walls of their container with more force—thus increasing the volume of gas, assuming that the container volume is able to expand. For a more specific example, consider warming the air within a balloon. As the temperature increases, the speed of the moving gas molecules increases inside the surface of the balloon. This in turn increases the rate at which the gas molecules bombard the surface. Because the surface of the balloon can stretch, the surface expands as a result of the increased internal bombardment. The volume of gas increases, and its density decreases. The act of cooling the gas inside a balloon will have the opposite effect and cause the pressure to be reduced and the balloon to shrink.

Charles's Law is sometimes expressed as

$$\frac{V_1}{T_1} = \frac{V_2}{T_2},$$

where the subscripts 1 and 2 refer to the volume and temperature of a gas before and after the volume or temperature has changed. Notice that we do not need to know the value of the constant k in order to make practical use of the law when comparing two volumes. For example, let us suppose we have a sample of gas at 15°C and at one atmosphere pressure, with a volume of 2.50 liters. We want to know what volume this gas will occupy at 40°C at the same atmospheric pressure. In other words, for this problem, the pressure remains the same, while the volume and temperature change. Our first step is to convert the temperature in degrees Celsius to degrees kelvin by adding 273 to both temperature values. Thus, we have 2.50 L $\div$ 288°K = $V_2 \div$ 313°K; thus, V_2 = 2.72 liters. The temperature has increased slightly, and the volume also increases slightly.

Jacques Charles (1746–1823), French mathematician, physicist, and inventor who studied the thermal expansion of gases and who was the first person to ascend in a hydrogen balloon (with Nicolas Robert).

CURIOSITY FILE: In the world of ballooning, Charles invented the nacelle, the basket suspended beneath the balloon by ropes and held in place by a hoop. This nacelle holds the passengers and their belongings. • Local peasants were so frightened by the landing of one of Charles's balloons that they tore it apart, believing it to be the work of the devil.

> Almost nothing is known of Charles's family or his upbringing, except that he received a liberal, nonscientific education. [Later] Charles published almost nothing of significance.
>
> —J. B. Gough, "Jacques Charles," in *Dictionary of Scientific Biography*

> Charles's view from the balloon was phenomenal, and as soon as they touched down, the ecstatic Jacques Charles jumped from the basket, crying out to the assembled onlookers his newly sworn creed: "I care not what may

> be the condition of the Earth—it is the sky that is for me now. What serenity! What a ravishing scene!"
>
> —Richard Hamblyn, *The Invention of Clouds*

Jacques Charles was born in Beaugency-sur-Loire, France, and later conducted research in many fields that ranged from electricity to gases to the flight of balloons. Very little is known about his family and early life, but we do know his education was liberal and did not focus on science. His first job was a low-level position in the bureau of finances in Paris. In 1779, at the same time that Ben Franklin was in Paris, Charles was fired from his job and became inspired to learn about those experimental sciences that did not require much mathematical sophistication.

Less than two years later, Charles had evolved into such a good science lecturer that his presentations in physics and chemistry attracted large audiences. He was named a resident member of the Académie des Sciences in 1795, and later he became professor of experimental physics at the Conservatoire des Arts et Metiers (Conservatory of Arts and Trades). In 1804, he married a woman described in the literature merely as an "attractive young lady."

Charles was most famous to his contemporaries for his various exploits and inventions pertaining to the science of ballooning and other practical sciences. For example, Charles

- promoted the idea of using hydrogen instead of hot air in balloons
- invented the *valve line*, which enables gas release for balloon descent
- invented the *appendix*, a tube that enables gas to escape to prevent balloon rupture
- invented the megascope, a device for projecting and magnifying objects
- invented a goniometer for measuring crystal angles
- developed ways to make balloons less porous in order to hold the hydrogen gas.

His first balloon journey took place in 1783, and his adoring audience of thousands watched as the balloon drifted. The balloon ascended to a height of nearly 3,000 feet (914 meters) and seems to have finally landed in a field outside of Paris, where it was destroyed by terrified peasants. In fact, the locals believed that the balloon was some kind of evil spirit or beast from which they heard sighs and groans, accompanied by a noxious odor.

The initial filling of this balloon had been quite a task. According to Charles M. Evans's *War of the Aeronauts: The History of Ballooning in the Civil War*,

> Charles used a crude hydrogen generator that consisted of a large wooden barrel containing iron filings.... Hinged doors were fashioned into the top of the barrel, which allowed Charles to pour sulfuric acid over the filings, causing the chemical reaction that produced hydrogen. The inflation process... took over four days to complete. More than 200 pints of acid and a ton of iron filings were consumed....

The balloon craze had thus begun. Here are some quick highlights: In 1784, John Jeffries and Jean-Pierre F. Blanchard flew a hydrogen balloon at various altitudes above England as they collected temperature and moisture data. In January of 1785, Jeffries and Blanchard crossed the English Channel from England to France. In June of 1785, Jean-François Pilâtre de Rozie and Pierre Romain attempted to cross the Channel from France to England in a double balloon, consisting of an upper gas balloon and a lower hot-air balloon. At an altitude of about 1,000 meters, the craft burst into flames, killing both de Rozier and Romain, who were the first to die in a balloon accident. The first flight made by a woman took place in 1894 when Marie Thible of Lyons, France, ascended to 8,500 feet in a 45-minute flight. The first balloon flight in America was undertaken by Blanchard in 1793 and witnessed by George Washington.

Charles had started a fad, but despite his initial fame, he rarely published his scientific findings, and his famous 1878 gas law was first made public by Gay-Lussac, who also improved upon Charles's experimental procedures.

Note that the relationship $P = kT$, where P stands for pressure, is a relationship that is usually not attributed in an eponymous fashion, although it is often seemingly mistakenly called Gay-Lussac's Law. Note also that French physicist Guillaume Amontons (1663–1705) also investigated the relationship between pressure and temperature in gases, although he did not have access to accurate thermometers. He did show that that the pressure of a gas increases as its temperature is raised while holding the volume and the amount of gas constant.

Amontons also discovered a relationship, which some have called Amontons's Law, that characterizes the friction between two surfaces. In particular, Amontons showed that the frictional force is directly proportional to the force normal (perpendicular) to the surfaces in contact, with a constant of proportionality (a frictional coefficient) that is constant and independent of the size of the contact area. These kinds of relationships were first suggested by Leonardo da Vinci and rediscovered by Amontons.

Several studies have been conducted in the early years of the twenty-first century to determine the extent to which Amontons's Law actually applies for materials at length scales from nanometers to millimeters. In particular, the validity of Amontons's Law is of concern today for researchers in the area of MEMS (micro-electromechanical systems), which make use of tiny devices such as those now used in inkjet printers and as accelerometers in car airbag systems. MEMS uses microfabrication technology to integrate mechanical elements, sensors, and electronics on a silicon substrate. Amontons's Law, which is often useful when studying traditional machines and moving parts, may not be applicable to larger machines, such as those the size of a pinhead and larger. [For additional principles in this book that relate to the field of tribology (friction), see Coulomb's Law of Friction, discussed in the entry for Coulomb.]

Returning to the life of Jacques Charles, his spirit of adventure and zest for life are exemplified by the joy he felt while traveling in his balloons. Richard Hamblyn, author of *The Invention of Clouds: How an Amateur Meteorologist Forged the Language of the Skies*, describes Charles's feelings while on one of his solo flights:

> Charles rose even higher on this second [flight of the evening], bringing the Sun back into view, where he stayed aloft until he watched its second setting, ravished by the sight, "hearing himself live."...When he finally relanded...he emerged from the basket more rhapsodic than ever, with the image of the twin sunsets, viewed from the vantage of a soaring balloon, scored indelibly onto his mind.

A lunar crater with a diameter of 1 kilometer was named after Charles and approved in 1976 by the International Astronomical Union General Assembly.

FURTHER READING

Evans, Charles, *War of the Aeronauts: The History of Ballooning in the Civil War* (Mechanicsburg, Pa.: Stackpole Books, 2002).

Gough, J. B., "Jacques Charles," in *Dictionary of Scientific Biography*, Charles Gillispie, editor-in-chief (New York: Charles Scribner's Sons, 1970).

Hamblyn, Richard, *The Invention of Clouds: How an Amateur Meteorologist Forged the Language of the Skies* (New York: Picador, 2001).

INTERLUDE: CONVERSATION STARTERS

We now know that there exist true propositions which we can never formally prove. What about propositions whose proofs require arguments beyond our capabilities? What about propositions whose proofs require millions of pages? Or a million, million pages? Are there proofs that are possible, but beyond us?

—Calvin Clawson, *Mathematical Mysteries*

For the religious, passivism [i.e., objects are obedient to the laws of nature] provides a clear role for God as the author of the laws of nature. If the laws of nature are God's commands for an essentially passive world . . . , God also has the power to suspend the laws of nature, and so perform miracles.

—Brian Ellis, *The Philosophy of Nature: A Guide to the New Essentialism*

The most beautiful thing we can experience is the mysterious. It is the source of all true art and science. He to whom this emotion is a stranger, who can no longer pause to wonder and stand rapt in awe, is as good as dead: his eyes are closed.

—Albert Einstein, "What I Believe," *Forum and Century*

One can argue that mathematics is a human activity deeply rooted in reality, and permanently returning to reality. From counting on one's fingers to moon-landing to Google, we are doing mathematics in order to understand, create, and handle things, and perhaps this understanding is mathematics rather than intangible murmur of accompanying abstractions. Mathematicians are thus more or less responsible actors of human history, like Archimedes helping to defend Syracuse (and to save a local tyrant), Alan Turing cryptanalyzing Marshal Rommel's intercepted military dispatches to Berlin, or John von Neumann suggesting high altitude detonation as an efficient tactics of bombing.

—Yuri I. Manin, "Mathematical Knowledge: Internal, Social, and Cultural Aspects," March 2007

1800–1900

Science will continue to surprise us with what it discovers and creates; then it will astound us by devising new methods to surprise us. At the core of science's self-modification is technology. New tools enable new structures of knowledge and new ways of discovery. The achievement of science is to know new things; the evolution of science is to know them in new ways. What evolves is less the body of what we know and more the nature of our knowing.

—Kevin Kelly, "Speculations on the Future of Science"

When a distinguished but elderly scientist states that something is possible, he is almost certainly right. When he states that something is impossible, he is very probably wrong.

—Arthur C. Clarke, *Profiles of the Future*, 1962

When, however, the lay public rallies round an idea that is denounced by distinguished but elderly scientists and supports that idea with great fervor and emotion—the distinguished but elderly scientists are then, after all, probably right.

—Isaac Asimov, *Quasar, Quasar, Burning Bright*, 1976

As the nineteenth century drew to a close, scientists could reflect with satisfaction that they had pinned down most of the mysteries of the physical world: electricity, magnetism, gases, optics, acoustics, kinetics, and statistical mechanics . . . all had fallen into order before them. They had discovered the X ray, the cathode ray, the electron, and radioactivity, invented the ohm, the watt, the Kelvin, the joule, the amp, and the little erg.

—Bill Bryson, *A Short History of Nearly Everything*

DALTON'S LAW OF PARTIAL PRESSURES

England, 1801. Each gas in a mixture of gases exerts a pressure as if the other gases were not present; the total pressure of the gases is the sum of the pressures created by each gas in the mixture.

CROSS REFERENCE: JAMES JOULE, CHARLES'S GAS LAW, THE LAW OF MULTIPLE PROPORTIONS, AND THE ATOMIC THEORY OF MATTER.

In 1801, Thomas Jefferson was inaugurated President of the United States. French inventor and silk weaver Joseph-Marie Jacquard developed a loom that used punched cards to control the weaving pattern. (Weavers destroyed his invention because they feared it would take away their jobs.) German chemist and physicist Johann Wilhelm Ritter discovered ultraviolet radiation. Ritter died at 33, perhaps, in part, because he continually used his own body in high-voltage experiments.

Dalton's Law of Partial Pressures states that the total pressure P_t exerted by a mixture of gases in a container is equal to the sum of the separate pressures that each of the gases would exert if just that single gas occupied the entire volume of the container. In formula form, Dalton's Law (also sometimes called Dalton's Law of Additive Pressures) may be expressed as

$$\boxed{P_t = p_\mathrm{a} + p_\mathrm{b} + p_\mathrm{c} + \ldots,}$$

where P_t is the pressure produced by a mixture of gases $a, b, c, \ldots,$ and p_a, p_b, and $p_\mathrm{c} \ldots$ are the partial pressures of the gases in the mixture. As suggested, the partial pressure of a single gas is the pressure that would be exerted by that gas if it were to occupy the whole container, at the same pressure and temperature, in the absence of other gases.

As an example, if we mix different quantities of argon, oxygen, and helium in a container so that they have partial pressures of 1, 2, and 3 atmospheres, respectively, the total pressure of the mixture is simply

$$P_t = p_\mathrm{argon} + p_\mathrm{oxygen} + p_\mathrm{helium} = 6\,\text{atmospheres}.$$

Dalton's Law implies that we can regard the atmospheric pressure of Earth as the sum of the partial pressures of its constituent gases, namely, the sum of a nitrogen pressure, an oxygen pressure, an argon pressure, a carbon dioxide pressure, a water vapor pressure, and a pressure from rare gases present in the atmosphere.

John Dalton assumed and suggested that a large amount of space exists between the gas molecules within a gas mixture, and therefore, one gas molecule has little influence on the motion of another. This led him to believe that the pressure of a gas sample is the same whether it is the only gas in a container or it is among other gases. If gas molecules begin to interact—as may be the case at very low temperatures and high pressures—deviations from Dalton's Law can occur. Also, Dalton's Law assumes that the gases do not chemically react with one another.

Although Dalton's Law may seem trivial, it is one of the more useful gas laws for scientists. For example, one laboratory method of gas collection involves displacing water from a bottle. Once all of the water has exited the bottle, we know that the gas occupies essentially the entire volume of the bottle, assuming that the gas solubility in the water is not significant in the experiment. In fact, you may recall from a high-school chemistry class that you used an inverted bottle filled with water as the bottle sat in a water bath. A tube from the reaction vessel transfers the gas into the upside-down bottle into which the gas bubbles to the top and displaces water. The water is forced out of the mouth of the bottle into the water bath. Note that the gas now trapped in the container is not composed entirely of the gas pumped into the bottle—rather, it is a mixture that contains a certain amount of water vapor. We are able to find the pressure of the dry gas alone, using Dalton's Law and by subtracting the pressure of the water vapor from the total pressure:

$$P_{(\text{dry gas})} = P_{\text{total}} - P_{(\text{water vapor})}$$

We can easily substitute an actual value for the partial pressure of the water vapor, $P_{(\text{water vapor})}$, because reference tables exist that show the pressure of water vapor at various temperatures. For example, the vapor pressure of water at 20°C is 2.3 kPa, and at 40°C the pressure is 7.4 kPa. (kPa is a unit of pressure and stands for kilopascals.) Thus, for example, if we have a sample of hydrogen gas collected over water that is at a temperature of 20°C and find the resultant pressure is 110 kPa, the pressure exerted by the dry hydrogen alone is 110 – 2.3 = 107.7 kPa.

Dalton's Law is of particular interest to scuba divers, because they want their gas tanks to deliver a partial pressure of 0.20 atmospheres of oxygen to ensure proper functioning of the respiratory systems of their bodies. The total pressure must account for the external pressure experienced by the diver in the water in order to prevent the lungs from collapsing. Thus, divers use a valve to equalize the pressure inside their lungs with the external pressure by adding helium gas. The valve, in effect, uses Dalton's Law to maintain appropriate pressures and oxygen levels.

John Dalton (1766–1844), English chemist, physicist, and meteorologist, famous for his contributions to the development of atomic theory and one of the fathers of modern physical science.

CURIOSITY FILE: Red-green color blindness has often been referred to as "Daltonism," after John Dalton, who suffered from and researched this affliction. • Dalton discovered butylene (a gas used today for making rubber) and determined the correct chemical formula for ether (formerly used as an anesthetic). • Dalton was exceeding versatile—in 1801, in addition to his important paper on the properties of gases, he published a book on English grammar. • In the 1990s, scientists studied one of his preserved eyes in order to understand the cause of his color blindness.

> Matter, though divisible in an extreme degree, is nevertheless not infinitely divisible. That is, there must be some point beyond which we cannot go in the division of matter.... I have chosen the word "atom" to signify these ultimate particles....
>
> —John Dalton, *A New System of Chemical Philosophy*, 1808

> Dalton transformed the atomic concept from a philosophical speculation into a scientific theory—framed to explain quantitative observations, suggesting new tests and experiments, and capable of being given quantitative form through the establishment of relative masses of atomic particles.
>
> —Arnold Arons, *Development of Concepts of Physics*

> Chemical analysis and synthesis go no farther than to the separation of particles one from another, and to their reunion. No new creation or destruction of matter is within reach of chemical agency. We might as well attempt to introduce a new planet into the Solar System, or to annihilate one already in existence, as to create or destroy a particle of hydrogen. All the changes we can produce consist in separating particles that are in a state of cohesion or combination, and joining those that were previously at a distance.
>
> —John Dalton, *A New System of Chemical Philosophy*, 1808

John Dalton was born in Eaglesfield in northern England. He was a quiet man who attained his professional success in spite of several hardships: He grew up in a family with little money, was a poor speaker, never had a wife to give him emotional support, was severely color blind, and was considered to be a fairly crude or simple experimentalist. Perhaps some of these challenges would have presented an insurmountable barrier to any budding chemist of his time, but Dalton persevered and made exceptional contributions to the development of *atomic theory*, which states that all matter is composed of atoms of differing weights that combine in simple ratios. During his time, atomic theory also suggested that these atoms could be considered to be indestructible.

Dalton was the son of a Quaker weaver who lived in the county of Cumberland, England. Dalton's childhood years were spent working in the fields and in his father's shop, where cloth was made. His sister sold paper, ink, and pens. When only 12, Dalton helped run the village Quaker school at which he also taught, and two years later he taught with his brother at a boarding school in the small town of Kendal. During these early years, Dalton acknowledged that his love of science was stimulated by John Gough, a blind natural philosopher. Dalton wrote in a 1783 letter,

> John Gough is . . . a perfect master of the Latin, Greek, and French tongues. . . . Under his tuition, I have since acquired a good knowledge of them. He knows by the touch, taste, and smell, almost every plant within twenty miles . . . he and I have been for a long time very intimate; as our pursuits are common—viz. mathematical and philosophical.

Starting in 1787, Dalton kept a daily meteorological diary, which he continued to update each day until the very day of his death. This amazing compendium eventually contained roughly 200,000 entries on meteorological observations for the variable climate of the lake district in which he lived. Some British scientists, such as John Frederic Daniell (1790–1845), have called Dalton "the father of meteorology." Dalton's obsession for note taking is also evidenced in his recreational life—even when he played the English lawn game of "bowls," he kept meticulous records of hits, misses, and other scores.

His mathematical prowess increased while at school, and even in his early years, he had a reputation for solving yearly puzzle competitions of the *Ladies' Diary* and *Gentleman's Diary*. His lecture topics at the Kendal school ranged from mechanics and optics to astronomy and pneumatics. According to Arnold Thackray's entry on Dalton in the *Dictionary of Scientific Biography*, Dalton soon became restless with his exclusive focus on teaching and argued that "very few people of middling genius or capacity for other business" become teachers. He wanted to become a physician, but this was difficult for a man of his meager finances. Also, his family thought that his awkward bedside manner would preclude him from being a great physician.

During his years in Kendal, Dalton collected plants and common insects. "Some of these," he wrote, "may be thought puerile; but nothing that enjoys animal life, or that vegetates is beneath the dignity of a naturalist to examine."

In 1792, Dalton was appointed a teacher of mathematics and natural philosophy at New College in Manchester. The Presbyterians had established this college in order to give an excellent education to those who could not attend Cambridge and Oxford, which had been open only to members of the Church of England.

In 1794, Dalton was elected to the Manchester Library and Philosophical Society, and within a month of his election, Dalton presented his first major paper, "Extraordinary Facts Relating to the Vision of Colours, with Observations." In this first paper ever published on color blindness, Dalton offered a systematic study of the affliction. He had discovered his own color blindness when he realized that flowers looked different to him than to his colleagues. In particular, when Dalton looked at a flower that most people saw as pink, he regarded the flower as being blue. (For some time after he presented his paper, color blindness was even known as "Daltonism.")

According to legends, Dalton once bought his mother special stockings for her birthday. The mother turned to Dalton and exclaimed, "Why did you buy me *scarlet* stockings?" Scarlet wouldn't have been suitable for a Quaker woman. Dalton had thought they were blue and turned to his brother to verify their suitable color. However, his brother *also* saw blue instead of scarlet, at which point John discovered that both he and his brother were color blind.

Even though Dalton's scientific theories on color blindness were not adequate to account for his inability to see the color red, his skill in giving a careful account of the phenomenon increased his growing prestige. In order to account for his color blindness, he wrote in "Extraordinary Facts" that it was "almost beyond doubt that one of the humours of my eye . . . is a colored medium." He asked that his own eye be cut open and studied after his death to confirm his hypotheses that blue fluids in his eye absorbed the color red. (The dissection indeed occurred, but did not support his theories.) About his own color blindness, he noted in "Extraordinary Facts,"

> that part of the image which others call red appears to me little more than a shade or defect of light. After that, the orange, yellow and green seem one colour which descends pretty uniformly from an

intense to a rare yellow, making what I should call different shades of yellow.

In 1800, Dalton opened his private "Mathematical Academy" that offered students classes in mathematics and chemistry—and made Dalton a good living. Close to this time, Dalton began to make strides in various areas that would make him famous:

1. He discovered Charles's Law, which described the expansion of gases at constant pressure. (As discussed under "Charles's Gas Law" in part II, the law is named after Jacques Charles, who discovered the law independently and earlier than Dalton.)
2. He formulated the law of additive partial pressures for gases, which was first published in *Meteorological Observations*. Dalton's Law, as it came to be called, states that every gas acts as an independent entity in a mixture of gases and that the total pressure of a mixture of gases equals the sum of the pressures of the gases in the mixture, each gas acting independently.
3. He promoted chemical atomic theory: All matter is atomic in nature. (He also calculated relative masses of atoms and elements such as hydrogen, oxygen, carbon, and nitrogen). According to Dalton, all elements are composed of tiny, indestructible particles called atoms that are all alike for a particular element and have the same atomic weight.

He also formulated the Law of Multiple Proportions that stated whenever two elements can combine to form different components, the masses of one element that combine with a fixed mass of the other are in a ratio of small whole numbers, such as 1:1, 2:1, and 1:2. These simple ratios provided evidence that atoms were the building blocks of compounds. Unfortunately, knowledge of such ratios was insufficient to determine the actual number of atoms in each compound. Nevertheless, Dalton's atomic theory set the stage for great advances in decades to come, leading many to call Dalton the "father of chemistry."

Dalton did encounter resistance to atomic theory. For example, the British chemist Sir Henry Enfield Roscoe (1833–1915) mocked Dalton in 1887, saying, "Atoms are round bits of wood invented by Mr. Dalton." Perhaps Roscoe was referring to the wood models that some scientists used in order to represent atoms of different sizes. Nonetheless, by 1850, the atomic theory of matter was accepted among a significant number of chemists, and most opposition disappeared.

Through the years, Dalton published essays on diverse subjects, including theories on the trade winds, dew points, heat, the aurora borealis, the solubility of gases in water, variations in barometric pressure, and evaporation. Despite prevailing contemporary views, he promoted the correct idea that the atmosphere was a physical mixture of about 80% nitrogen and 20% oxygen instead of being a compound of elements. He published his idea that the air was not a vast chemical solvent in *Meteorological Observations*. Neither this publication nor his forthcoming law of additive partial pressures brought much immediate scientific reaction.

In 1801, he expressed part of his famous law of partial pressures in his paper "New Theory of the Constitution of Mixed Aeriform Fluids, and Particularly of the Atmosphere" in the *Journal of Natural Philosophy, Chemistry and the Arts*:

> When two elastic fluids, denoted by A and B, are mixed together, there is no mutual repulsion amongst their particles; that is, the particles of A do not repel those of B, as they do one another. Consequently, the pressure or whole weight upon any one particle arises solely from those of its own kind.

Although we now know that Dalton was inaccurate when he said that only like atoms in a mixture of gases repel and unlike atoms are indifferent toward each other, Dalton's basic ideas did point him in the correct direction, causing him and his followers to reject a commonly held theory that all atoms in matter were alike. Dalton believed that atoms of different elements had different sizes and masses and that each element had its own unique and identical kind of atoms—all key points of his atomic theory. He maintained that any two molecules of the same chemical compound are composed of the same combination of atoms. This atomic hypothesis is essential to the field of chemistry today.

He formally claimed the special status of chemical atoms when he wrote in *A New System of Chemical Philosophy*, "We might as well attempt to introduce a new planet into the Solar System, or to annihilate one already in existence, as to create or destroy a particle of hydrogen." Further, he wrote, "I should apprehend there are a considerable number of what may be called elementary principles, which can never be metamorphosed, one into another, by any power we can control."

Although the total number of atoms in the world was very large, he suggested that the number of different types of atoms is quite small. His original writings listed about twenty different elements, which he thought of as species of atoms. Today, we know of more than one hundred naturally occurring and manmade elements.

In 1816, Dalton was elected to the position of corresponding member of the French Académie des Sciences, and in 1822 he visited Paris, where he met other famous scientists of his time, such as Pierre-Simon Laplace (1749–1827), Joseph Louis Gay-Lussac (1778–1850), and André-Marie

Ampère (1775–1836). In 1817, Dalton became president of the Manchester Literary and Philosophical Society, which he presided over for the remaining 27 years of his life. He was elected to the Royal Society in 1922, and received the Royal Medal in 1826 in recognition of his chemical atomic theory. In 1831, he chaired various scientific committees of the British Association for the Advancement of Science. In 1836, he became vice president-elect of the association, but his participation was cut short by two severe paralytic attacks in 1837. He was a partial invalid for the rest of his life. Thackray writes in the *Dictionary of Scientific Biography* of society's growing respect for Dalton during his final years:

> Dalton's later life also illustrates the growing recognition that society was beginning to offer the man of science. Impeccable scientific credentials, a blameless personal life, and in old age a calm and equable temperament all combine to make Dalton a peculiarly suitable recipient of civil honor.

In 1794, Dalton explained why he had never married and had no progeny: "My head is too full of triangles, chymical processes, and electrical experiments, etc., to think much of marriage." His needs were always simple, which was a reflection of his Quaker faith.

In 1844, he had another stroke. On July 26 of that year, he recorded with a shaking hand his final meteorological observation. A day later, he fell from his bed and was found dead. More than 40,000 people filed past his coffin in Manchester Town Hall. Stores and offices closed for a day as a mark of respect. Dalton's funeral procession stretched for two miles. According to Bill Bryson's *A Short History of Nearly Everything*, Dalton's entry in Britain's 1885 *Dictionary of National Biography* is one of the longest, "rivaled in length only by those of Darwin and Lyell among nineteenth-century men of science."

As Dalton had requested, his eye was cut open, and the liquids of his eye were found to be normal. One of his eyes was preserved at the Royal Institution, and in the 1990s, cellular analysis revealed that the eye lacked the pigment that provides sensitivity to green. Today, we call this form of color blindness "deuteranope." Roughly five out of every 100 males is deuteranomalous to at least some degree.

As another curious aside, in 2006, researchers at Cambridge University and the University of Newcastle upon Tyne in England discovered that people afflicted with red-green color blindness actually have a special sensitivity to *other* hues. For example, the researchers found that color-blind subjects could actually distinguish between very subtly different tones of khaki, whereas people with normal vision could not. Elise Kleeman noted, "The findings lend credence to the theory that people with red-green color blindness make good hunters or soldiers because they are not easily fooled by camouflage." The researchers suggest that red-green color blindness could have been retained for evolutionary reasons because it helped early humans locate predators and food in forests.

Returning our attention to the legacy of Dalton, a lunar crater with a diameter of 60 kilometers was named after Dalton and approved in 1964 by the International Astronomical Union General Assembly. Dalton's contribution to humankind has been considered so great by Michael H. Hart, author of *The 100: A Ranking of the Most Influential Persons in History*, that Hart ranks Dalton as the thirty-second most influential person in all of history. Hart concludes:

> So convincingly did Dalton present his [atomic] theory that within twenty years it was adopted by the majority of scientists. Furthermore, chemists followed the program that his book suggested: determine exactly the relative atomic weights; analyze chemical compounds by weight; determine the exact combination of atoms which constitutes each species of molecule. The success of that program has, of course, been overwhelming. It is difficult to overstate the importance of the atomic hypotheses. It is the central notion in our understanding of chemistry.

Within just a few decades of his death, many Englishmen thought of Dalton with profound reverence. In 1874, Henry Lonsdale wrote in *The Worthies of Cumberland*:

> As pilgrims to the shrines of saints draw thousands of English Catholics to the Continent, there may be some persons in the British Islands sufficiently in love with science, not only to revere the memory of its founders, but to wish for a description of the locality and birth-place of a great master of knowledge—John Dalton—who did more for the world's civilization than all the reputed saints in Christendom.

In 1895, Henry E. Roscoe's *John Dalton and the Rise of Modern Chemistry* forever immortalized Dalton and his fellow great scientist from Manchester, James Joule, whom I discuss in a separate entry:

> In the vestibule of the Manchester Town Hall are placed two life-sized marble statues facing each other. One of these is that of John Dalton . . . the other that of James Prescott Joule. . . . Thus the honour is done to Manchester's two greatest sons—to Dalton, the founder of modern Chemistry and of the Atomic Theory, and the

laws of chemical-combining proportions; to Joule, the founder of modern physics and the discoverer of the law of Conservation of Energy.

The one gave to the world the final proof...that in every kind of chemical change no loss of matter occurs; the other proved that in all the varied modes of physical change, no loss of energy takes place.

FURTHER READING

"Biography of John Dalton," Salt Lake Community College; from ww2. slcc.edu/ schools/hum_sci/physics/whatis/biography/dalton.html.

Bryson, Bill, *A Short History of Nearly Everything* (New York: Random House, 2003).

Dalton, John, "Extraordinary Facts Relating to the Vision of Colours, with Observations," in *Memoirs Of The Literary And Philosophical Society Of Manchester*, Volume 5 (London: Cadell and Davies, 1798).

Dalton, John, "New Theory of the Constitution of Mixed Aeriform Fluids, and Particularly of the Atmosphere," *Journal of Natural Philosophy, Chemistry and the Arts*, 5: 241–244, 1801.

Dalton, John, *A New System of Chemical Philosophy* (Manchester, U.K., 1808).

Cardwell, D., *John Dalton and the Progress of Science* (New York: Manchester University Press/Barnes & Noble Inc., 1968).

Greenway, Frank, *John Dalton and the Atom* (Ithaca, N.Y.: Cornell University Press, 1966).

Hart, Michael H., *The 100: A Ranking of the Most Influential Persons in History* (New York: Citadel Press, 1992).

Kleeman, Elise, "In Combat, Stick with the Color-Blind," *Discover*, 27(3): 11, March 2006.

Lonsdale, Henry, *The Worthies of Cumberland* (London: George Routledge & Sons, 1874).

Roscoe, Henry, *John Dalton and the Rise of Modern Chemistry* (London: Cassell & Company, 1895).

Thackray, Arnold, "Dalton," in *Dictionary of Scientific Biography*, Charles Gillispie, editor-in-chief (New York: Charles Scribner's Sons, 1970).

INTERLUDE: CONVERSATION STARTERS

> Why do the laws that govern [the universe] seem constant in time? One can imagine a Universe in which laws are not truly lawful. Talk of miracles does just this, invoking God to make things work. Physics aims to find the laws

instead, and hopes that they will be uniquely constrained, as when Einstein wondered whether God had any choice when He made the Universe.

—Gregory Benford, in John Brockman's *What We Believe but Cannot Prove*

Until now, physical theories have been regarded as merely models which approximately describe the reality of nature. As the models improve, so the fit between theory and reality gets closer. Some physicists are now claiming that supergravity *is* the reality, that the model and the real world are in mathematically perfect accord.

—Paul Davies, *Superforce*

Physical concepts are free creations of the human mind, and are not, however it may seem, uniquely determined by the external world. In our endeavor to understand reality, we are somewhat like a man trying to understand the mechanism of a closed watch. He sees the face and the moving hands, even hears it ticking, but he has no way of opening the case. If he is ingenious, he may form some picture of the mechanism which could be responsible for all the things he observes, but he may never be quite sure his picture is the only one which could explain his observations. He will never be able to compare his picture with the real mechanism, and he cannot even imagine the possibility of the meaning of such a comparison.

—Albert Einstein, *The Evolution of Physics*

The burgeoning field of computer science has shifted our view of the physical world from that of a collection of interacting material particles to one of a seething network of information. In this way of looking at nature, the laws of physics are a form of software, or algorithm, while the material world—the hardware—plays the role of a gigantic computer.

—Paul Davies, "Laying Down the Laws," *New Scientist*

HENRY'S GAS LAW

England, 1802. The amount of a gas dissolved in a liquid is proportional to the pressure of the gas above the liquid, provided that no chemical reaction takes place.

In 1802, astronomer William Herschel discovered binary stars (two stars both orbiting around their center of mass). Today, we know that a large percentage of stars are part of binary (or multiple) star systems. Also in 1802, German naturalist Gottfried Treviranus coined the term "biology." The U.S. Military Academy, also known as West Point, was established. British experimenter Thomas Wedgwood produced the first photograph. Ludwig van Beethoven performed "Moonlight Sonata" for the first time.

Henry's Law is one of the many gas laws discussed in this book. In short, it states that the amount of a gas—usually taken to mean the mass of the gas—that is dissolved in a liquid is directly proportional to the pressure of the gas above the solution. This assumes that the system under study has reached a state of equilibrium and that the gas does not chemically react with the liquid. A common formula used today for Henry's Law is

$$\boxed{P = kC,}$$

where P is the partial pressure of the particular gas above the solution, C is the concentration of the dissolved gas, and k is the Henry's Law constant. As just two examples, for dissolved oxygen, $k = 4.34 \times 10^4$ L·atm/mol, and for dissolved carbon dioxide, $k = 1.64 \times 10^3$ L·atm/mol, when these gases are dissolved in water at 299°K. Sometimes, Henry's Law constants are given in units of (mL gas)/(mL solvent · atm). Henry's Law is most accurate for dilute solutions and low gas pressures.

We can visualize why Henry's Law works in several ways. For example, consider a scenario in which the partial pressure of a gas above a liquid increases by a factor of 2. Then, on the average, twice as many molecules will collide with the liquid surface in a given time interval, and thus, twice as many gas molecules will enter the solution. Note that different gases have different solubilities, and this also affects the process, as well as the value of Henry's constant. Many Henry's constants have been published at 25°C, and they decrease with increasing temperature.

Let's work on a small problem that involves Henry's Law. In particular, let's determine what the oxygen (O_2) content of pure water is in the presence of air at 20°C and 1 atmospheric pressure—given a Henry Law's constant for pure oxygen at 20°C, which can be expressed as 0.031 mL O_2/mL

H_2O at 1 atmosphere total pressure. Because air consists of approximately 20% O_2, the O_2 content of air is only 0.2 atmospheres. Using Henry's Law, we find that the solubility of O_2 in the presence of air to be 0.2 atm × [0.031 mL O_2/(mL H_2O · 1 atm)] = 0.0062 mL O_2/mL H_2O.

Henry's Law has been used by researchers to better understand the noise associated with "cracking" of finger knuckles. Gases that are dissolved in the synovial fluid in joints rapidly come out of solution as the joint is stretched and pressure is decreased. This *cavitation*—the sudden formation and collapse of low-pressure bubbles in liquids by means of mechanical forces—produces a characteristic noise.

In scuba diving, the pressure of the air breathed is roughly the same as the pressure of the surrounding water. The deeper one dives, the higher the air pressure and the more air that dissolves in the blood. When a diver ascends rapidly, the dissolved air may come out of solution too quickly in the blood, and the bubbles in the blood cause a painful and dangerous disorder known as decompression sickness.

Researchers at Like-A-Fish Technologies are currently trying to use Henry's Law to extract breathable oxygen from seawater for divers. Perhaps someday we can all return to the sea without the traditional scuba tank—or at least that is the dream of the Israel-based company. The process works by placing seawater under low pressure, which according to Henry's Law will cause dissolved gas to be released from the liquid for the diver to breathe. The company holds patents in Europe and a pending patent in the United States. A laboratory model of the device was developed and successfully tested.

William Henry (c. 1774–1836), British chemist famous for his law that relates the amount of gas dissolved in a liquid to the pressure of the gas above the liquid.

CURIOSITY FILE: Gas bubbles leave a soda drink once you open the soda can due to Henry's Law. When the can is closed, the carbon dioxide gas in the can is under pressure and remains dissolved. When opened, the pressure is removed, and the dissolved gas rapidly leaves the liquid in the form of bubbles. • Henry's father, Thomas Henry, was an apothecary who discovered a new way of making magnesium carbonate, which he used as an antacid and soon became a popular medicine known as Henry's Magnesia. • William Henry killed himself because he could no longer endure his physical afflictions.

> Henry was somewhat shy and reserved with strangers, who often thought him cold. This is perhaps not surprising in a man with such physical infirmities since his youth. He appears to be a man who could not be induced to relax. He regretted that his lifelong struggle with pain, and digestive disorders, had reduced his capacity for scientific and literary creativity.
>
> —Craig Thornber, "Thomas Henry, FRS and his son, William Henry, MD, FRS, GS"

William Henry was born in Manchester, England. At age 10, Henry suffered an injury from a falling beam, which caused him to suffer from pain throughout his life. The injury also limited his physical activities; thus, he focused his attention on reading and studying. His son William Charles Henry later wrote in *A Biographical Account of the Late Dr. Henry*:

> His fortitude, while yet a child, in supporting the sudden paroxysms of pain, which were often so intense as to oblige him to rest in the streets, was most remarkable. In his efforts to banish the perception of his physical sufferings by an absorbing mental occupation, he manifested that energy of resolution and purpose, which throughout life compelled a feeble bodily frame to keep pace with the exertions of an ardent and unfatigued spirit.

William Charles says nothing about of his father's marriage or his family life. We do know that Henry was close friends with John Dalton, the English chemist, famous for his contributions to the development of atomic theory. Henry entered Edinburgh University in 1795, received his M.D. in 1807, and later specialized in urinary diseases. Henry worked as a physician at the Manchester Infirmary, studying bladder stones and writing an essay on diabetes. He was elected fellow of the Royal Society in 1808, and in 1809, the Royal Society awarded him the Copley Medal.

Henry devoted much of his research time to chemistry, with an emphasis on the behavior of gases. He read one of his best-known papers to the Royal Society in 1802 and published it in 1803. The paper described his experiments on the amount of a gas absorbed by water at different temperatures and under different pressures. For example, he demonstrated that if a gas was compressed to twice the normal atmospheric pressure, twice as much was dissolved. These kinds of observation led to Henry's Law, which Linus Pauling rigorously defined in modern terms in *General Chemistry*:

> At constant temperature, the partial pressure in the gas phase of one component of a solution is, at equilibrium, proportional to the concentration of the component in the solution, in the region of low concentration. This is equivalent to saying that the solubility of a gas in a liquid is proportional to the partial pressure of the gas.

Henry published papers that dealt with the composition of hydrochloric acid and of ammonia, the flammability of mixtures of gases, and the disinfecting powers of heat. His *Elements of Experimental Chemistry*, based on his lectures given from 1798–1799 at Manchester, was published in 1801, and it went through 11 editions in 30 years. During Henry's life, the book was the most popular and successful chemistry text in English.

Sometime after 1824, Henry underwent surgical operations on his hands. Through the years, he suffered from chronic ill health and severe pains that still persisted from his original injury. His problems became so debilitating that he could no longer sleep, and he killed himself in 1836. Craig Thornber, a biographer of Henry, writes:

> When we look at the summation of William Henry's interests in chemistry, botany, geology, medicine, literature and business, together with his role as one of the founders of the British Association for the Advancement of Science, and as Vice Chairman of both the Literary and Philosophical Society and the Natural History Society of Manchester, we perceive that he was at the forefront of intellectual life in Manchester in a period when it becoming the first industrial city in the world.

FURTHER READING

Henry, Charles, *A Biographical Account of the Late Dr. Henry* (Manchester, U.K.: F. Looney, 1837).

Kimbrough, Doris R., "Henry's Law and Noisy Knuckles," *Journal of Chemical Education*, 76(11): 1509, 1999.

Odian, George, and Ira Blei, *Schaum's Outline of General, Organic and Biological Chemistry* (New York: McGraw-Hill Professional, 1994).

Pauling, Linus, *General Chemistry* (New York: Dover, 1988).

Scott, E. L., "William Henry," in *Dictionary of Scientific Biography*, Charles Gillispie, editor-in-chief (New York: Charles Scribner's Sons, 1970).

Thornber, Craig, "Thomas Henry, FRS and his son, William Henry, MD, FRS, GS"; see www.thornber.net/cheshire/ideasmen/henry.html.

INTERLUDE: CONVERSATION STARTERS

The supreme task of the physicist is to arrive at those universal elementary laws from which the cosmos can be built up by pure deduction.

—Albert Einstein, 1949 interview with Alfred Werner, *Liberal Judaism*

When one sees a mathematical truth, one's consciousness breaks through into this world of ideas.... One may take the view that in such cases the mathematicians have stumbled upon works of God.

—Roger Penrose, *The Emperor's New Mind*

The two greatest logical geniuses of the last century both killed themselves. Alan Turing died after taking a bite of an apple that was laced with cyanide. Kurd Gödel...refused to eat and consequently starved to death. Did these two men self-destruct because of their logical prowess, or in spite of it?...And was it just a coincidence that both were inordinately fond of the Walt Disney movie "Snow White"?

—Jim Holt, "Obsessive-Genius Disorder," *New York Times Book Review*, September 3, 2006

There is a noble vision of the great Castle of Mathematics, towering somewhere in the Platonic World of Ideas, which we humbly and devotedly discover (rather than invent). The greatest mathematicians manage to grasp outlines of the Grand Design, but even those to whom only a pattern on a small kitchen tile is revealed, can be blissfully happy....Mathematics is a proto-text whose existence is only postulated but which nevertheless underlies all corrupted and fragmentary copies we are bound to deal with. The identity of the writer of this proto-text (or of the builder of the Castle) is anybody's guess....

—Yuri I. Manin, "Mathematical Knowledge: Internal, Social, and Cultural Aspects"

Symmetry doesn't so much control as it does describe or account for nature....I give Einstein credit for introducing symmetry into modern physics... with his special theory of relativity—$e = mc^2$. Wow! The big increase in knowledge is the statement that the laws of physics apply to any system that you want; the laws are invariant to a change in the velocity of the system—that's relativity.... Symmetry in fact makes things much more simple... it is the overriding basis of the mathematics of physics... symmetry produces an elegance and a beauty to the description of nature.

—Leon Lederman, interview, September 22, 2005, in Siobhan Roberts's *King of Infinite Space*

GAY-LUSSAC'S LAW OF COMBINING GAS VOLUMES

France, 1808. The volumes of gases that chemically react with each other, or are produced in reactions, are expressed in ratios of small, whole numbers.

CROSS REFERENCE: CHARLES'S GAS LAW, AVOGADRO'S GAS LAW, JOHN DALTON, JEAN-BAPTISTE BIOT, ALEXANDER VON HUMBOLDT, AND LOUIS JACQUES THENARD.

In 1808, the United States prohibited importation of slaves from Africa, and Napoleon abolished the Inquisition in Spain and Italy. Ludwig van Beethoven conducted and performed in a concert that featured the premiere of his *Fifth Symphony*, *Sixth Symphony*, and *Fourth Piano Concerto*.

Gay-Lussac's Law of Combining Gas Volumes states that the ratio between reacting gas volumes and the volume of gaseous product can be expressed in small whole numbers. In 1811, this law and related experimental data stimulated Italian chemist Amedeo Avogadro (1776–1856) to state his famous hypothesis that equal volumes of gases, at the same temperature and pressure, contain the same number of particles, or molecules (see "Avogadro's Gas Law," below).

In Gay-Lussac's 1809 paper, "Memoir on the Combination of Gaseous Substances with Each Other," he discusses the law that would someday bear his name:

> It is my intention to make known some new properties in gases, the effects of which are regular, by showing that these substances combine amongst themselves in very simple proportions, and that the contraction of volume which they experience on combination also follows a regular law. I hope by this means to give a proof of an idea... that we are perhaps not far removed from the time when we shall be able to submit the bulk of chemical phenomena to calculation.

Gay-Lussac had performed a number of experiments with Prussian naturalist Alexander von Humboldt (1769–1859) on the creation of water vapor by passing sparks through mixtures of hydrogen and oxygen. Gay-Lussac observed that, for any given volume of oxygen completely converted in the reaction, exactly twice this volume of hydrogen was required.

The measurements were quite precise and pointed to the 2:1 ratio with an accuracy of about 0.1%. In particular, they found that

2 volumes hydrogen + 1 volume oxygen → 2 volumes water vapor,

with all volumes being measured at the same temperature and pressure. The reacting volumes and resulting volumes are small whole-number ratios to each other. He also investigated volume relations for other gaseous substances and found, for example,

3 volumes hydrogen + 1 volume nitrogen → 2 volumes ammonia gas,

and

1 volume nitrogen + 1 volume oxygen → 2 volumes nitric oxide gas.

Gay-Lussac went on to explain:

> Thus it appears evident to me that gases always combine in the simplest proportions when they act on one another; and we have seen in reality in all the preceding examples that the ratio of combinations is 1 to 1, 1 to 2, or 1 to 3. It is very important to observe that in considering weights there is no [integral] relation between the elements of any one compound; it is only when there is a second compound between the same elements that the new proportion of the element that has been added is a multiple of the first quantity. Gases, on the contrary, in whatever proportions they may combine, always give rise to compounds whose [constituents] by volume are [integral] multiples of each other.
>
> Not only, however, do gases combine in very simple proportions, as we have just seen, but the apparent contraction of volume which they experience on combination has also a simple relation to the volume of the gases....
>
> These ratios by volume are not observed with solid or liquid substances, nor when we consider weights, and they form a new proof that it is only in the gaseous state that substances are in the same circumstances and obey regular laws.

Gay-Lussac's Law seemed to support the idea that equal volumes of gases contain equal numbers of particles, which could account for the simple proportions of reacting volumes. Moreover, Gay-Lussac said that his results were very favorable to English chemist John Dalton's (1766–1844) "ingenious idea" about the composition of molecules and atomic-molecular theory. However, strangely, Dalton never accepted the "round numbers" of Gay-Lussac.

Simply by carefully studying gas volumes, it became possible for scientists to understand the underlying structure of matter. Maurice Crosland writes in *Gay-Lussac: Scientist and Bourgeois*:

> The volumetric approach to matter of Gay-Lussac and his successors could easily be overlooked today. Yet as one of the principal methods of investigation of the basic problems of chemical composition and reactions, it influenced much of the chemistry of the first half of the nineteenth century. When there were so few keys to the understanding of physical and chemical units, it provided a valuable means of approach and one which could claim to be solidly based on experimental evidence.

In 1802, Gay-Lussac also formulated the law that stated that the volume V of a fixed amount of gas at fixed pressure is directly proportional to its temperature T in degrees kelvin, that is, $V = kT$. This law, which was first published by Gay-Lussac, is today usually known as Charles's Law because Gay-Lussac referenced unpublished work by French chemist and physicist Jacques Charles (1746–1823) from around 1787 (see "Charles's Gas Law" in part II).

Joseph Louis Gay-Lussac (1778–1850), French chemist and physicist, famous for contributions to the physical chemistry of gases.

CURIOSITY FILE: Gay-Lussac invented a portable barometer. • While working at the Paris Mint, he invented a device for quickly estimating the purity of silver, which was the only legal measure in France until 1881. • He went on a balloon flight to a record-setting height of 23,000 feet to test hypotheses on the magnetic field of Earth and the composition of the air. • When he mixed metallic potassium with another element, it exploded, destroying his laboratory and temporarily blinding him.

> Probably Gay-Lussac's greatest single achievement is based on the law of combining volume of gases, which he announced at a meeting... in 1808. For Gay-Lussac himself, the law provided a vindication of his belief in regularities in the physical world, which it was the business of the scientist to discover.
>
> —Maurice P. Crosland, "Joseph Gay-Lussac," in *Dictionary of Scientific Biography*

I have not chosen a career that will lead me to a great fortune, but that is not my principal ambition.

—Joseph Gay-Lussac, quoted in Maurice Pierre Crosland's *Gay-Lussac: Scientist and Bourgeois*

Joseph Louis Gay-Lussac was born in Saint-Léonard-de-Noblat in central France. His father, Antoine Gay, was a lawyer who used the surname Gay-Lussac in order to distinguish himself from other people with the last name of Gay. The father created "Gay-Lussac" by appending the name of the hamlet of Lussac, in which resided some family property.

Gay-Lussac received his early education at home and went to Paris in 1794 to prepare for studies at the École Polytechnique. His father was arrested during the French Revolution.

In 1809, Gay-Lussac became professor of chemistry at the École Polytechnique. During the same year, he married Geneviève-Marie-Joseph Rojot. (He had met the beautiful Geneviève earlier when she worked as a linen shop assistant and while she was surreptitiously studying a chemistry book.) Gay-Lussac and Rojot had five children.

In 1832, he became professor of chemistry at the Jardin des Plantes, the main botanical garden in France. Here are some brief highlights of his accomplishments and interests:

- In 1802, Gay-Lussac formulated the law that a gas expands linearly with a fixed pressure and rising temperature (which is today usually known as Charles's Gas Law, although in 1787 Charles did not measure the coefficient of expansion). Gay-Lussac concluded that equal volumes of all gases expanded equally with the same increase in temperature. He determined that the expansion of gases was 1/266.66 of the volume, at 0°C, for each degree rise in temperature. (Today, we know that at a constant pressure a gas expands 1/273.15 of its volume at 0°C, for each degree Celsius of rise in temperature.)
- In 1804, Gay-Lussac embarked on a hydrogen balloon ascent to a height of five kilometers with French physicist Jean-Baptiste Biot (1774–1862) in order to investigate the atmosphere of Earth. They concluded that the magnetic intensity decreased with increasing altitude. During a second trip that year, Gay-Lussac went alone and repeated his observations of pressure, temperature, humidity, and magnetism. He reached a height of 7,106 meters above sea level, a record not broken for another fifty years.
- In 1805, with Prussian naturalist and explorer Alexander von Humboldt (1769–1859), he discovered that the basic composition of

the atmosphere does not change with increasing altitude—for the range of altitudes they were able to explore.

- In 1808, Gay-Lussac and French Chemist Louis Jacques Thenard (1777–1857) discovered boron. At about the same time, Gay-Lussac and Thenard suggested that the rate of decomposition of an electrolyte depends only on the electrical current strength and not on the size of the electrodes.
- In 1815, Gay-Lussac made cyanogen (C_2N_2), a toxic chemical used for the production of insecticides.
- In 1816, Gay-Lussac recognized five oxides of nitrogen, whose modern formulas are N_2O, NO, N_2O_3, NO_2, and N_2O_5.

It is interesting to note that Gay-Lussac's first publication on his gas law was important not only because of its scientific value but also because almost identical research was carried out simultaneously and independently by the English chemist, physicist, and meteorologist John Dalton (1766–1844). In their near-simultaneous publication in 1802 of research on the thermal expansion of gases, Dalton and Gay-Lussac both concluded that all gases expand by the same proportion for a particular temperature rise at constant pressure.

Perhaps Gay-Lussac's most important contribution to industry was his 1827 refinement of the lead-chamber process used for the production of sulfuric acid. The tall absorption towers were known as Gay-Lussac towers and facilitated the following chemical reaction:

$$SO_{2(g)} + NO_{2(g)} \rightarrow SO_{3(g)} + NO_{(g)}$$

This reaction was carried out in a lead-lined chamber in which the sulfur trioxide (SO_3) was subsequently dissolved in water to produce sulfuric acid. Gay-Lussac provided a means by which the nitrogen monoxide (NO) could be recycled after oxidizing to NO_2. Sulfuric acid was produced this way well into the twentieth century.

Crosland reminds us in *Gay-Lussac: Scientist and Bourgeois* that Gay-Lussac lived in a special time for France:

> Gay-Lussac was in many ways typical of the new men of science who emerged from post-Revolutionary France. His was the first generation which could receive a full training in science and go on to earn a living as a scientist. Before his time, one was fortunate if one could follow a single course of lectures on some branch of science to supplement what could be learned from books.

Throughout his life, Gay-Lussac remained passionate about trying to understand the laws of science, and he wrote, as quoted by Crosland,

"If one were not animated with the desire to discover laws, they would often escape the most enlightened attention." He also emphasized the interdependence of scientists through history: "A discovery is the product of a previous discovery, and in its turn it will rise to a further discovery."

Gay-Lussac's German chemist friend Justus von Liebig (1803–1873) recalled the happiest years of his life, which were spent in Gay-Lussac's private laboratory:

> Never shall I forget the years passed in the laboratory of Gay-Lussac. When we had finished a successful analysis (you know without my telling you that the method and the apparatus discovered in our joint memoir were entirely his), he would say to me, "Now you must dance with me just as [the scientist] Louis Thénard and I always danced together when we discovered something." And then we would dance.

Gay-Lussac died in 1850 in Paris, France. A lunar crater with a diameter of 26 kilometers was named after Gay-Lussac and approved in 1935 by the International Astronomical Union General Assembly. His name is one of 72 names of prominent French scientists whom Gustave Eiffel placed on the Eiffel Tower (see "Coulomb's Law of Electrostatics" in part II).

FURTHER READING

Brock, William, *Justus von Leibig: The Chemical Gatekeeper* (Cambridge, U.K.: Cambridge University Press, 1997).

Crosland, Maurice P., *Gay-Lussac: Scientist and Bourgeois* (Cambridge, U.K.: Cambridge University Press, 1978).

Crosland, Maurice P., "Joseph Gay-Lussac," in *Dictionary of Scientific Biography*, Charles Gillispie, editor-in-chief (New York: Charles Scribner's Sons, 1970).

Gay-Lussac, Joseph, "Memoir on the Combination of Gaseous Substances with Each Other," *Mémoires de la Société d'Arcueil*, 2(207), 1809; translation (Alembic Club Reprint No. 4) reprinted in Henry A. Boorse and Lloyd Motz, editors, *The World of the Atom*, volume 1 (New York: Basic Books, 1966).

INTERLUDE: CONVERSATION STARTERS

I believe in science. Unlike mathematical theorems, scientific results can't be proved. They can only be tested again and again until only a fool would refuse to believe them. I cannot prove that electrons exist, but I believe fervently in their existence. And if you don't believe in them, I have a high-voltage cattle prod I'm willing to apply as an argument on their behalf. Electrons speak for themselves.

—Seth Lloyd, in John Brockman's *What We Believe but Cannot Prove*

When you discover mathematical structures that you believe correspond to the world around you... you are communicating with the universe, seeing beautiful and deep structures and patterns that no one without your training can see. The mathematics is there, it's leading you, and you are discovering it. Mathematics is a profound language, an awesomely beautiful language. For some, like Leibniz, it is the language of God. I'm not religious, but I do believe that the universe is organized mathematically.

—Anthony Tromba, July 2003 UC Santa Cruz press release

This property of human languages—their resistance to algorithmic processing—is perhaps the ultimate reason why only mathematics can furnish an adequate language for physics. It is not that we lack words for expressing all this $E = mc^2$ and $\int e^{iS}(\phi) D\phi$ stuff..., the point is that we still would not be able to do anything with these great discoveries if we had only words for them.... Miraculously... even very high level abstractions can somehow reflect reality: knowledge of the world discovered by physicists can be expressed only in the language of mathematics.

—Yuri I. Manin, "Mathematical Knowledge: Internal, Social, and Cultural Aspects," March 2007

AVOGADRO'S GAS LAW

Italy, 1811. Equal volumes of gases contain the same number of molecules.

CROSS REFERENCE: DALTON'S ATOMIC THEORY AND GAY-LUSSAC'S LAW OF COMBINING GAS VOLUMES.

In 1811, Jane Austen published *Sense and Sensibility*, and "Luddites" destroyed factory machines in northern England. Scottish anatomist Sir Charles Bell published his *New Idea of the Anatomy of the Brain*, in which he discusses his discovery of nerve functions and their relationships to different parts of the brain.

Avogadro's Law, named after physicist Amedeo Avogadro, who proposed it in 1811, states that equal volumes of gases at the same temperature and pressure contain the same number of molecules, regardless of the molecular makeup of the gas. This law may be stated simply as

$$\boxed{N_1 = N_2,}$$

where N_1 is the number of molecules in one gas, and N_2 is the number of molecules in another gas. The law assumes that the gas particles are acting in an "ideal" manner, which is a valid assumption for most gases at pressures at or below a few atmospheres near room temperature.

A variant of the law, also attributed to Avogadro, states that the volume of a gas is directly proportional to the number of molecules of the gas. This is represented by the formula $V = a \times N$, where a is a constant, V is the volume of the gas, and N is the number of gas molecules. Other contemporary scientists believed such proportionality should be true, but Avogadro's Law went further than competing theories because Avogadro essentially defined a molecule as the smallest characteristic particle of a substance, and the molecule could consist of several atoms. For example, he proposed that a water molecule consisted of two hydrogen atoms and one oxygen atom.

Avogadro's number, 6.0221367×10^{23}, is the number of atoms found in 1 mole of an element, and more particularly, today we define Avogadro's number as the number of carbon-12 atoms in 12 grams of unbound carbon-12. A mole is the amount of an element that contains precisely the same number of grams as the value of the atomic weight of the substance. For example, because nickel has an atomic weight of 58.6934, there are 58.6934 grams in a mole of nickel. Note that the actual number of molecules in 1 mole—now called Avogadro's number in honor of Avogadro's contributions to the theory of gases and molecular weights—was never actually determined by Avogadro but rather was computed later by Austrian physicist and chemist Johann Josef Loschmidt (1821–1895).

Because atoms and molecules are so small, the magnitude of Avogadro's number is difficult to visualize. If an alien were to descend from the sky to deposit an Avogadro's number of unpopped popcorn kernels on the surface of the United States, the surface would be covered with the kernels to a depth of more than nine miles.

To give you another feel for the significance of this number, recall that an Avogadro's number of carbon atoms would exist in a 12-gram sample of carbon, which is about the mass of two American quarters. Another interesting bit of trivia—if you placed 24 numbered balls, numbered 1 through 24, into a hat and drew them out randomly one at a time, the probability of drawing them out in numerical order is about 1 chance in Avogadro's number—a very slim chance indeed!

Although Avogadro proposed his now-famous law in 1811 while a professor of physics at the University of Turin, it was not readily used until about 1858, when Italian chemist Stanislao Cannizzaro (1826–1910) provided additional chemical explanations to support the law. In particular, Cannizzaro showed that molecular weights of gases could be determined by weighing 22.4 liters of each gas, and he presented a coherent system of atomic weights based on Avogadro's hypothesis.

When Avogadro published his work on gases, only hydrogen, oxygen, nitrogen, and chlorine were known to be gaseous under the temperatures and pressures that were readily available in the laboratory. Therefore, scientists had a very limited set of materials with which to test this law.

Amedeo Avogadro (1776–1856), Italian physicist and chemist famous for his studies on gases.

CURIOSITY FILE: The Avogadro family chapel and grave in Quarengra, Italy, were spared damage by a large flood in November 1969. Today, you can visit his grave. • An Avogadro's number of soda cans would cover the entire surface of Earth to a depth of 200 miles. • Today, "Avogadro's Number" is the name of a popular restaurant with live musical entertainment, located in Fort Collins, Colorado. • "Avogadro's Number" is also the name of a folk-rock band based in the Susquehanna Valley of Pennsylvania.

The hypothesis we have just proposed is based on that simplicity of relation between the volumes of gases on combination, which would appear to be otherwise inexplicable.

—Amedeo Avogadro, "Essay on a Manner of Determining the Relative Masses of the Elementary Molecules of Bodies, and the Proportions in Which They Enter into These Compounds," 1811

Amedeo Avogadro was born in Turin, Italy. His father was a famous lawyer who became senator of Piedmont, Italy, and who encouraged his son to study law. In 1769, Avogadro obtained his doctorate in ecclesiastical law and began his practice. As time permitted, he also privately studied physics and mathematics. In 1809, he became a professor of natural philosophy at the College of Vercelli, and in 1820 he was appointed "first chair" of mathematical physics at the University of Turin.

Little is known of his personal life except that he married and had several children. According to Mario Morselli, author of *Amedeo Avogadro, a Scientific Biography*, "As is often the case with men of his disposition, Avogadro found more satisfaction and enjoyment in the serene atmosphere of his family—they had seven children—than in the pursuit of social and professional success." He enjoyed reciting poetry in several languages and oversaw a family newspaper that reported on the events in his family's lives.

Through much of his life, Avogadro was a modest man, which may have contributed to some of his obscurity outside of Italy. Most of his early work was done in isolation. He first published on the topic of electricity and charged plates before he studied the physical properties of gases.

Avogadro's Law has far-reaching implications. For example, we can infer directly from the law that the relative weights of the molecules of any two gases are the same as the ratios of the densities of these gases at the same temperature and pressure. He also correctly postulated the existence of diatomic gases whose molecules were composed of two atoms, such as N_2, O_2, and H_2 (written in today's notations).

In 1811, he stated the correct formulas for carbon dioxide, sulfur dioxide, and hydrogen sulfide, and a decade later he gave the correct ratios of atoms in organic compounds such as turpentine and ether. He also gave correct molecular weights for numerous elements, such as mercury, iron, silver, lead, copper, and calcium. Later, he correctly theorized that the average distance between the molecules of all gases is the same under the same conditions.

Aaron Ihde reiterates and summarizes the key aspects of Avogadro's Law in *The Development of Modern Chemistry*:

> Avogadro's paper of 1811, based upon Gay-Lussac's Law and Dalton's atomic theory... started with the assumption that equal volumes of all gases contained equal numbers of molecules under similar conditions.... He supposed the existence of molecules of elemental gases which contained more than a single atom.... He saw how the theory of combining volumes should be applied in determining [chemical] formulas.... Despite the soundness of Avogadro's reasoning, his hypothesis was generally rejected or ignored.

English chemist John Dalton (1766–1844) never appreciated Avogadro's work, partly because Dalton did not accept Gay-Lussac's Law of Combining Gas Volumes. Ihde suggests that Avogadro's work was ignored by the scientific community because the concepts were too new, noting that many chemists of the time even rejected Dalton's concept of atoms.

Other scientists who worked around the time of Avogadro considered similar laws but either soon rejected them or did not apply the principles to the whole field of chemistry. Avogadro's Law was important because by comparing the masses of equal volumes of different gases under the same conditions, the ratios of weights of the gas molecules could be measured, and this was the basis for the first correct measurements of atomic weights.

Today, scholars debate why many years passed before Avogadro's formulation of his law was recognized. Some historians suggest that the hot spots of European science tended to be in France, England, and Germany. Thus, Avogadro's sequestration in Turin, coupled with his modesty, tended to isolate him from the mainstream science of his day. Also, Avogadro did not really support his law with a comprehensive set of experimental results. Finally, his idea of diatomic gases was considered to be heresy by those who believed that two atoms of the same element would have similar charges and repel each other.

When Avogadro died in 1856, the editors of the scientific journal *Nuovo Cimento* wrote a short obituary in which they remarked on his "retiring disposition" and the "simplicity" of his life. According to Morselli, the obituary noted that Avogadro's

> studies on atomic volumes and their relationship with the chemical affinities and the electrochemical series, and the electrochemical researches conducted with Michelotti on the chemical theory of the voltaic pile would be remembered. However, no mention at all appeared [in the obituary] on this occasion of the 1811 gas generalization, whose full significance would remain unrecognized in Italy, as well as in the rest of the scientific world, until the late 1860s.

The mausoleum of the Avogadro family, marked with the family coat of arms, still stands in Quaregna, Italy. Amedeo Avogadro, his wife Felicita, and their children are buried there. A lunar crater with a diameter of 139 kilometers was named after Avogadro and approved in 1970 by the International Astronomical Union General Assembly.

FURTHER READING

Crosland, Maurice P., "Amedeo Avogadro," in *Dictionary of Scientific Biography*, Charles Gillispie, editor-in-chief (New York: Charles Scribner's Sons, 1970).

Ihde, Aaron, *The Development of Modern Chemistry* (New York: Dover, 1964).

Morselli, Mario, *Amedeo Avogadro, a Scientific Biography* (Hingham, Mass.: Kluwer, 1984).

INTERLUDE: CONVERSATION STARTERS

> It has sometimes been suggested that the laws of nature are not real—that they are entirely inventions of the human mind, attempting to make sense of the universe. This is very strongly argued against by the spectacular efficacy of science: a) its power to solve otherwise intractable problems, and make accurate predictions, and b) by the fact that newly-discovered laws have typically suggested the existence of previously unknown or undiscovered phenomena, which have then been confirmed to exist.
>
> —"Physical Law," *Wikipedia*

> A rock pile ceases to be a rock pile the moment a single man contemplates it, bearing within him the image of a cathedral.
>
> —Antoine de Saint-Exupery, *Flight to Arras*

> [Some claim that] we do not understand why the universe has rules; therefore, God must have done it.... This ignores the possibility that the universe has to have rules, or could not exist. Or the fact that, had it no rules, we would not be able to exist.
>
> —Ben Hoskin, "God of the Gaps," Letter to *New Scientist*, March 24, 2007

> The case of Loschmidt's [actual calculation of Avogadro's number] is a prime example of the zeroth theorem, which states that a discovery, rule or insight

named after an individual often does not originate with that person. Others include the Dirac delta function, a mathematical trick used by engineer Oliver Heaviside 30 years before the English physicist Paul Dirac... and Olbers' paradox, that the night sky is dark even though the endless succession of stars in an infinite universe should fill the entire sky. German astronomer Heinrich Olbers discussed it in 1823, but it was well known to Johannes Kepler more than 200 years before.

— "Zeroth Theorem," *New Scientist* (unsigned article)

BREWSTER'S LAW OF LIGHT POLARIZATION

Scotland, 1815. The amount of the polarization of light reflected from a transparent surface is a maximum when the reflected ray is at right angles to the refracted ray.

CROSS REFERENCE: SNELL'S LAW AND WILLIAM LAWRENCE BRAGG.

In 1815, the first commercial cheese factory was founded in Switzerland. Napoleon abdicated and Louis XVIII returned to Paris. Brazil declared itself a kingdom "equal" to Portugal.

A light wave consists of an electric field and a magnetic field that oscillate perpendicular to each other and to the direction of travel. Usually, the electric vector of light vibrates in all directions. However, it is possible to restrict the vibrations of the electric field by plane-polarizing the light beam. For example, one may pass the light through oriented dichroic crystals in a plastic film so that the electric field in one direction is almost completely absorbed, while a large fraction of the electric field in a direction perpendicular to the absorbed component is transmitted. (More generally, "dichroic" often refers to a material in which light in different polarization states experiences a varying absorption as it travels through the material.)

Another approach for obtaining plane-polarized light is via the reflection of light from a surface between two media such as air and glass. The component of the electric field parallel to the surface is most strongly reflected. At one particular angle of incidence on the surface, called the *Brewster angle*, the reflected beam consists entirely of light whose electric vector is parallel to the surface—and the reflected and refracted beams are at right angles. The Brewster angle can be found by

$$\boxed{\theta_B = \arctan\left(\frac{n_2}{n_1}\right),}$$

where n_1 and n_2 are the refractive indices of the two media. This equation is one way to express Brewster's Law. (See "Snell's Law of Refraction" in part I for an explanation of refraction, which is exemplified by the bending of light when it passes from one material into another.)

For the example of shining light from air onto glass, n_2 is approximately equal to 1.5 for glass, and n_1 is approximately equal to 1 for air. We find that Brewster's angle for visible light is approximately 56° to the normal of the glass surface, where the term "normal" refers to an imaginary line perpendicular to the surface. Because the refractive index for a given medium depends on the wavelength of light, Brewster's angle also varies with wavelength. This also means that a beam of white light does not have a unique polarizing angle—hence one reason that 56° is mentioned as a mere approximation.

Because the incident beam is often traveling through air in many practical experiments, n_1 can be set to 1, and the equation simplifies to $\tan \theta_B = n_2$. For another typical problem, we may solve for the Brewster angle for a beam of light traveling in air when it is reflected by a pool of water with a refractive index of 1.33. Using Brewster's Law, we find $\tan \theta_B = 1.33$, and thus the polarizing angle is 53.1°.

One simplified but useful model of light, which allows us to visualize the notion of polarization, involves our thinking of each photon as having electric and magnetic field components that oscillate perpendicularly to each other, and both directions are mutually perpendicular to the direction of travel. However, individual photons can be rotated by varying amounts, about the direction of motion, relative to other individual photons. Given this model, unpolarized light is a disorderly jumble of photons with a variety of orientations. Polarized light may arise when this jumble encounters a polarizing filter that only allows photons with particular orientations to pass.

Note that light is electromagnetic radiation for which the amplitude is perceived as brightness and frequency (or wavelength) is perceived as color. The phenomenon of polarization (or angle of vibration) is not usually perceptible to humans or is only very weakly perceptible. However, all light that reflects from a flat surface is usually at least partially polarized. To test this, you can use a polarizing filter (e.g., Polaroid sunglasses) and hold the filter at 90° to the reflection, and the reflected light will be reduced or eliminated. In fact, polarization is commonly achieved today using a sheet of commercial material called Polaroid, invented by Edwin H. Land in 1938. During the manufacturing process, the sheet is stretched in one direction so that long-chain hydrocarbon molecules in the sheet become aligned in one direction. These chains conduct at optical frequencies when the sheet is treated with iodine. If light is incident with its electric field vector E parallel to the chains, electric currents are set up along the chains, and the light energy is absorbed. If the electric field E is perpendicular to the chains, the light is transmitted.

Polarization by light scattering in our atmosphere sometimes produces a glare in the skies. Photographers can reduce this partial polarization using special materials so that the glare does not produce an image of a washed-out sky.

Many animals, such as pigeons and bees, are quite capable of perceiving the polarization of light, which they can use for navigation because the linear polarization of sunlight is perpendicular to the direction of the sun.

In 2000, several researchers at the 3M Film/Light Management Technology Center in St. Paul, Minnesota, constructed complex materials from multilayer mirrors that they say "generalizes" Brewster's Law and increases the reflectivity of a mirror in interesting ways as the incidence angle of light is increased. Researchers suggest that these materials may have applications that range from brighter computer screens to cosmetics to decorative packaging. Details of the work can be obtained from Michael Weber and colleague's "Giant Birefringent Optics in Multilayer Polymer Mirrors," which is listed under Further Reading at the end of this entry.

David Brewster (1781–1868), Scottish physicist and science writer famous for his experimental work in optics and polarized light and for his invention of the kaleidoscope.

CURIOSITY FILE: Brewster named his kaleidoscope invention using the Greek words *kalos* (beautiful), *eidos* (form), and *scopos* (watcher). • Brewster was a famous debunker of fraudulent psychics. • In 1826, British scientist Charles Wheatstone (1802–1875) invented the kaleidophone, naming it in honor of Brewster's device. The kaleidophone involved either a vibrating piano string wrapped in silver wire or a vibrating metal rod that produced beautiful patterns when placed near a light source.

> The power of a theory . . . to explain and predict facts is by no means a test of its truth. . . .
>
> —David Brewster, "Observations on the Absorption of Specific Rays, in Reference to the Undulatory Theory of Light"

> As a devout evangelical Presbyterian who believed in the unity of truth, he felt that such unbridled speculation in physics had profoundly serious implications for religion. To Brewster, "Speculation engenders doubt, and doubt is frequently the parent either of apathy or impiety."
>
> —Edgar W. Morse, "David Brewster," in *Dictionary of Scientific Biography*

> Brewster's Angle is useful in all kinds of practical applications, from adjusting radio signals to building microscopes capable of examining objects on a molecular scale. It is central to the development of fiber optics, lasers, and to the study of meteorology [and] cosmology.
>
> —Cozy Baker, *Kaleidoscopes: Wonders of Wonder*

David Brewster was born in Jedburgh, Scotland, and educated for the ministry at the University of Edinburgh. He was awarded an honorary M.A. in 1800, and in 1804 he was licensed to preach in the Church of Scotland. Of Brewster's brief time at the pulpit, James Hogg, a colleague, wrote in a letter to publisher James Fraser:

> He was licensed, but the first day he mounted the pulpit was the last, for he had then, if he has not still, a nervous something about him that made him swither when he heard his own voice and saw a congregation eyeing him; so he stacked his discourse, and vowed never to try that job again. It was a pity for Kirk, [the National Church of Scotland] . . . but it was a good day for Science . . . for if the doctor had gotten a manse [cleric's house], he might most likely have taken to his toddy [drink] like other folk.

Brewster's fascination with science grew in parallel with his religious interests. In 1799, he began his investigations into the polarization, reflection, and absorption of light. Brewster enjoyed making devices of various kinds, including sundials, microscopes, and telescopes. His income depended mostly on his skillful editorial abilities, and he edited the *Edinburgh Magazine* (1802–1806), *Scots Magazine* (1802–1806), and the *Edinburgh Encyclopedia* (1807–1830). Throughout his life, he authored many popular books and articles. Like British physicist William Lawrence Bragg (1890–1971; see "Bragg's Law of Crystal Diffraction" in part IV), Brewster enthusiastically promoted scientific education for the general public.

In 1810, Brewster married Juliet McPherson. Their marriage, which produced five children, was happy and lasted forty years until Juliet's death. When he was 74, Brewster married his second wife, Jane Purnell, with whom he had a daughter.

In 1815, Brewster found that for any dielectric reflector (an electrical insulator substance such as glass in which an electric field may be maintained with near-zero power dissipation), a simple relationship exists between the polarization angle for the reflected light of a particular wavelength and the refractive index of the substance for the same wavelength. The law may be used to determine the refractive index of a solid, even if we only have a very small sample of the substance, because only a small reflecting surface is needed. The "Brewster angle," or polarizing angle, of the dielectric substance may be thought of as the angle of incidence for which a wave polarized parallel to the plane of incidence is entirely transmitted (no reflection). This means that an unpolarized wave that is incident with this angle is resolved into a transmitted, partly polarized component and a reflected, perpendicularly polarized component.

Brewster's first major publication appeared in *A Treatise on New Philosophical Instruments* (1813), in which he describes his work on the optical properties of hundreds of substances that he made during his attempts to improve telescopes.

He was elected fellow of the Royal Society in 1815 and invented the kaleidoscope in 1816. Cozy Baker, founder of the Brewster Kaleidoscope Society and author of *Kaleidoscopes: Wonders of Wonder*, writes of the kaleidoscope fad that occurred after news of Brewster's invention spread:

> His kaleidoscope created unprecedented clamor. A universal mania for the instrument seized all classes, from the lowest to the highest, from the most ignorant to the most learned, and every person not only felt, but expressed the feeling that a new pleasure had been added to their existence.

American inventor Edwin Land (1909–1991) wrote in the *Journal of the Optical Society of America*, "The kaleidoscope was the television of the 1850s, and no respectable home would be without a kaleidoscope in the middle of the library." In 1818, Dr. Peter M. Roget (1779–1869, famous for *Roget's Thesaurus*) wrote of Brewster's kaleidoscope, "In the memory of man, no invention, and no work, whether addressed to the imagination or to the understanding, ever produced such an effect." Brewster produced his initial design using a tube in which he placed pairs of mirrors at one end, and pairs of translucent disks at the other end. Between the two ends, he placed beads.

Brewster was granted a patent for his kaleidoscope, but due to a flaw in the patent registration process, he did not realize any remuneration for his invention. When the person he employed to manufacture the devices showed them to London opticians so that he could take orders from them, the basic idea was no longer a secret, and a frenzy of interest ensued. Before Brewster could derive financial rewards, kaleidoscopes were suddenly manufactured by zealous entrepreneurs who made large sums of money by selling hundreds of the devices. Within three months, more than 200,000 kaleidoscopes were sold in England and France.

In 1818, Brewster wrote to his wife:

> You can form no conception of the effect which the instrument excited London.... No book and no instrument in the memory of man ever produced such a singular effect.... Thousands of poor people make their bread by making and selling them.... It will create, in a single hour, what a thousand artists could not invent in the course of a year.

Brewster described his invention and the principles behind it in *The Kaleidoscope*, published in 1819. Marjorie Senechal writes in "Reflections of Kaleidoscope":

> It was the first book on the mathematics of symmetry for a general readership.... No one (until M. C. Escher over a century later) has done more to popularize the science of symmetry.... Brewster's little book is almost as interesting as the kaleidoscope itself.

The lenticular stereoscope was another of Brewster's inventions that quickly became a very popular toy. This device was used to produce the illusion of a three-dimensional object, and it resulted from his interest in early photography. This lenticular stereoscope was a closed box, which could be opened on the sides to admit light, and it had two adjustable lenses. When examples were sent to the Great Exhibition at Crystal Palace in 1851, Queen Victoria was apparently fascinated by it, triggering another fad.

Brewster's optical research led to advances in the British lighthouse system. Although Augustin Fresnel (1788–1827) had also worked on special dioptric lenses, Brewster described the apparatus in 1812, and he pressed for its adoption in British lighthouses. These kinds of lenses could be large and lightweight and produce a beam of light that could be seen over much longer distances than could be produced by traditional lenses.

In the 1820s, Brewster became fascinated by the ongoing debate over the number of colors in the spectrum. Isaac Newton, for example, has suggested that seven colors existed. Other researchers felt that there were fewer colors because yellow was simply a combination of red and green. Using various color-absorbing glasses, Brewster established the separate existence of yellow.

Brewster was knighted in 1831. In the 1830s, he began to write biographies, including those for Newton, Galileo, Tycho Brahe, and Johannes Kepler—as well as numerous entries for the *Encyclopaedia Britannica*. In 1859, he was elected principal of the University of Edinburgh.

Brewster's 1854 book, *More Worlds Than One: The Creed of the Philosopher and the Hope of the Christian*, suggested that every star has a planetary system similar to ours and that every planet, sun, and moon were inhabited by life forms. In this book, Brewster writes:

> Neither in the Old nor in the New Testament is there a single expression incompatible with the great truth, that there are other worlds than our own which are the seats of life and intelligence. Many passages, on the contrary, are favorable to the

> doctrine.... The beautiful text, for example, in which the Psalmist expresses his surprise that the Being who fashioned the heavens and ordained the moon and the stars, should be mindful of so insignificant a being as man, is, we think, a positive argument for the plurality of worlds.

In 1855, Brewster was asked to study the Scottish psychic medium Daniel Dunglas Home (1833–1886), who was alleged to have been able to exhibit all kinds of odd phenomena and sounds during a séance. Brewster exposed the scam by publishing a letter in the *Morning Advertiser*, in which he denounced spiritualism and added, "I saw enough to satisfy myself that [the séance effects] could all be produced by human hands and feet."

A heated newspaper controversy over the séance ensued. In response to an irate newspaper letter that contradicted Brewster, he replied that he had not been allowed to look under the table and that "rather than believe that spirits made the noise, I will conjecture that the raps were produced by Mr. Home's toes, and rather than believe that spirits raised the table, I will conjecture that it was done by the agency of Mr. Home's feet, which were always below it."

At age 87, David Brewster contracted pneumonia. He knew he was dying when he said, "I shall see Jesus and that will be grand. I shall see Him who made the worlds." Shortly after Brewster's death, his daughter, Mrs. Margaret M. Gordon, published a biography titled *The Home Life of Sir David Brewster*—a large book in which she mentions his publication of more than 2,000 scientific papers. A lunar crater with a diameter of 10 kilometers was named after Brewster and approved in 1976 by the International Astronomical Union General Assembly.

During his life, Brewster's mind had ranged far—from such topics as the laws of polarization and the effects of heat and pressure on polarization, to the discovery of crystals with two axes of double refraction and the laws of metallic reflection.

FURTHER READING

Baker, Cozy, *Kaleidoscopes: Wonders of Wonder* (Concord, Calif.: C&T Publishing, 1999).

Brewster, David, *The Kaleidoscope* (Edinburgh: Constable & Company, 1819; reprint edition, Holyoke, Massachusetts: Van Cort Publishers, 1987).

Brewster, David. *More Worlds Than One: The Creed of the Philosopher and the Hope of the Christian* (New York: Robert Carter & Brothers, 1854).

Land, Edwin, "Some Aspects of the Development of Sheet Polarizers," *Journal of the Optical Society of America*, 41(12): 957–963, 1951.

Morse, Edgar, W., "David Brewster," in *Dictionary of Scientific Biography*, Charles Gillispie, editor-in-chief (New York: Charles Scribner's Sons, 1970).

Pendergrast, Mark, *Mirror Mirror: A History of the Human Love Affair with Reflection* (Basic Books: New York, 2003).

Senechal, Marjorie, "Reflections of Kaleidoscope," in *Symmetry 2000* (Proceedings from a symposium held in Stockholm, September 2000), edited by I. Hargittai and T. C. Laurent (London: Portland Press, 2002).

Weber, Michael F., Carl A. Stover, Larry R. Gilbert, Timothy J. Nevitt, and Andrew J. Ouderkirk, "Giant Birefringent Optics in Multilayer Polymer Mirrors," *Science*, 287(5462): 2451–2456, March 31, 2000.

INTERLUDE: CONVERSATION STARTERS

Selection [of universes] arises because only firm laws can yield constant, benign conditions to form new life. Once life-forms realize this, they could intentionally make more smart universes with the right fixed laws to produce ever grander structures.

—Gregory Benford, in John Brockman's *What We Believe but Cannot Prove*

The mathematics involved in string theory... in subtlety and sophistication... vastly exceeds previous uses of mathematics in physical theories.... String theory has led to a whole host of amazing results in mathematics in areas that seem far removed from physics. To many this indicates that string theory must be on the right track....

—Michael Atiyah, "Pulling the Strings," *Nature*

Superstring theory has been absorbed into membrane theory, or M-theory, as they call it. There is not a scintilla of empirical evidence to support it. Although I have only a partial understanding of M-theory, it strikes me as comparable to Ptolemy's epicycles. It's getting more and more baroque.

—Martin Gardner in "Interview with Martin Gardner," *Notices of the American Mathematical Society*, 2005

When we look back at the scientific revolution from our vantage point of three centuries and attempt to understand the momentous transformation of Western thought by isolating its central characteristic, the ever greater role of mathematics and of quantitative modes of thought insistently catches our eye—what Alexandre Koyré dubbed the geometrization of nature. Initiated in the sixteenth and seventeenth centuries, the geometrization of nature has proceeded with gathering momentum ever since. To be a scientist today is to understand and to do mathematics; such is perhaps our most distinctive legacy from the scientific revolution.

—Richard S. Westfall, "Newton's Scientific Personality," *Journal of the History of Ideas*

We've managed to push the program of understanding the universe to small scales and large scales, by pursuing this approach of looking for simplicity. Particularly when we look at the cosmos; now we can see out to the farthest observable edges of the cosmos, we can see that the laws of physics are the same, and that the physical conditions are also remarkably similar throughout observable universe.

—Paul Steinhardt, "Einstein: An *Edge* Symposium"

THE DULONG-PETIT LAW OF SPECIFIC HEATS

France, 1819. The specific heats of elements are in inverse proportion to their atomic weights.

CROSS REFERENCE: HUMPHRY DAVY, ANDRÉ-MARIE AMPÈRE, JÖNS BERZELIUS, LOUIS-JACQUES THÉNARD, FRANÇOIS ARAGO, ALBERT EINSTEIN, DEBYE'S T^3 LAW, AND DULONG-PETIT'S FIVE-FOURTHS POWER LAW.

In 1819, the United States purchased Florida from Spain for $5 million. (To be more precise, Spain ceded East Florida to the United States and renounced all claims to West Florida. In return, the United States assumed $5 million of liability for damages done by American citizens who rebelled against Spain.) The *SS Savannah* became the first steamship to cross the Atlantic Ocean.

According to the Dulong-Petit Law, proposed in 1819 by French chemists Pierre Louis Dulong and Alexis Thérèse Petit, the specific heat capacity C of a crystal is

$$\boxed{C = 3\frac{R}{M}.}$$

C is often measured in joules per kelvin per kilogram. R is the gas constant (8.314472 joules per kelvin per mole), and M is the molar mass (measured in kilograms per mole). This law is fairly accurate at high temperatures for solids with relatively simple crystal structures. At low temperatures, the law fails because quantum mechanical effects become more important. The law also assumes that the materials under observation do not melt, boil, or change their crystal structure in the temperature range being studied.

Diamond has a special place in the history of modern physics because it exhibits the largest departure from the Dulong-Petit Law, even at room temperatures, and this observation stimulated Einstein to consider the possible quantum effects of materials on specific heats. These effects are quite strong in diamond due to its particular tetrahedral-lattice atomic structure. Note that one reason diamond is so hard is that the chemical bonds between its carbon atoms are extremely strong. Another is that the atoms form a rigid structure—each atom is connected to four others, forming a regular network.

To best understand specific heat, consider that materials with low specific heats, such as metals, require less input of energy to increase their temperature than those with high specific heats such as water. In some sense, specific heat measures how well a substance holds its temperature, or how well it "stores" heat, which is one reason we often use the phrase "heat capacity."

More precisely, materials differ from one another in the quantity of heat needed to produce a given rise of temperature in a given mass. The ratio of the heat supplied to a body to its corresponding temperature rise is the heat capacity, or specific heat, of the body. This value is a characteristic of the material under study.

Note that the *gram-atomic capacity*, which is the specific heat multiplied by the atomic weight, is approximately a constant for solid elements. This implies that if we can measure the specific heat of an element, we can derive its atomic weight using this law.

Some authors write the Dulong-Petit Law in the following manner, which emphasizes the fact that the specific heat at constant volume is the rate of change with temperature (i.e., the temperature derivative) of that energy:

$$\boxed{C = \frac{\partial}{\partial T}(3kTN_A) = 3kN_A/\text{mole} = 24.94\ \text{J/mole}}$$

Here, k is Boltzmann's constant, T is the temperature in degrees kelvin, N_A is Avogadro's number, and the quantity $3kTN_A$ is the energy per mole. As an example, we can observe that the specific heats of copper and lead turn out to be quite similar when expressed in units of J/mole °C:

Copper	0.386 J/g °C × 63.6 g/mole = 24.6 J/mole °C
Lead	0.128 J/g °C × 207 g/mole = 26.5 J/mole °C

In 1819, Dulong and Petit used different units, and they showed that molar heat capacities of almost all substances have values close to 6 calories/mole °K. (Note that the scientific literature requires us to become accustomed to different units for heat capacities; however, the results are all compatible.) Molar heat capacities for various metals at room temperature are as follows:

Aluminum	5.82 calories/mole °K
Copper	5.85 calories/mole °K
Gold	6.11 calories/mole °K
Lead	6.32 calories/mole °K

At higher temperatures, the heat capacity becomes very close to 6 calories/mole °K.

A related famous law of Dulong and Petit, known as the Five-Fourths Power Law, states that when a body is cooling in still air, the loss of heat is proportional to $(T - T_S)^{5/4}$, where T is the temperature of the body and T_s is the temperature of the surroundings.

Pierre Louis Dulong (1785–1838) and Aléxis Thérèse Petit (1791–1820), French chemists and physicists who discovered that the specific heat multiplied by the atomic weight of an element is a constant.

CURIOSITY FILE: Dulong and Petit both had their lives marred by tragedy. Dulong was an orphan who later blew off his fingers and lost an eye during a chemistry experiment. Petit's wife died shortly after he married her, and he died before age 30.

> Six months after their marriage, Petit's wife became ill and died. . . . This was too much for him. He suffered physical and mental lassitude and exhibited the symptoms of premature senescence. When Petit was no longer able to speak in public, Dulong and Arago took over his lectures so that he might continue to draw a salary.
>
> —Jaime Wisniak, "Alexis-Thérèse Petit," *Educacion Quimica*

> The heat capacities of all solids turn out to depend on the temperature at which they are measured. If Dulong and Petit had lived in a very much colder world, they might never have discovered their law.
>
> —Alan Holden, *The Nature of Solids*

Pierre Louis Dulong was born in Rouen in northwestern France—the same town in which the English burned Joan of Arc in 1431. Both of his parents died before he was 5 years old, and an aunt took care of him. He entered the École Polytechnique in Paris in 1801 at the minimum admittance age of 16; however, excessive study demands appeared to have taken a toll on his health, and he left the Polytechnique in his second year.

Years later, he studied medicine and treated the poor and indigent, even if they could not pay for his services. He left medicine when it was clear that he could not make a good living in this area, given his predilection for offering free treatments and paying for his patients' prescriptions.

Dulong married in 1803 and, over the course of his life, had four children, one of whom died in infancy. Dulong's next and most successful career was in chemistry, although this career brought him much hardship. For example, in order to continue his studies in chemistry, he spent nearly all of his money on equipment. Moreover, during his studies of highly explosive nitrogen trichloride, which he discovered in 1811, he lost an eye and nearly lost a hand.

The explosion had occurred when Dulong was handling containers of chlorine gas and a solution of ammonium chloride. The oil nitrogen trichloride (NCl_3) formed and violently exploded. In addition to his eye, the explosion cost Dulong between one and three fingers, a quantity that varies according to different biographers. However, even after this horrendous accident, Dulong continued working with NCl_3. He found that this dangerous yellow oil boiled at 71°C, and the foul-smelling vapor irritated the eyes and mucous membranes. Today, we know that the oil explodes on contact with many substances or even when exposed to bright light.

British chemist and physicist Humphry Davy (1778–1829) heard about Dulong's work and repeated the experiment so that he, too, could understand the power of nitrogen trichloride. He prepared a sample "scarcely as large as a grain of mustard seed," and it exploded, causing glass shards to enter his cornea. Later, Davy wrote of the incident to French physicist André-Marie Ampère (1775–1836), "The fulminating oil which you mentioned roused my curiosity and nearly deprived me of an eye. After some months of confinement I am again well."

One would think that the next experiments would be performed so carefully that no additional mayhem would ensue. However, British chemist and physicist Michael Faraday (1791–1867), who was Davy's assistant at the time, collaborated with Davy to produce several more detonations until, he writes in an 1813 letter:

> The experiment was repeated again with a larger portion of the substance. It stood for a moment or two and then exploded with a fearful noise; both Sir H. and I had masks on, but I escaped this time the best. Sir H. had his face cut in two places about the chin, and a violent blow on the forehead struck through a considerable thickness of silk and leather; and with this experiment he has for the present concluded.

After receiving additional injuries, Dulong himself abandoned this particular line of research.

Dulong was a professor of physics in Paris from 1820 to 1830, and then director of studies at the École Polytechnique. He collaborated with the French chemist Alexis Thérèse Petit from 1815 to 1820, and upon Petit's death continued his research alone on specific heat capacities, publishing his findings in 1829.

In 1819 he and Petit discovered the chemistry law that now bears their names. Their law states that, for many elements that are solid at room temperature, the specific heats of the elements are in inverse proportion to their atomic weights—and the law was also useful in determining atomic weights. In their 1819 paper, *Recherches sur quelques points importants de la Théorie de la Chaleur*, they write:

> The simple inspection of these [specific heat and relative atomic weight] numbers exhibits an approximation too remarkable by its simplicity not to immediately recognize in it the existence of a physical law capable of being generalized and extended to all elementary substances. These products, which express the capacities of the different atoms, approach so near equality . . . we are authorized to deduce from them the following law: The atoms of all simple bodies have exactly the same capacity for heat.

When Dulong was appointed professor of physics at the École Polytechnique in 1820, he wrote of his dear colleague Petit who had just died: "Through a weakness of character for which I reproach myself incessantly, I have consented to accept the professorship of physics at the École Polytechnique, which the death of my unfortunate friend [Petit] has left vacant."

In 1823, Dulong was elected to the physics section of the Académie des Sciences. In 1829, Dulong determined that equal volumes of all gases evolve or absorb the same quantity of heat when they are suddenly expanded or compressed to the same fraction of their original volumes. This relationship assumes that the experiment is performed under the same conditions of gas temperature and pressure. He also discovered that the accompanying temperature changes are inversely proportional to the specific heat capacities of the gases at constant volume.

Some related highlights of Dulong's interests and publications:

- 1811, published a paper on the reversibility of chemical reactions
- 1815, investigated the properties of mercury thermometers as well as the laws of cooling in a vacuum
- 1816, investigated color changes of dinitrogen tetroxide, which is a colorless solid at –20°C and a red gas when heated
- 1820, published a paper with Swedish chemist Jöns Berzelius (1779–1848) on fluid densities and investigated with French chemist Louis-Jacques Thénard (1777–1857) the use of metals to facilitate combinations of gases
- 1826, investigated the refracting power of gases

- 1829, investigated the specific heat of gases by measuring the effect of temperature changes on the tones produced when gases were passed through a flute
- 1830, published paper with French physicist François Arago (1786–1853) on the elasticity of steam at high temperatures
- 1838, published a paper on heat produced in chemical reactions

Although Dulong had personally experienced the devastating effects of dangerous chemical experiments when he lost his eye in an explosion, this did not seem to deter him when he worked with Arago on a long and risky study of the pressure of steam at high temperatures. The French government was worried about the safety of boilers and asked for Dulong's help in understanding boilers under high pressure. Dulong decided to help the government and performed many experiments on pressures as great as 27 atmospheres.

His name is one of 72 names of prominent French scientists whom Gustave Eiffel placed on the Eiffel Tower (see "Coulomb's Law of Electrostatics" in part II).

Aléxis Petit was a child prodigy who was born in Vesoul, France. He completed the entrance requirements of the École Polytechnique in Paris before he was 11 years of age, and surpassed the entrance exam scores of all other candidates at the time. Although his life was tragically short, he made strides in many scientific fields. He is best remembered for his collaborations with Dulong regarding the law that states that atoms of simple materials have the same capacity for heat, but he also made important contributions to the determination of heat capacities and to our understanding of the refractive power of materials and the conversion of kinetic energy to mechanical power.

Little is known about the young life of Petit. He obtained his doctorate in 1811 for his thesis "Mathematical Theory of Capillary Action." His objective was to determine the laws that describe the movement of liquids in capillary spaces, for example, the laws that describe the ability of a narrow glass tube to draw a liquid upward. Capillarity happens to be one of the reasons why water flows upward in plants.

Capillary action refers to the interaction between a liquid and the solid it touches. In the process, the surface of the liquid becomes elevated or depressed at the point where it contacts the solid. For example, if you were to carefully observe the surface of water in a drinking glass, you would find that the surface is slightly higher at the edges, where the water touches the glass, than in the middle. Capillarity is the result of competing forces: adhesion (the force between the molecules of a liquid and those of the container) and cohesion (the force between the molecules of the liquid).

More specifically, Petit found that the force H, which the capillary wall exerts on the liquid, and the force H', which the liquid exerts on itself, is described by $H = H' \cos^2(\bar{\omega}/2)$, where $\bar{\omega}$ represents the angle between the capillary wall and the surface of the liquid at the point that the liquid touches the wall. (The angle opens downward into the liquid.)

In 1814, Petit married; unfortunately, six months later his wife became sick, and she died in 1817. Petit had become a full professor of physics at the Polytechnique in 1815. For much of his early research, he collaborated with his brother-in-law, the French physicist François Arago (1786–1853). Together, they examined the effect of temperature on the refractive indices of gases. Their results led Petit to become an early supporter of the wave theory of light. Petit's first major paper was published jointly with Arago in 1816, which investigated the variation of the refractive power of a substance in different states of aggregation.

In 1818, Petit won the Paris Academy Prize for work on the law of cooling, and in the same year, he published a paper on the general principles of machine theory and another paper on the theory of heat. Petit collaborated with Dulong on several research papers related to the theory of heat. They also studied the laws that govern the cooling of materials in a vacuum, in air, and in other gases. His first joint paper with Dulong involved laws that described the expansions of solids and liquids. The 1818 annual French prize in physics went to both Dulong and Petit, and it consisted of a gold medal valued at 3,000 francs. Also working with Dulong, in 1819 he formulated the famous empirical law concerning the specific heat of elements.

Petit also investigated water wheel and cannon efficiency. His cannon equations allowed him to calculate the amount of charge powder need to produce the maximum effect for a given bullet and recoil velocity.

Possibly due to the death of his wife, Petit himself began to suffer from periods of extreme fatigue and depression. He no longer spoke in public. Petit's life and scientific contributions were cut short when he died of tuberculosis at age 29. The law of constant molar heat capacities was his most famous work.

Petit was buried in the Cimetiere de l'Est, where Dulong would also be buried years later. A lunar crater with a diameter of 5 kilometers was named after Petit and approved in 1976 by the International Astronomical Union General Assembly.

As I noted above, the Dulong-Petit Law for specific heat C is not accurate at low temperatures, when quantum effects must be considered. According to Donald W. Rogers, author of *Einstein's "Other" Theory: The Planck-Bose-Einstein Theory of Heat Capacity*, this inaccuracy attracted the attention of Albert Einstein (1879–1955):

Einstein noticed that the law of Dulong and Petit fails badly for diamond. Later low-temperature studies showed that it always fails, provided the temperature is low enough... and later cryoscopic studies showed that the Dulong and Petit constant... approaches zero near 0°K.... Einstein set out only to remedy problems in predicting the heat capacity of diamond, but in so doing he developed a general theory of the variation of C_v with T for all solids at all temperatures, even down to 0°K.

In order to create a more useful formulation of the law, Einstein visualized matter as a collection of harmonic oscillators that moved in three dimensions while connected to regularly spaced lattice points in a crystal. Using this quantized approach, both Einstein and Dutch-American physicist Peter Debye (1884–1966) determined more accurate expressions that worked at high and low temperatures, one version of which can be expressed as

$$C = \frac{\pi^2 N_A k^2}{2E_F} T + \frac{12\pi^4 N_A k}{5T_D^3} T^3,$$

where k is Boltzmann's constant, T is the temperature in degrees kelvin, N_A is Avogadro's number, and E_F is the Fermi energy. T_D is called the Debye temperature and is equal to $h\nu_D/k$, where ν_D is the maximum allowed phonon frequency (now called the Debye frequency), and h is Planck's constant. Phonons refer to vibrations that propagate through a material at the speed of sound and affect the specific heat. Debye showed that there is a certain characteristic temperature for each crystalline solid at which its atomic heat should equal 5.67 calories per degree. Einstein's theory expressed this temperature as $h\nu_D/k$, and ν_D can be thought of as a frequency that is characteristic of the atom vibrating in the crystal lattice.

A shorter version of this expression is sometimes called Debye's T^3 Law because specific heat varies as T^3. This formulation was given by Debye in 1912:

$$C = \frac{12\pi^4 N_A k}{5T_D^3} T^3$$

FURTHER READING

Crosland, Maurice P., "Pierre Dulong," in *Dictionary of Scientific Biography*, Charles Gillispie, editor-in-chief (New York: Charles Scribner's Sons, 1970).

Curry, Roger; "Fulminating Oils—Sweat of the Devil—Nitrogen Trichloride and Nitroglycerine," in *Lateral Science*; see lateralscience.co.uk/oil/.

Faraday, Michael, personal letter to Benajmin Abbott, April 8, 1813.

Fox, Robert, "Alexis Petit," in *Dictionary of Scientific Biography*, Charles Gillispie, editor-in-chief (New York: Charles Scribner's Sons, 1970).

Nave, Carl R., "HyperPhysics: Law of Dulong and Petit," Department of Physics and Astronomy, Georgia State University; see hyperphysics.phy-astr.gsu.edu/HBASE/hframe.html.

Rogers, Donald W., *Einstein's "Other" Theory: The Planck-Bose-Einstein Theory of Heat Capacity* (Princeton, N.J.: Princeton University Press, 2005).

Wisniak, Jaime, "Alexis-Thérèse Petit," *Educacion Quimica*, 13(1): 55–60, 2002.

INTERLUDE: CONVERSATION STARTERS

Science does have some metaphysical assumptions, not the least of which is that the universe follows laws. But science leaves open the question of whether those laws were designed. That is a metaphysical question. Believing the universe or some part of it was designed or not does not help understand *how* it works.

—Robert Todd Carroll, "Intelligent Design," *The Skeptic's Dictionary*

The universe seems to operate by several sets of rules that act in layers, independently of each other. The most apparent of these basic rules of nature, gravity, controls the biggest objects in the universe: the stars, the planets, you and me. The other three that scientists have uncovered operate at the subatomic level.

—John Boslough, *Stephen Hawking's Universe*

Imagination is more important than knowledge. For while knowledge defines all we currently know and understand, imagination points to all we might yet discover and create.

—Albert Einstein, "On Science"

Every theoretical physicist who is any good knows six or seven different theoretical representations for exactly the same physics. He knows that they are all equivalent, and that nobody is ever going to be able to decide which one is right at that level, but he keeps them in his head, hoping that they will give him different ideas for guessing.

—Richard Feynman, *The Character of Physical Law*

It's become hackneyed, but if a theory is so simple that its deep equation can be put across a T-shirt in 20-point type, then we generally view that as fairly simple. Certainly that is the case of both general relativity and quantum mechanics.

—Brian Greene, "Einstein: An *Edge* Symposium," edge.org

The essence of the scientific method is rationality and logic: we suppose that things are the way they are for a reason. Yet when it comes to the laws of physics themselves, well, we are asked to accept that they exist "reasonlessly." If that were correct, then the entire edifice of science would ultimately be founded on absurdity.

—Paul Davies, "Laying Down the Laws," *New Scientist*

THE BIOT-SAVART LAW OF MAGNETIC FORCE

π *France, 1820.* Current through a wire causes magnetic field lines to form concentric circles around the wire. The magnitude of the field is inversely proportional to the square of the distance from the wire. Many other aspects of the law become clear in the detailed description.

CROSS REFERENCE: AMPÈRE'S CIRCUITAL LAW OF ELECTROMAGNETISM, BIOT'S ABSORPTION LAW, BIOT'S LAW OF ROTARY DISPERSION, HANS ØRSTED, DANIEL BERNOULLI, AND JOSEPH GAY-LUSSAC.

The Missouri Compromise, which regulated the practice of slavery in the western territories of the United States, became law. Joseph Smith, Jr., founder of the Latter Day Saint movement that gives rise to Mormonism, claimed to be visited in a vision by God and Jesus. Maine is admitted as the twenty-third U.S. state.

The Biot-Savart Law states that the magnetic flux density (or magnetic induction) near a long, straight conductor is directly proportional to the current in the conductor and inversely proportional to the square of the distance from the conductor. We often see the Biot-Savart Law expressed as

$$d\mathbf{B} = \frac{\mu_0 I}{4\pi} \frac{d\mathbf{s} \times \hat{\mathbf{r}}}{r^2},$$

where the boldface letters are vectors. (A vector is a quantity specified by a magnitude and a direction.) The formula shows that the magnetic field **B** produced by a short segment of wire d**s** is directly related to the steady current I. In more technical terms, d**s** may be considered the differential length vector of the current element. The direction of d**s** is the same as the direction of the current. The unit vector $\hat{\mathbf{r}}$ points from the current element of a wire to a field point somewhere in the space around the wire. In experimental terms, $\hat{\mathbf{r}}$ points from a short segment of current to a probe point where we desire to compute the magnetic field.

The r in the denominator is the distance from the current element to the field point. This means that the value of the magnetic field depends on the location of a particular point with respect to the segment of wire, and in particular, the magnetic field is inversely proportional to the square of the distance from the current element that produces it. When the current is measured in amps, distance in meters, and magnetic field in units of teslas, the value of μ_0 is $4\pi \times 10^{-7}\,\mathrm{T \cdot m/A}$. μ_0 is known as the permeability of free space.

The value of **B** at some point in space around the wire may be thought of as the sum of all the contributions for each small segment of the wire. To compute the numerator in Biot-Savart's formula, the vector (cross) product $d\mathbf{s} \times \hat{\mathbf{r}}$ means we calculate the product of the magnitudes of vectors $d\mathbf{s}$ and $\hat{\mathbf{r}}$ and multiply by $\sin \theta$, where θ is the angle between the two vectors.

The Biot-Savart Law can be derived from Ampère's Circuital Law of Electromagnetism (see entry below), so in some sense it is not truly a separate principle. However, although Ampère's Law is a general statement of the behavior of steady currents, its application can have some practical disadvantages. In order to use Ampère's Law effectively, the magnetic field must be sufficiently simple so that **B** can be removed from within the integral sign in Ampère's Law.

For certain particular configurations of wires, we may use simplifications of the Biot-Savart formula in order to determine an approximation for **B**. For example, we can often use the simplified formula for the magnetic field around an infinitely long wire whenever we want to estimate the field near a segment of wire. Similarly, we may use the formula for the magnetic field at the center of a circular loop of wire whenever we want to estimate the magnetic field near the center of any wire loop. The magnitude of the magnetic field at a point that is a distance r from an infinitely long wire carrying current I is

$$B = \frac{\mu_0 I}{2\pi r}.$$

The magnetic flux density **B** near the straight wire is at every point perpendicular to the plane determined by the point and the line of the conductor. This means that the lines of inductions are circles, with their centers on the wire. The direction of the magnetic field is given by the right-hand rule, which you may recall from introductory physics classes: Lift your right hand. Point the thumb of your right hand in the direction of the current. Your fingers curl around the wire and indicate the direction of the circular magnetic field lines around the wire.

The Biot-Savart formulation is useful for studying other configurations. Consider current flowing through a circular loop of wire of radius R. A simple formula also exists for the magnitude of the magnetic field at the center of such a loop:

$$B = \frac{\mu_0 I}{2R}$$

Again, the direction of the magnetic field is given by the right-hand rule. If you curl the fingers of your right hand in the direction of the current flow, your thumb points in the direction of the magnetic field inside the loop. Other simple formulas exist for the magnetic field within a long, thick wire and for the magnetic field inside a long solenoid made with N turns of wires. (A solenoid is a current-carrying coil of wire.)

Jean-Baptiste Biot (1774–1862) and Félix Savart (1791–1841), French physicists famous for their Biot-Savart Law concerning magnetic fields near conductors.

CURIOSITY FILE: The idea that meteorites fell from the sky was considered to be a superstition until Jean-Baptiste Biot convinced the French that the rocks had an extraterrestrial origin. • The green mineral biotite, named after Biot, occurs in the lava of Mount Vesuvius. • The Biot-Savart law has additional applications in the field of fluid dynamics, particularly in aerodynamics. • Savart conducted pioneering studies of the turbulent sounds in blood vessels.

> One of the most important acoustical physicists of the first half of the nineteenth century, Félix Savart, offered a singular explanation of the voice.... Savart compared the larynx to a hunter's birdcall—a short cylinder, each end of which was covered by a thick plate with a small hole in the center.... To support his suggestions, Savart referred to a plaster cast of the inside of a cadaver's throat.
>
> —Thomas Hankins and Robert Silverman, *Instruments and the Imagination*

> Isaac Newton's conceptions seem to have surpassed the limits of thought of mortal man.... Words fail to convey the profound impression of astonishment and respect which one experiences in studying the work of this admirable observer of nature.
>
> —Jean-Baptiste Biot, *Journal de Physique*

> Jean-Baptiste Biot (1774–1862), a steady contributor to many fields, reverted to his childhood faith after an audience with the Pope.
>
> —Dan Graves, *Scientists of Faith: Forty-Eight Biographies of Historic Scientists and Their Christian Faith*

Shortly after Danish physicist and chemist Hans Christian Ørsted's (1777–1851) discovery in 1819 that a compass needle moves when placed near a current-carrying wire, French scientists Jean-Baptiste Biot and Félix Savart reported that a conductor carrying a steady current exerts a force on a magnet. From their experimental results, Biot and Savart derived a formula for the magnetic field at a point in space in terms of the current that produces the field. They found that the intensity of the magnetic field produced by a current flowing through a wire varies inversely with the distance from the wire.

Jean-Baptiste Biot was born in Paris, France. His father had always planned for his son to enter the world of commerce, and Biot was provided with a private math tutor in his youth. Just as in the life of physicist and mathematician Daniel Bernoulli (1700–1782; see "Bernoulli's Law of Fluid Dynamics" in part II), young Biot rebelled against his father's desire for him to go into business.

After his education at the college of Louis-le-Grand, in 1793 Biot joined the French army. After serving briefly in the artillery, he attended the École Polytechnique in Paris. He then spent a brief time in jail for his part in an antigovernment insurrection. After his release, he became professor of mathematics at the École Centrale at Beauvais in 1797—the same year that he married the 16-year-old Antoine Brisson, whom he taught science and mathematics. Three years later, he became professor of mathematical physics at the Collège de France.

Biot is famous for insisting upon the reality of meteorites. His 1803 report on the fall of a meteorite convinced scientists that rocks fall from the sky and have an extraterrestrial origin. Prior to Biot's paper, this idea was dismissed as mere superstition. After Biot assured scientists that these objects did come from the sky, U.S. President Thomas Jefferson wrote to the American naturalist Andrew Elliot:

> The exuberant imagination of a Frenchman...runs away with his judgment. It even creates facts for him which never happened, and he tells them with good faith....The evidence of nature, derived from experience, must be put into one scale, and in the other the testimony of man, his ignorance, the deception of his senses, his lying disposition.

In 1804, Biot sailed with French chemist Joseph Gay-Lussac (1778–1850) on the first balloon flight. The balloon was loaded with scientific equipment to assist them in their studies. The two men ascended to a height of approximately 13,000 feet, and the research performed in the balloon demonstrated that Earth's magnetic field does not vary appreciably with altitudes accessible to balloons. They also studied the chemical composition of the atmosphere at various elevations.

In 1835, Biot showed that sugar solutions rotate the plane of polarized light and that the angle of rotation can be used to measure the concentration of the sugar solution. He also showed that polarized light that passed through an organic substance can be rotated clockwise or counterclockwise depending on the optical axis of the substance.

Over the course of his life, Biot made advances in a range of subject areas, including applied mathematics, astronomy, elasticity, electricity, magnetism, optics, and mineralogy. In 1847, he discovered unique optical properties of mica—and in honor of his research, the dark green, mica-based mineral biotite $[K_2(Mg,Fe,Al)_6(Si,Al)_8O_{20}(OH)_4]$ is named after him.

Biot was a prolific author, having written more than 250 works on a variety of subjects before he died at the age of 88. His most famous work was his *Elementary Treatise on Physical Astronomy* (1805).

Biot's research covered many areas of science, but we can get a flavor of the diversity of topics from the following brief highlights: Biot

- wrote a biography of Isaac Newton
- studied refraction as well as polarization of light and sound
- studied mirages
- performed a comparative study of rhombic aragonite and hexagonal calcite
- determined the meridian of Paris
- studied the composition of air contained in swim bladders of fish that lived on the shores of Ibiza and the Formentera islands
- studied heat flow in bars and the expansion of liquids
- derived a relationship in 1818, sometimes referred to as Biot's Absorption Law, that showed how the intensity of sunlight depended on the thickness of the atmosphere; the law can be expressed as $I' = Ie^{-kt}$, where I is the intensity of incident radiation, I' is the intensity of radiation transmitted through thickness t, e is Euler's number, and k is the absorption coefficient
- gave us Biot's Law of Rotary Dispersion, $\alpha = k/\lambda^2$, in which α refers to the rotation of polarized light, and λ is the wavelength; for example, the rotation angle of polarized light produced by a quartz plate decreases with change of color from violet to red

Before he died, English mathematician and proto-computer scientist Charles Babbage (1791–1871) paid Biot a visit. Babbage wrote in *Passages from the Life of a Philosopher*,

The last time during M. Biot's life that I visited Paris, I went, as usual, to the Collège de France. I inquired of the servant who opened the door after the state of M. Biot's health, which was admitted to be feeble. I then asked whether he was well enough to see an old friend. Biot himself had heard the latter part of this conversation. Coming into the passage, he seized my hand and said: "My dear friend, I would see you even if I were dying."

A lunar crater with a diameter of 12 kilometers was named after Biot and approved in 1935 by the International Astronomical Union General Assembly.

Félix Savart was born in Mézières, France, and as an adult embarked on a medical career. However, during those times when he had few patients, he conducted experiments on the violin and began to devote himself to the study of the acoustics of air, bird songs, and vibrating solids. One of his goals was to enhance the tone of violins and also to produce louder instruments that could be heard in the larger orchestras and concert halls.

He began teaching at the Collège de France in 1828. Aside from his work with Biot on magnetic fields, he is well known today for his acoustical experiments, his explanations of the working of violins, the creation of a trapezoid-shaped fiddle, and the "Savart disk," which is a serrated rotating wheel that produces a sound wave of known frequency. The Savart cup or bell is a related device that is set into oscillation with a violin bow to produce a sound of definite pitch.

The *savart* is a unit, named in his honor, used in music to describe the frequency ratio between notes. For example, there are approximately 301 savarts in an octave. If one note is only 1 savart higher than another note, the higher note has a frequency equal to $2^{1/301} = 1.002305$ times the frequency of the lower note.

FURTHER READING

Babbage, Charles, *Passages from the Life of a Philosopher* (New Brunswick, N.J.: Rutgers University Press, 1994).

Bueche, Frederick, *Introduction to Physics for Scientists and Engineers* (New York: McGraw-Hill, 1975).

Consolmagno, Guy, and Martha Schaefer, *Worlds Apart: A Textbook in Planetary Sciences* (Englewood Cliffs, N.J.: Prentice Hall, 1994).

Crosland, Maurice P., "Jean-Baptiste Biot," in *Dictionary of Scientific Biography*, Charles Gillispie, editor-in-chief (New York: Charles Scribner's Sons, 1970).

Dostrovsky, Sigalia, "Félix Savart," in *Dictionary of Scientific Biography*, Charles Gillispie, editor-in-chief (New York: Charles Scribner's Sons, 1970).

Graves, Dan, *Scientists of Faith: Forty-Eight Biographies of Historic Scientists and Their Christian Faith* (Grand Rapids, Mich.: Kregel Publications, 1996).

INTERLUDE: CONVERSATION STARTERS

This solution to the problem of induction involves accepting the existence of laws of nature, and it involves recognizing these laws not just as regularities in the behavior of things (consistencies in how the world works in different places and times), but as forms of natural necessity—as laws whose obtaining *ensures* that things behave and interact in certain regular ways.

—John Foster, *The Divine Lawmaker: Lectures on Induction, Laws of Nature, and the Existence of God*

Much of the history of science, like the history of religion, is a history of struggles driven by power and money. And yet, this is not the whole story. Genuine saints occasionally play an important role, both in religion and science. For many scientists, the reward for being a scientist is not the power and the money but the chance of catching a glimpse of the transcendent beauty of nature.

—Freeman Dyson, introduction to John Cornwell's *Nature's Imagination: The Frontiers of Scientific Vision*

We have an unfortunate tendency to oversimplify invention to lists of names, dates, and other statistics. Look more closely and you'll find a rich and fascinating ecosystem. It's not just the idea that counts—the way it is implemented and the context are equally important.

—Jeff Hecht, "More Than the Sum of Their Parts," *New Scientist*, August 12, 2006

It is the quest of this special classic beauty, the sense of harmony of the cosmos, which makes us choose the facts most fitting to contribute to this harmony. It is not the facts but the relation of things that results in the universal harmony that is the sole objective reality.

—Robert M. Pirsig, *Zen and the Art of Motorcycle Maintenance*

FOURIER'S LAW OF HEAT CONDUCTION

 π *France, 1822.* The rate of heat flow between two points in a material is proportional to the difference in the temperatures of the points and inversely proportional to the distance between the two points.

CROSS REFERENCE: RENÉ DESCARTES, LORD KELVIN, NICOLAS DE CARITAT CONDORCET, ANTOINE LAVOISIER, AND OLIVER HEAVISIDE.

In 1822, Jean-François Champollion revealed his first successful attempts at deciphering Egyptian hieroglyphs through the use of the Rosetta Stone. English mathematician and mechanical engineer Charles Babbage proposed the construction of a Difference Engine—a special-purpose mechanical digital calculator. Alas, this first model would have required nearly 25,000 parts, and he did not complete the construction.

If we place one end of a metal spoon into a hot cup of tea, the temperature at the end of the handle of the spoon begins to rise. This heat transfer is caused by molecules at the hot end exchanging their kinetic and vibrational energies with adjacent regions of the spoon through random motions. The energy is always transported from the tea to the spoon tip, that is, from high to low temperature, through a process known as conduction.

Fourier's Law of Heat Conduction is concerned with the transmission of heat in materials. The law states that the heat flux, Q, which is the flow of heat per unit area and per unit time, is proportional to the gradient of temperature difference:

$$Q = -KA\frac{\Delta T}{\Delta x}$$

The law is often applied to objects such as a slab of material, a body of water, or insulated wires. Here, A is the surface area for heat transfer; Δx is the thickness of the matter through which the heat is passing; K is a conductivity constant, which is dependent on the nature of the material and its temperature; and ΔT is the temperature difference through which the heat is being transferred. The minus sign is placed before the conductivity constant to indicate that heat flows in the direction of decreasing temperature. Note that although the heat conduction equation refers to one-dimensional conduction, the formula can be generalized to three dimensions by observing that heat flow may be a vector quantity with x, y, and z components.

A concrete example may help us to visualize the operation of Fourier's Law. Consider an insulated metal rod with ends A and B. The rate of flow of energy, which might be thought of as a "heat current," is proportional to the difference in temperatures at A and B and inversely proportional to the distance between A and B. This means that the heat current is doubled if the temperature difference is doubled or length of the rod is halved.

If we let U be the conductance of the material, that is, the measure of the ability of a material to conduct heat, we may write

$$U = \frac{K}{\Delta x}$$

and rewrite Fourier's law as

$$\boxed{Q = -UA\Delta T.}$$

A material with a high value for thermal conductivity often also has a high value for electrical conductivity, and metals are good conductors of heat. One exception to this rule is diamond, which has a very high thermal conductivity but low electrical conductivity. Among the best thermal conductors, in order of thermal conductivity values, are diamond, carbon nanotubes, silver, copper, and gold. Examples of thermal conductors that are much weaker conductors than these materials are glass, water, and air.

By using simple instruments, the high thermal conductivity of diamonds is sometimes used to help experts distinguish real diamonds from fakes. An informal, but less reliable, test of diamond authenticity is known as the "breath test." Because diamond has the highest thermal conductivity of any known material and is cool to the touch, moisture from a person's breath evaporates from a diamond more rapidly than from a fake—due to the high rate of heat transfer from the diamond to the water vapor.

Note that the high thermal conductivity of diamond could make it an ideal substrate for electronic chips, because the conductivity would distribute and help dissipate the heat of a chip very effectively. As computer chips become physically smaller, heat dissipation problems become more and more important. Of course, diamond is not used as a substrate for chips for reasons of cost, but if diamonds were cheap, it is possible to imagine technology in which chips were made of diamond in order to dissipate heat.

Some technologists suggest that if certain materials challenges can be overcome, diamond could make an excellent semiconducting material itself, because it will operate at an extremely high temperature, operate at higher microprocessor frequencies, and will allow the chip to cool faster than chips that use conventional materials. In fact, experimental diamond transistors have been run at 81 gigahertz, compared to a typical personal computer processor speed of around 2 to 3 gigahertz, in 2008. Silicon exhibits significant thermal stress around 100°C, but diamond can withstand several times this temperature without problems. When a small amount of boron is added to diamond, it becomes a semiconductor.

Today, Fourier's Law is used in many diverse areas of science. As just one example, the law is used in the study of "soil heat flux," which is a crucial research area for scientists interested in understanding the energy balance for the surface of Earth.

In 1872, Irish-Scottish physicist William Thomson (1824–1907), better known as Lord Kelvin, used Fourier's Law of Heat Conduction in order to date the age of Earth. However, due to several incorrect assumptions, he calculated that between 20 and 400 million years would be required for Earth to cool to its present temperature—and thus he suggested that Earth was 20 to 400 million years old. Today, we know that Earth is much older, and Kelvin's young Earth would not have allowed sufficient time for evolution to occur. Kelvin's computation was inaccurate partly because he did not realize that the radioactive elements in Earth serve as an internal heating mechanism that opposes and slows down the cooling.

On April 28, 1862, Lord Kelvin presented his "On the Secular Cooling of the Earth," which in 1864 was printed in the *Transactions of the Royal Society of Edinburgh*:

> Fourier's mathematical theory of the conduction of heat is a beautiful working out of a particular case belonging to the general doctrine of the "Dissipation of Energy." A characteristic of the practical solutions it presents is, that in each case a distribution of temperature, becoming gradually equalised through an unlimited future, is expressed as a function of the time, which is infinitely divergent for all times longer past than a definite determinable epoch....
>
> [In the past] I suggested, as an application of these principles, that a perfectly complete geothermic survey would give us data for determining an initial epoch in the problem of terrestrial conduction. The chief object of the present communication is to estimate from the known general increase of temperature in the Earth downwards, the date of the first establishment of that consistentior status, which, according to Leibnitz's theory, is the initial date of all geological history....

Jean Baptiste Joseph Fourier (1768–1830), French mathematician and Egyptologist, famous for his influence in many areas of mathematical physics and for his formulas on the conduction of heat in solid materials.

CURIOSITY FILE: Diamonds of any size are cool to the touch because of their high thermal conductivity, perhaps contributing to the common use of the word "ice" when referring to diamonds. • Even though Fourier became an expert on heat transfer, he was never good at regulating his own heat. He was always so cold, even in the summer, that he wore several large overcoats. • A seemingly inordinate number of busts and statues of Fourier have been destroyed. For example, one bust by French artist Pierre-Alphonse Fesard was shattered in World War II. Similarly, a bronze statue of Fourier, erected in his hometown of Auxerre, was melted by the Nazis in order to build armaments. • During his last months, Fourier often spent his time in a box to support his weak body. • Fourier invented the notation $\int_a^b$ for the integral from *a* to *b*.

> Profound study of nature is the most fertile source of mathematical discoveries.
> —Joseph Fourier, *The Analytic Theory of Heat*

> Where it is a duty to worship the sun, it is pretty sure to be a crime to examine the laws of heat.
> —John Morley, "Voltaire," in *Critical Miscellanies, 1872*

> The great shock caused by his trigonometric expansions was due to his demonstration of a paradoxical property of equality over a finite interval between algebraic expressions of totally different form.... So powerful was his approach that a full century passed before nonlinear differential equations regained prominence in mathematical physics.
> —Jerome Ravetz and I. Grattan-Guiness, "Fourier," in *Dictionary of Scientific Biography*

Fourier was born in Auxerre, France, a cathedral town that overlooked the river Yonne. He was the ninth of twelve children of his father's second marriage. His father was a master tailor named Joseph, who had fifteen children distributed between his two marriages. At age 9, Fourier's father and mother died. The town's archbishop placed him in military school run by Benedictine monks. Here, he developed his love for mathematics. At night, he would often spend hours solving equations by candlelight, after everyone else was asleep.

By the age of 14, Fourier had completed a study of the six volumes of mathematician Charles Bézout's (1730–1814) *Cours de mathématiques*.

When he was only 16, Fourier discovered a new proof of Descartes's rule of signs, which states that the number of positive real roots of a polynomial is bounded by the number of changes of sign in its coefficients. (E.g., given a polynomial such as $x^4 + 9x^3 - 4x^2 - x - 5$, Descartes, in 1637, would have said that there is at most one positive real root.) Fourier's teenage achievement quickly became the standard proof. He also generalized Descartes's rule that is used to estimate the number of real roots within a given interval.

In his late teens, Fourier considered training for the priesthood, but instead mathematics became his main love. However, he wondered if he could make a significant contribution to mathematics and wrote to a professor, "Yesterday was my 21st birthday, at that age Newton and Pascal had already acquired many claims to immortality." In 1787, he arrived at the Benedictine abbey of St. Benoit-sur-Loire to prepare for his vows while at the same time teaching mathematics to other novices. However, Fourier never did take his vows and in 1789 left the abbey. His mind had always been on mathematics.

During the French Revolution, Fourier was arrested and imprisoned after he attempted to defend many of the victims of the Reign of Terror, in which thousands were guillotined after accusations of counterrevolutionary activities. One of Fourier's heroes, the mathematician Nicolas de Caritat Condorcet (1743–1794), defied the radical Jacobins and died in prison. Another hero, Antoine Lavoisier (1743–1794), the founder of modern chemistry, was guillotined. Fourier himself feared the he would go to the guillotine, but political changes eventually enabled Fourier to be freed.

In 1795, he joined the faculty of the École Polytechnique. In 1797, he succeeded French-Italian mathematician Joseph-Louis Lagrange as the chair of analysis and mechanics. Fourier was renowned as an outstanding lecturer.

Fourier accompanied Napoleon on his 1789 expedition of Egypt. During the next few years, Fourier spent much of his time studying Egyptian artifacts, participated in diplomatic undertakings, and oversaw the massive *Description de l'Egypte* (*Description of Egypt*), which discussed the Egyptian materials found during his expedition and gave an account of the history of ancient Egypt.

After his return to France in 1801, Napoleon sent Fourier to Grenoble, where Fourier held a government position overseeing the drainage of swamps and the construction of roads. At this time, he continued some of his mathematical work.

His research on the mathematical theory of heat began around 1804, and in 1807 he had completed his important memoir *On the Propagation of Heat in Solid Bodies*. One of his interests was heat diffusion in various different shapes such as rectangles, rings, spheres, cylinders, and prisms. His work enabled him to mathematically express the conduction of heat in thin sheets of material, which may be thought of as being two-dimensional objects. His formulation, in terms of a differential equation, may be expressed as

$$\frac{\partial u}{\partial t} = k\left(\frac{\partial^2 u}{\partial x^2} + \frac{\partial^2 u}{\partial y^2}\right),$$

where u is the temperature at time t at a point (x, y) of the plane, and k is a constant called the diffusivity of the material. For these problems, researchers and mathematicians are usually given the temperatures at points on the surface as well as at its edges at time $t = 0$. Fourier introduced a series with sine and cosine terms in order to find solutions to these kinds of problems. In fact, Fourier's theorem helps researchers today to analyze a wide range of function in terms of sines and cosines. These so-called "Fourier series" play an important role in many branches of modern mathematics and physics.

Jerome Ravetz and I. Grattan-Guiness note in the *Dictionary of Scientific Biography*,

> Fourier's achievement can be understood by [considering] the powerful mathematical tools he invented for the solutions of the equations, which yielded a long series of descendents and raised problems in mathematical analysis that motivated much of the leading work in that field for the rest of the century and beyond.

Any differentiable function can be represented to arbitrary accuracy by a sum of sine and cosine functions, no matter how bizarre the function may look when graphed. Consider the application of Fourier series in acoustics in which some periodic function $y(t)$ represents the displacement of air particles near a clarinet or drum. Fourier's theorem tells us that this function can be written as

$$y(t) = \sum_n A_n \sin \omega_n t + B_n \cos \omega_n t.$$

The lowest angular frequency ω_1 corresponds to the actual period of the waveform $y(t)$. In other words, $\omega_1 = 2\pi/T$, where T is the period. The relative values of A and B depend on the waveform shape. As another example, a saw-tooth waveform, which looks like the edge of some serrated handsaws, can be thought of as the sum of an infinite number of sinusoidal waves as follows:

$$y = 2y_0\left[\sin\left(\frac{2\pi t}{T}\right) - \frac{1}{2}\sin\left(\frac{2\cdot 2\pi t}{T}\right) + \frac{1}{3}\sin\left(\frac{3\cdot 2\pi t}{T}\right) - \cdots\right]$$

Here, y_0 is the amplitude of the wave, and T is the period. British physicist Sir James Jeans (1877–1946) wrote in *Science and Music*,

> Fourier's theorem tells us that every curve, no matter what its nature may be, or in what way it was originally obtained, can be exactly reproduced by superposing a sufficient number of simple harmonic curves—in brief, every curve can be built up by piling up waves.

Some of Fourier's other research activities included investigations of thermometers, heating in houses, estimating the age of Earth, various approaches to distinguishing between real and imaginary roots of equations, and estimating the errors of measurements.

In 1808, Napoleon conferred a barony on him, and later Napoleon made him a count. In 1817, Fourier was elected to the Académie des Sciences. Later he was also elected a foreign member of the Royal Society.

In 1822, Fourier published his mathematical theory of heat conduction in solids based on a differential equation that indicates that the rate of flow of heat through a unit area perpendicular to an x-axis is proportional to the temperature gradient (rate of change of temperature, dT/dx) in the x-direction. It is interesting that Fourier wrote and developed his theory in terms of "caloric theory," an incorrect theory that held that changes in temperature are due to the transfer of an invisible and weightless fluid called caloric. Nevertheless, Fourier's Law of Heat Conduction is correct and in agreement with experiments, even if Fourier's idea of the *nature* of heat was not. As discussed in the introduction of this book, a law can explain how the universe works, even if the researcher who discovered the law is not quite sure *why* it works.

In the early 1820s, Fourier wondered how Earth stays sufficiently warm to support life. For example, some researchers felt that heat generated by the rays of the Sun should reflect off the land and oceans and be lost in outer space. Fourier proposed that although some heat does escape, the atmosphere acts as a translucent dome, like a glass lid of a pot, that absorbs some of the heat of the Sun and reradiates it downward to Earth. Thus, Fourier's ideas were forerunners of today's proposed mechanism of global warming.

Ever since his return from the heat of Egypt, Fourier's bodily thermostat had never seemed to readjust itself, because he was always cold. Some have suggested that he was a victim of myxedema (caused by decreased thyroid activity), which lowers the body's metabolic rate. Whatever Fourier's affliction, he rarely went outside without an overcoat and a servant bearing another in reserve, even in the middle of summer. Eventually, he confined himself to his own heated quarters until his death by heart attack in 1830. A little before he died, he wrote a friend of having already seen "the other bank where one is healed of life."

During his last years, Fourier often lived in a box to support his aging body. According to Gale E. Christianson, author of *Greenhouse: The 200-Year Story of Global Warming*:

> [Fourier], who had originated the idea of global warming, found himself back inside a wooden box, a device he used because he was so weakened by chronic rheumatism that to bend over was to risk a fatal attack of breathlessness.... It kept his body upright by allowing only his head and arms to protrude, thus enabling him to work on his scientific papers to the last, even as he doggedly engaged in the voluminous correspondence required of the permanent secretary of the Académie des Sciences.

Fourier's thermophilia is reminiscent of the heat-loving characteristics of other great mathematical physicists such as English electrical engineer, mathematician, and physicist Oliver Heaviside (1850–1924). A final contender for the 1912 Nobel Prize, Heaviside established mathematical foundations for modern electric-circuit design and vector analysis for electromagnetics. His electrical theories allow us today to enjoy long-distance telephony. Heaviside loved working in swelteringly hot rooms by the light of smoky oil lamps. His friends called his private work area "hotter than hell." Heaviside's thermophilia reached new highs as he required ever increasing quantities of gas to run both his lights and fires. So strong were Heaviside's cravings for heat that he constantly fought with the local gas company about not paying his gas bills. In fact, he used gas at the prodigious rate of 800,000 cubic feet per year.

Many of Fourier's groundbreaking mathematical theories were not accepted by his colleagues, partly because he did not provide rigorous proofs for his ideas. S. F. Sun writes in *Physical Chemistry of Macromolecules*:

> Fourier must have died a sad man, having never earned distinction in his lifetime among his peers in mathematics. However, Fourier's position in the history of mathematics was gradually recognized.... Today, the Fourier series is developed in modern analysis alongside the rapid growth of automatic computing. The Fourier integral and Fourier transform, which are derived directly form the Fourier series, are involved in all technical fields, such as engineering, physics, chemistry, biology, and medicine.

Ioan James aptly sums up Fourier's accomplishments in *Remarkable Physicists*:

> Throughout his career, Fourier won the loyalty of younger friends by his unselfish support. . . . His scientific achievements lie mainly in the study of the diffusion of heat and in the mathematical techniques he introduced to further that study. . . . He had a superb mastery of analytical technique, and this power, guided by physical intuition, brought him [lasting] success.

A lunar crater with a diameter of 51 kilometers was named after Fourier and approved in 1935 by the International Astronomical Union General Assembly. His name is one of 72 names of prominent French scientists whom Gustave Eiffel placed on the Eiffel Tower (see "Coulomb's Law of Electrostatics" in part II).

FURTHER READING

Christianson, Gale, *Greenhouse: The 200-Year Story of Global Warming* (New York: Walker & Company, 1999); see www.nytimes.com/books/first/c/christianson-greenhouse.html.

James, Ioan, *Remarkable Physicists: From Galileo to Yukawa* (New York: Cambridge University Press, 2004).

Jeans, James, *Science and Music* (New York: Dover, 1968).

Jiji, Latif, *Heat Transfer Essentials: A Textbook* (New York: Begell House, 1998).

Ravetz, Jerome, and I. Grattan-Guiness, "Fourier," in *Dictionary of Scientific Biography*, Charles Gillispie, editor-in-chief (New York: Charles Scribner's Sons, 1970).

Smith, Eric J., "81GHz Diamond Semiconductor Created," *Geek News*, August 27, 2003; see www.geek.com/news/geeknews/2003Aug/gee20030827021485.htm.

Sun, S. F., *Physical Chemistry of Macromolecules: Basic Principles and Issues* (Hoboken, N.J.: Wiley, 2004).

Thomson, William (Lord Kelvin), "On the Secular Cooling of the Earth," *Transactions of the Royal Society of Edinburgh*, 23: 167–169, 1864; read April 28, 1862; see zapatopi.net/kelvin/papers/on_the_secular_cooling_of_the_earth.html.

INTERLUDE: CONVERSATION STARTERS

What makes the planets go around the sun? At the time of Kepler, some people answered this problem by saying that there were angels behind them beating their wings and pushing the planets around an orbit. As you will see, the answer is not very far from the truth. The only difference is that the angels sit in a different direction and their wings push inwards.

—Richard Feynman, *The Character of Physical Law*

The laws of the universe are cunningly contrived to coax life into being.... If life follows from [primordial] soup with causal dependability, the laws of nature encode a hidden subtext... which tells them: Make life!... It means the laws of the universe have engineered their own comprehension.

—Paul Davies, *The Fifth Miracle*

Somewhere in that great ocean of truth, the answers to questions about life in the universe are hidden.... Beyond these questions are others that we cannot even ask, questions about the universe as it may be perceived in the future by minds whose thoughts and feelings are as inaccessible to us as our thoughts and feelings are to earthworms.

—Freeman Dyson, "Science & Religion: No Ends in Sight," *New York Review of Books*

AMPÈRE'S CIRCUITAL LAW OF ELECTROMAGNETISM

π *France, 1825.* The magnetic circulation in free space is proportional to the total current through the surface bounding the path over which the circulation is computed. This circulation along concentric paths around a straight wire carrying a current is proportional to the current.

CROSS REFERENCE: BIOT-SAVART LAW, MAXWELL'S EQUATIONS, HANS ØRSTED, JOSEPH HENRY, MICHAEL FARADAY, JEAN-BAPTISTE BIOT, AND FÉLIX SAVART.

> In 1825, Bolivia gained independence from Peru. The Erie Canal provided a passage from Albany, New York, to Lake Erie. Thus, the canal connected the Great Lakes with the Atlantic Ocean. French law made sacrilege a capital offense.

The connection between electricity and magnetism was largely unknown until 1819, when Danish physicist Hans Christian Ørsted (1777–1851) discovered that a compass needle moves when an electric current is switched on or off in a nearby wire. Although not fully understood at the time, this simple demonstration suggested that electricity and magnetism were related phenomena, a finding that led to various applications of electromagnetism and eventually culminated in telegraphs, radios, televisions, and computers. In 1820, Ørsted published a pamphlet on his findings, and his observations caused a sensation, particularly in France, where there was great interest in electric and magnetic phenomena.

Subsequent experiments during a period from 1820 to 1825 by French physicist André-Marie Ampère (1775–1836) and others showed that any conductor that carries an electric current I produces a magnetic field around it. This basic finding, and its various consequences for conducting wires, is sometimes referred to as Ampère's Law of Electromagnetism. For example, a current-carrying wire produces a magnetic field **B** that circles the wire. (The use of bold signifies a vector quantity.) The magnitude of **B** has a constant value, which is proportional to I, along an imaginary circle of radius r centered on the axis of the wire. Ampère and others showed that electric currents attract small bits of iron, and Ampère proposed a theory that electric currents are the source of magnetism.

Readers who have experimented with electromagnets, which can be created by wrapping an insulated wire around a nail and connecting the ends of the wire to a battery, have experienced Ampère's Law firsthand. In short, Ampère's Law expresses the relationship between the magnetic field and the electric current that produces it. This law, like most laws described in this book, has practical applications and is useful in the building and understanding of electromagnets, motors, generators, and transformers.

Ampère's Law is expressed in many forms, perhaps most famously with the integral calculus equation:

$$\oint_s \mathbf{B} \cdot \mathrm{d}\mathbf{s} = \mu_0 I_{\mathrm{enc}},$$

where **B** is the magnetic field. The integral is along the closed loop **s**. μ_0 is a magnetic constant known as the permeability of free space and equals $1.2566 \times 10^{-6}\ \mathrm{Wb} \cdot \mathrm{A}^{-1} \cdot \mathrm{m}^{-1}$. (Wb is an abbreviation for webers, a unit of magnetic flux, and A for amps.) I_{enc} is the current enclosed by the curve **s**. The formulation indicates that the line integral of the magnetic field around an arbitrarily chosen path is proportional to the net electric current enclosed by the path. Ampère's Law may be used to determine the magnetic field both inside and outside a long straight wire.

Notice that in Ampère's Law (also called Ampère's Circuital Law), the quantity $\oint_s \mathbf{B} \cdot d\mathbf{s}$ is independent of the radius of the closed path around the wire and is constant for the path **s** as long as the current is constant.

Scottish physicist James Clerk Maxwell (1831–1879) refined the law to better describe the relationship between magnetic fields and current in charging capacitors, an expression that is part of a set of equations known as Maxwell's Equations:

$$\oint_s \mathbf{B} \cdot \mathrm{d}\mathbf{s} = \mu_0 I_{\mathrm{enc}} + \frac{\mathrm{d}\Phi_{\mathrm{e}}}{\mathrm{d}t},$$

where Φ_{e} is the flux of the electric field through the surface. This Ampère-Maxwell Law can also be expressed in differential calculus form:

$$\nabla \times \vec{B} = \mu_0 \vec{J} + \mu_0 \varepsilon_0 \frac{\partial \vec{E}}{\partial t}$$

$\vec{J}$ is referred to as the current density. Maxwell also showed that a changing electric field is accompanied by a changing magnetic field, even in empty space.

Additional connections between magnetism and electricity were demonstrated by the experiments of American scientist Joseph Henry (1797–1878), British scientist Michael Faraday (1791–1867), and Maxwell. About a month after Ørsted described his findings on the effect of electric current on compass needles, French physicists Jean-Baptiste Biot (1774–1862) and Félix Savart (1791–1841) also studied the relationship between electrical current in wires and magnetism. The Biot-Savart Law, described in its own entry in this book, was also deduced by Ampère. Although Ampère's Law is completely general for steady currents and can be easily applied for simple current paths, such as for the field at the center of a single circular wire path, Ampère's Law is sometimes difficult to apply in practical computations for more complicated examples, such as those involving coils of wire. The Biot-Savart Law is related to Ampère's Circuital Law and can be used in situations in which alternative methods are desired for computing the magnetic field **B** that results from a current.

One week after Ampère became aware of Ørsted's discovery, he showed that two parallel currents attract each other if the currents are in the same direction and repel each other if the currents are in opposite directions.

André-Marie Ampère (1775–1836), French physicist who by 1825 had established the foundation of electromagnetic theory.

CURIOSITY FILE: Ampère believed that he had proven the existence of the soul and of God. • Ampère's father was executed by guillotine.

> [My father] never required me to study anything, but he knew how to inspire in me a desire to know.
>
> —André-Marie Ampère, quoted in James R. Hofmann's *André-Marie Ampère*

> I am going to take up mathematics again. I have some troubles at first, but when I have overcome the initial repugnance, I no longer want to leave the calculations. I still experience a great charm there when I can eliminate every other thought and occupy myself with it alone, absolutely alone.
>
> —André-Marie Ampère, quoted in James R. Hofmann's *André-Marie Ampère*

Ampère was born in Poleymieux-au-Mont-d'Or, near Lyon, France. Many scientific lawmakers in this book showed signs of genius at an early age, Ampère among them. According to the 1911 edition of the *Encyclopaedia Britannica*, "he took a passionate delight in the pursuit of knowledge from his very infancy, and is reported to have worked out long arithmetical sums by means of pebbles and biscuit crumbs before he knew the figures."

Michael O'Reilly and James Walsh write of Ampère's childhood in their 1909 book *Makers of Electricity*:

> The first marvelous faculty that began to develop in him was an uncontrollable tendency to arithmetical expression. Before he knew how to make figures, he had invented for himself a method of doing even rather complicated problems in arithmetic by the aid of a number of pebbles or peas. During an illness that overtook him as a child, his mother, anxious because of the possible evil effects upon his health of mental work, took his pebbles away from him.

Ampère is also said to have mastered most of all known mathematics by the time he was 12 years old. This may be an exaggeration, but he did start writing geometry treatises at this age, and he eventually became a professor of physics and chemistry at Bourg École Centrale at age 26, and a professor of mathematics at the École Polytechnique in Paris eight years later. His interests had always been varied, and when he was about 15 years old, he discovered a 20-volume French encyclopedia, which he read from start to finish.

Ampère was not a careful experimenter but had flashes of insight and was quick to understand the implications of observations made by others. Only a week after physicist Hans Christian Ørsted discovered that an electric current passing through a wire affects the motion of a nearby compass needle, Ampère wrote the first of many papers that gave a rather complete theory of these observations. In 1826, Ampère published *Memoir on the Mathematical Theory of Electrodynamic Phenomena, Uniquely Deduced from Experience*, his most famous paper on electricity and magnetism. The memoir described experiments as well as his mathematical derivations of the electrodynamic force law. The speed with which Ampère laid the foundation of electromagnetism led Maxwell to write in *A Treatise on Electricity and Magnetism*:

> We can scarcely believe that Ampère really discovered the law of action by means of the experiments which he describes.... He tells us himself that he discovered the law by some process which he has not shown us, and that when he had afterwards built up a perfect demonstration, he removed all traces of the scaffolding by which he had raised it.

Ampère's Law goes beyond the circuital formulation emphasized in the first part of this book entry, and Ampère also mathematically described the magnetic force between two electric currents. For example, he established that two electric currents attract one another when they move parallel to one another in the same direction, and they repel each other when they move in opposite directions. As early as 1820, he formulated a law of force between two current elements and gave a formula that related this force to the value of the currents and the relative orientations of the current-carrying wires. In particular, the force of the electric current between two wires will exhibit the inverse square law, which states that the force decreases with the square of the distance between two conductors. The force is proportional to the product of the two currents. These attractions and repulsions are different in nature from the attractions and repulsions of static electricity.

Ampère owes some of his early success to his father, a wealthy merchant who exposed young Ampère to a large library and encouraged his son to learn whatever he wished. Ampère promptly memorized entire encyclopedia articles, taught himself number theory, and worked through the early books of Euclid with no teacher. He learned Latin just so he could read the mathematical works of Leonhard Euler (1707–1783), whose papers were often in Latin.

James Hofmann writes in *André-Marie Ampère: Enlightenment and Electrodynamics*:

> With no formal education, Ampère's protective family circle encouraged him to adopt both the optimistic scientific outlook of the Enlightenment and a devotion to the Catholic faith. This combination of intellectual expectation and emotional spirituality produced a tension that became his most definitive characteristic.

When Ampère was in his late teens, his father was executed by guillotine during the French Revolution, which caused Ampère to withdraw from the world. For a year, he had little contact with friends. He later recalled that two factors helped lift him from his depression: a renewed interest in botany and *Corpus poetarum latinorum*, a book containing the works of Roman poets.

During this low point of his life, he also met Catherine-Antoinette Carron, who became his wife in 1799. The next four years were the happiest in his life. Sadly, Catherine-Antoinette died suddenly in 1803, a short time after the birth of their son Jean-Jacques, who eventually became a famous historian and philologist (historical linguist). His daughter, who was born later during Ampère's second marriage that ended in divorce, married one of Napoleon's lieutenants in 1827. Her husband's alcoholism, violence, and gambling led to police intervention and was a source of stress for Ampère.

After his great theories in the 1820s, unifying the fields of electricity and magnetism, his interest in creative science diminished. He spent the remainder of his life focusing on philosophy and how best to classify science and human knowledge. Through his life, Ampère was a religious man and believed that he had proven the existence of the soul and of God.

In his old age, financial problems became a major concern for Ampère. His sister incurred large debts in her attempts to maintain his household.

Ampère's health declined quickly in 1829, and he had been afflicted with severe bouts of bronchitis through much of the 1820s. Physicians treated him by applying leeches to his body. He finally died in Marseille, France, in 1836. In 1883, William Thomson (Lord Kelvin) honored him by proposing that the unit for current be called "amperes." Ampère's name is one of 72 names of prominent French scientists whom Gustave Eiffel placed on the Eiffel Tower (see "Coulomb's Law of Electrostatics" in part II).

O'Reilly and Walsh write of the importance of Ampère's genius:

> Few men of the nineteenth century are so interesting as André-Marie Ampère, who is... deservedly spoken of as the founder of the science of electrodynamics. Extremely precocious as a boy... he grew up to be a young man of the widest possible interests.... [Dominique] Arago has said of Ampère's discovery identifying magnetism and electricity that "the vast field of physical science perhaps never presented so brilliant a discovery, conceived, verified, and complete with such rapidity."

O'Reilly and Walsh also note that during the terrible period of the French Revolution, Ampère had some doubts with respect to religious truth, but later he became "one of the most faithful practical Catholics of his generation." Ampère seldom passed a day "without finding his way into a church, and his favorite form of prayer was the rosary."

FURTHER READING

Darrigol, Olivier, *Electrodynamics from Ampere to Einstein* (New York: Oxford University Press, 2000).

Hofmann, James, *André-Marie Ampère: Enlightenment and Electrodynamics* (New York: Cambridge University Press, 1996).

James, Ioan, *Remarkable Physicists: From Galileo to Yukawa* (New York: Cambridge University Press, 2004).

Maxwell, James Clerk, *A Treatise on Electricity and Magnetism*, (London: Macmillan, 1873),

O'Reilly, Michael, and James Walsh, *Makers of Electricity* (New York: Fordham University Press, 1909).

Williams, L. Pearce, "André-Marie Ampère," in *Dictionary of Scientific Biography*, Charles Gillispie, editor-in-chief (New York: Charles Scribner's Sons, 1970).

INTERLUDE: CONVERSATION STARTERS

A law is not a cause; yet it is more than merely a description. It is true because it is beautiful and simple; yet it is never quite true at all. "That is the same with all of our laws—they are not exact. There is always an edge of mystery, always a place where we have some fiddling around to do yet."

—James Gleick, *Genius: Life & Science of Richard Feynman*, quoting Richard Feynman's *The Character of Physical Law*

Science must be testable in principle, but that is not necessarily the same thing as testable in practice, given current technological limitations.... It is not uncommon for decades to go by before theories in physics are confirmed. In some cases, such as the atomic theory, it has taken centuries.

—Tom Siegfried, "A Great Unraveling," *New York Times Book Review*, September 17, 2006

There are many people ... who would be perfectly able to argue the value of having read Shakespeare but would see no usefulness at all in being aware of chemical laws.... While it's true that such laws might not make it possible to increase your IRA earnings, they ... describe the universe we live in and reveal the mysteries still contained in it.... If you are familiar with both the First and Second Law of Thermodynamics, you will be much less likely to waste money investing in a perpetual motion machine.

—Jay Ingram, *The Barmaid's Brain and Other Strange Tales from Science*

We have no reason to suppose any physical law can be more accurate than 1 part in 10^{120}. Beyond that we can expect the law to break down and become fuzzy.

—Paul Davies, "Laying Down the Laws," *New Scientist*

OHM'S LAW OF ELECTRICITY

Germany, 1827. The current flow through a conductor is proportional to the voltage and inversely proportional to the resistance.

CROSS REFERENCE: ROWLAND'S LAW, POISEUILLE'S LAW OF FLUID FLOW, FOURIER'S LAW OF HEAT CONDUCTION, JAMES CLERK MAXWELL, HUMPHRY DAVY, ISAAC NEWTON, JOSEPH HENRY, AND THE BERNOULLI FAMILY.

In 1827, German composer Ludwig van Beethoven and English poet William Blake died. The Baltimore & Ohio (B&O) Railroad was incorporated and become the first railroad in America to offer commercial transportation of both freight and people.

Ohm's Law may be represented in various forms. One familiar expression of the law states that the steady electric current I in a circuit is proportional to the constant voltage V (or total electromotive force) across a resistance and inversely proportional to the value R of the resistance:

$$\boxed{I = \frac{V}{R}}$$

Ohm's experimental discovery of the law in 1827 suggested that the law held for a number of different materials. As made obvious from the equation, if the potential difference V (in units of volts) between the two ends of a wire is doubled, then the current I in amperes also doubles. For a given voltage, if the resistance doubles, the current is decreased by a factor of 2. Resistance is given in units of ohms. The resistance of the material is usually constant over large ranges of voltage and current at a fixed temperature.

The equation variables may be modified so that it can also apply to alternating-current (AC) circuits. For example, Ohm's Law for an AC circuit that consists of an alternating voltage source and a resistor may also be expressed as $I = V/R$, where V and I are now the root mean square (rms), or effective, values of voltage and current. The potential difference across the resistor varies sinusoidally and is in phase with the current. In other words, the current and voltage reach a maximum and minimum value at the same time.

Also with additional modifications, Ohm's Law may be reformulated for application to magnetomotive forces, which involve phenomena that give rise to magnetic fields. More particularly, the law can be modified so that it applies to the constant ratio of the magnetomotive force (mmf) to magnetic flux in magnetic circuits. For example, we can express the magnetomotive version of Ohm's Law as mmf = $\Phi\mathfrak{R}$, where mmf is the magnetomotive field force, Φ is the field flux, and Я is the reluctance. Magnetic reluctance may be thought of as the resistance of a material to a magnetic field. However, unlike the case for Ohm's Law applied to electrical circuits, the reluctance of a material to a magnetic flux changes with the concentration of flux going through it, which makes mmf = ΦЯ nonlinear. (An analogous situation would be one in which an electrical resistor changed resistance as the current through it varied.) The equation mmf = ΦЯ is sometimes referred to as Rowland's Law after Henry Augustus Rowland (1848–1901), the first physics professor at Johns Hopkins University and a brilliant experimentalist who conducted important work on electricity and magnetism.

An "Ohm's Law" for acoustics exists that involves sound pressure, acoustic impedance of air, particle velocity, and sound intensity. When applied to a sound wave traveling through air, particle velocity refers to the speed of an air molecule as it moves back and forth in the direction that the sound wave is traveling as the wave passes.

When the term "conductor" is used in discussions of Ohm's Law, it often refers to the circuit element across which a voltage is to be measured. Resistors are conductors that limit the passage of electricity by some amount. For example, a resistor with a high value of resistance, such as a resistance above 20 megaohms, is a poor conductor. In modern times, resistors are often manufactured from nonmetals that obey Ohm's Law. Both metallic and nonmetallic resistors are sometimes called ohmic devices, because they obey Ohm's Law within a range of voltage, currents, and temperatures. Scottish physicist James Clerk Maxwell (1831–1879) and Scottish mathematician George Chrystal (1851–1911) later showed that Ohm's Law is valid even when the currents are so powerful as to almost fuse a conducting wire.

An ohmic material is generally defined as a material for which the resistance in the expression $I = V/R$ is independent of I and V. In other words, a conductor obeys Ohm's Law only if a plot of V versus I is linear. Thus, the relationship $R = V/I$ by itself is not a statement of Ohm's Law and is generally true as a definition of the resistance in a conductor whether or not the conductor obeys Ohm's Law. A lightbulb filament is generally considered to be non-ohmic because the V versus I plot is not linear.

Although not part of Ohm's initial law, we may state the temperature dependence of resistance that scientists often find in conductors. As background, when the temperature of a conductor increases, the collisions of atoms increase. According to the classical model of electric conduction, resistivity is inversely proportional to the mean free path of the electrons traveling in the conductor between collisions with atoms. As temperature increases, the atoms vibrate more, and the mean free path decreases, thus increasing resistivity. This means that the electrons have a decreased ability to flow without meeting any interference as a result of various collisional processes within the wire as temperature increases. Today, we know that although the classical theory of conduction is successful for predicting Ohm's Law, the theory is usually replaced by more modern theories of conduction based on quantum mechanics, which more accurately explain the temperature dependence of resistivity.

The resistance of an ohmic substance depends on temperature in the following way:

$$R = (L/A) \times \rho = (L/A) \times \rho_0[\alpha(T - T_0) + 1],$$

where L is the constant length of the conductor, A is the cross-sectional area of the conductor, T is its temperature, T_0 is a reference temperature, ρ is the resistivity, and ρ_0 and α are constants that are characteristic of the material. Table 7 gives some sample values for resistivity ρ and temperature coefficient α.

Many metals exist for which the resistivity is zero below a critical temperature T_c. This phenomenon, called superconductivity, was discovered in 1911 by Dutch physicist and Nobel laureate Heike Kamerlingh Onnes (1853–1926). For example, $T_c = 1.2$°K for aluminum. The phenomenon of superconductivity can be understood with the aid of quantum mechanics. Modern research with more practical, higher temperature superconductors has led Antony Anderson in *New Scientist* to humorously predict the demise of the ohm in the future when superconductivity may be commonplace:

> Soon we will banish resistance from our machines, every electrical contact will be perfect, and the standard ohm will lose pride of place beside the standard volt and ampere.... Superconductivity may usher in a wattless wonderland, but I do hope that we leave room for resistance here and there. The occasional ohm might come in useful during cold weather, especially if wrapped up in an [electric] blanket!

Ohm's Law has relevance in determining the dangers of electrical shocks on the human body. Generally, the higher the current flow, the more dangerous the shock is. The amount of current is equal to the voltage applied between two points on the body, divided by the electrical resistance of the body. Precisely how much voltage a person can experience and survive depends on the total resistance of the body, which varies from person to person and may depend on such parameters as body fat, fluid intake, skin sweatiness, and how and where contact is made with the skin. Death

TABLE 7 Resistivity and Temperature

Material	Resistivity ρ at 20°C (Ω·m)	Temperature Coefficient α at 20°C (per °C)
Silver	1.6×10^{-8}	3.8×10^{-3}
Copper	1.7×10^{-8}	3.9×10^{-3}
Aluminum	2.8×10^{-8}	3.9×10^{-3}
Tungsten	5.5×10^{-8}	4.5×10^{-3}
Iron	10×10^{-8}	5.0×10^{-3}
Lead	22×10^{-8}	4.3×10^{-3}
Mercury	96×10^{-8}	0.9×10^{-3}
Carbon	$3{,}500 \times 10^{-8}$	-0.5×10^{-3}
Germanium	0.45	-48×10^{-3}
Silicon	640	-75×10^{-3}
Wood	10^{8} to 10^{14}	
Glass	10^{10} to 10^{14}	
Hard rubber	10^{13} to 10^{16}	

can result from heat failure or asphyxia, which is caused by the electrical effect on either the muscles of the chest or the respiratory center in the brain.

Because wet skin can have a thousand-ohm resistance or even lower, being wet increases the chance of electrocution. For comparison, dry skin has an electrical resistance of about 500,000 ohms. This kind of information is of interest to executioners, as indicated in Michael S. Morse's paper titled, "Report on Findings and Recommendations, Prepared Following visit to Florida State Penitentiary at Starke, FL":

> Mr. Wiechert and I conducted several tests on the electrocution equipment. Initially, tests were run to measure voltage and current produced by the system. A bucket filled with water into which

> electrodes were placed was used to provide an appropriate resistive load. . . . The resistive load was calculated to be between 200 and 250 ohms which is in the range of typical values for a human as observed during an execution.

Electrical resistance is used today to monitor corrosion and material loss in pipelines. For example, net change in the resistance in a metal wall may be attributable to metal loss. A corrosion-detection device may be permanently installed to provide continuous information, or the device may be portable to gather information as needed.

See "Poiseuille's Law of Fluid Flow" below for a description of a law similar to Ohm's Law but with application to the flow of liquids through a cylindrical tube of constant cross section. With Poiseuille's Law, a pressure drop corresponds to voltage, and liquid flow rate corresponds to current.

Georg Ohm (1787–1854), German physicist famous for his work on voltage and resistance in circuits.

CURIOSITY FILE: Although Georg Ohm discovered one of the most fundamental laws in the field of electricity, his work was ignored by his colleagues, and he lived in poverty for much of his life. • His harsh critics called his work a "web of naked fancies."

> I feel clearly that only that which is simple can be great.
> —Georg Ohm, quoted in Kenneth Caneva's "Georg Ohm"

> Ohm's work stands alone, and, reading it at the present time, one is filled with wonder at his prescience, respect for his patience and prophetic soul, and admiration at the immensity and variety of ground covered by his little book, which is indeed his best monument.
> —Thomas Lockwood, 1891 preface to Ohm's "The Galvanic Circuit Investigated Mathematically"

Georg Ohm was born in Erlangen, Bavaria, which is now part of Germany. He was the son of Johann Ohm, a locksmith by trade, but also highly intellectual and self-educated in various scientific fields. Of Ohm's seven siblings, only two survived—brother Martin, who eventually became a famous mathematician, and his sister Elizabeth. Johann taught his two sons mathematics, science, and philosophy. Because Johann was such a good educator in mathematics, various professors compared the Ohms to the Bernoulli family, an extraordinary Swiss family that contained eight outstanding mathematicians within three generations.

In 1805, Georg Ohm enrolled in the University of Erlangen, where he focused more on having a good time with friends than on serious study. Johann was so angry at his son that he forced Ohm to leave the university, where he was wasting his time, and sent him to Switzerland, where the young Ohm became a mathematics teacher in a school in Gottstadt bei Nydau. Ohm continued private studies of mathematics, which allowed him to receive a doctorate from Erlangen in 1811, where he also began to teach mathematics. However, the teaching position paid so poorly that Ohm was in poverty. In 1813, he took a post at a lower quality but higher paying school in Bamberg. He worked unhappily at this school until it closed in 1816.

In 1817, Ohm started to teach mathematics and physics at Jesuit Gymnasium of Cologne. The Gymnasium was a relatively good school at which to teach, but its quality gradually deteriorated during Ohm's period of employment. Ohm would eventually reach his scientific pinnacle mostly as result of his private studies, reading the texts of the leading French mathematicians and physicists, and experiments he conducted in isolation to satisfy his curiosity. As Kenneth Caneva says in the *Dictionary of Scientific Biography*,

> Overburdened with students, finding little appreciation for his conscientious efforts, and realizing that he would never marry, he turned to science both to prove himself to the world and to have something solid on which to base his petition for a position in a more stimulating environment.

In 1825, Ohm published his first paper describing his experiments that showed how the electromagnetic force produced by a wire decreases as the length of the wire increases. Around this time, he also came to believe that the current through a conductor is proportional to the potential difference applied across the material. In 1826, Ohm published two papers that provided a mathematical description of conduction in circuits modeled on Fourier's Law of Heat Conduction (see entry above).

In 1827, Ohm's famous law appeared in his book *Die galvanische Kette, mathematisch bearbeitet* (The Galvanic Circuit Investigated Mathematically). Here, he discussed his theory of electricity and provided a mathematical introduction to the entire field. As described above, the equation

$I = V/R$ is known today as Ohm's Law, and it states that the amount of steady current I through a material is proportional to the voltage V across the material divided by the electrical resistance R of the material. (Humphry Davy had also investigated the "conducting" powers of wires for various wire lengths, but it was Ohm who clarified the relationship with a mathematical statement.)

Ohm had believed that his publications would entice a prestigious university to offer him a position, but no offers were made. In fact, many of Germany's physicists did not appreciate his work, which appeared to them to represent an overly mathematical approach to physics and to the expression of the laws of nature.

Ohm's Law was so poorly received and Ohm's emotions scraped so raw by the scientific community's general lack of enthusiasm that in 1828 he resigned his post at Jesuit's College of Cologne, where he was still professor of mathematics. One critic reviewed Ohm's book, saying that its "sole effort is to detract from the dignity of nature." The German Minister of Education said that Ohm was "a professor who preached such heresies was unworthy to teach science." This is reminiscent of the time Isaac Newton went nearly mad in an exchange on his theory of colors with several English Jesuits who had criticized Newton's experiments. The correspondence between Newton and his critics lasted for quite some time until Newton had a nervous breakdown.

In 1833, Ohm accepted a professorship position at the Polytechnic School of Nüremberg. His research started to gain broad acceptance outside of Germany, and American physicist Joseph Henry (1797–1878) suggested that Ohm had added great clarity and insight with respect to electrical circuits. Ohm's work was finally recognized by the Royal Society with its award of the Copley Medal in 1841.

In 1843, Ohm asserted his fundamental principle of physiological acoustics, which focused on the way in which the human ear perceives combinations of tones. However, Ohm had made certain mathematical assumptions that were not justified, which triggered an angry dispute with the physicist August Seebeck (1805–1849).

In 1852, two years before his death, Ohm finally achieved his lifelong goal of being appointed to the chair of physics at the University of Munich. Caneva sums up Ohm's life:

> The resulting inwardness of Ohm's character and the highly intellectualized nature of his ideals of personal worth were an essential aspect of the man who would bring the abstractness of mathematics into the hitherto physical and chemical domain of galvanic electricity.

Today, Ohm is honored by the ohm, a unit of electrical resistance. The ohm corresponds to a conductor in which a current of 1 ampere is produced by a potential of 1 volt across its terminals. The symbol for the ohm is the Greek letter omega, written as Ω. Units of ohms, kilohms (10^3 Ω), and megaohms (10^6 Ω) are all commonly used by electrical engineers. As an additional honor, a lunar crater with a diameter of 64 kilometers was named after Ohm and approved in 1970 by the International Astronomical Union General Assembly.

In 1930, biographer Rollo Appleyard gave a fitting eulogy for Ohm in *Pioneers of Electrical Communication*:

> A century ago, the science and practice of electrical measurement...hardly existed. With a few exceptions, ill-defined expressions relating to quantity and intensity, combined with immature ideas of conductivity and derived circuits, retarded the progress of quantitative electrical investigations. Yet, amidst this confusion, a discovery had been made that was destined to convert order out of chaos, to convert electrical measurement into the most precise of all physical operations, and to aid almost every other branch of quantitative research. This discovery resulted from the arduous labours of Georg Simon Ohm.

Anderson leaves us with an upbeat tribute to resistance and Dr. Ohm:

> Electrical resistance in cables and conductors can lead to burnt varnish, smoke, sudden short circuits and melted metal; but without the benefit of the damping provided by resistance, without even the vestige of Joule heating...our machines might be superefficient, but they would be afflicted with the mechanical equivalent of Parkinson's disease.... Without resistance, our electric blankets, kettles, and incandescent lamp bulbs would be useless.

FURTHER READING

Anderson, Antony, "Spare a Thought for the Ohm," *New Scientist*, May 7, 1987; see www.antony-anderson.com/ohm.htm.

Appleyard, Rollo, *Pioneers of Electrical Communication* (London: Macmillan & Company, 1930).

Bueche, Frederick, *Introduction to Physics for Scientists and Engineers* (New York: McGraw-Hill, 1975).

Caneva, Kenneth, "Georg Ohm," in *Dictionary of Scientific Biography*, Charles Gillispie, editor-in-chief (New York: Charles Scribner's Sons, 1970).

Lockwood, Thomas, preface to "The Galvanic Circuit Investigated Mathematically" by Georg Ohm, Berlin, 1827; translated by William Francis (New York: D. Van Nostrand Company, 1891).

Morse, Michael S., "Report on Findings and Recommendations, Prepared Following Visit to Florida State Penitentiary at Starke, FL," Florida Corrections Commission, April 8, 1997; from www.fcc.state.fl.us/fcc/reports/methods/emappa.html (website no longer accessible).

INTERLUDE: CONVERSATION STARTERS

Ohm never described his experimental results in the compact and simple form $i \propto V$, or $V = iR$. This was not done until 1849, when Gustav Kirchhoff (1824–1887) "saw through" the experimental complications and understood the macroscopic phenomenon of electric conduction in essentially modern terms. Nevertheless, any equation which relates current to voltage in linear fashion is called Ohm's law.

—Robert M. Eisberg and Lawrence S. Lerner, *Physics*

Consider Andrei Linde's suggestion that, rather than there being only one universally valid set of physical laws, there are many different universes, each with its own laws of nature, each randomly different from the other. . . . Is the assumption that there is any unique universal physical law another childish dream from which we must awaken? . . . If random, they cannot be God's thoughts, because they are not the product of any thought, much less that of God.

—Peter Pesic, "Bell & the Buzzer: On the Meaning of Science," *Daedalus*, Fall 2003

A "law of nature" is one of the concepts that slips through your fingers the more you try to grasp it. The most that can be said about a physical law is that it is a hypothesis that has been confirmed by experiment so many times that it becomes universally accepted. There is nothing natural about it, however: it is a wholly human construct.

—*New Scientist*, "Editorial: Breaking the Laws." April 29, 2006 (unsigned)

Science, [Freeman] Dyson says, is an inherently subversive act. Whether overturning a longstanding idea (Heisenberg upending causality with quantum mechanics, Gödel smashing the pure platonic notion of mathematical decidability) or marshaling the same disdain for received political wisdom (Galileo, Andrei Sakharov), the scientific ethic—stubbornly following your nose where it leads you—is a threat to establishments of all kinds.

—George Johnson, "Dancing with the Stars," *New York Times Book Review*

He studied scientific truths, then became upset even more by the apparent cause of their temporal condition.... The time spans of scientific truths are an inverse function of the intensity of scientific effort. Thus the scientific truths of the twentieth century seem to have a much shorter life-span than those of the last century because scientific activity is now much greater.... What shortens the life-span of the existing truth is the volume of hypotheses offered to replace it.... And what seems to be causing the number of hypotheses to grow in recent decades seems to be nothing other than scientific method itself.

—Robert Pirsig, *Zen and the Art of Motorcycle Maintenance*

GRAHAM'S LAW OF EFFUSION

Scotland, 1829. The rates of effusion of two gases are inversely proportional to the square roots of the gases' densities. At equal pressure and temperature, less massive gases will effuse more rapidly than more massive gases.

CROSS REFERENCE: JOHN DALTON, MICHAEL FARADAY, JOHANN DÖBEREINER, AND AVOGADRO'S GAS LAW.

> In 1829, slavery was abolished in Mexico, and the first U.S. patent on a typewriter was granted to William Burt of Detroit. The device was called a "typographer." Greece received its autonomy from the Ottoman Empire.

Graham's Law of Effusion states that the rate of effusion of a gas is inversely proportional to the square root of the mass of its particles. (The term "effusion" is defined shortly.) This formula can be written as

$$\frac{R_1}{R_2} = \sqrt{\frac{M_2}{M_1}},$$

where R_1 is the rate of effusion of one gas, R_2 is the rate of effusion for a second gas, M_1 is the molar mass of the first gas, and M_2 is the molar mass of the second gas. Graham's Law applies for both effusion and diffusion. Note that because equal volumes of different gases contain the same number of particles (see "Avogadro's Gas Law," above), the number of moles per liter at a given temperature and pressure is constant. Therefore, the density of a gas is directly proportional to its molar mass. (The term "mole" is also defined in "Avogadro's Gas Law.")

Effusion is a process in which individual particles flow through a hole so small that the particles go through one at a time. The rate of effusion depends on the molecular weight of the gas. For example, gases like hydrogen with a low molecular weight effuse more quickly than do heavier particles because the low-weight particles are generally moving at higher speeds. We can understand this phenomenon by imagining two different gas particles with the same kinetic energy, $E = \frac{1}{2}mv^2$. If two gases having the same energy, the light particle is moving faster. For this reason, a balloon filled with oxygen will deflate more slowly than one filled with hydrogen.

Diffusion is a term usually used to describe the spread of one substance through a second substance, such as perfume molecules diffusing through the atmosphere. As is the case for effusion, diffusion is faster for lighter molecules than for heavier ones.

Thomas Graham's first major paper in 1829 was actually concerned with the diffusion of gases, and he reported that the relative rates of effusions of gases are comparable to the relative rates of diffusion. Although his 1829 paper "A Short Account of Experimental Researches on the Diffusion of Gases Through Each Other, and Their Separation by Mechanical Means" contained the essentials of Graham's Law, a subsequent paper, "On the Law of Diffusion of Gases," published in 1833 established the principle with greater clarity. Graham wrote in the 1833 paper:

> The diffusion or spontaneous intermixture of two gases in contact is effected by an interchange in position of indefinitely minute volumes of gases, which volumes are not necessarily of equal magnitude, being, in the case of each gas, inversely proportional to the square root of the density of that gas; [that is] diffusion takes place between the ultimate particles of gases, and not between sensible masses.

Graham suggested that his law could be used to more accurately determine the specific gravity of gases than by other available means. When Graham measured the effusion of gases through a tiny hole in a metal plate, he found that the velocities of flow were inversely proportional to the square roots of the densities.

As an example, let us determine the relative rates of effusion of the gases hydrogen (H_2, molecular weight 2) and nitrogen (N_2, molecular weight 28):

$$\frac{R_{H_2}}{R_{N_2}} = \sqrt{\frac{28}{2}} = 3.74$$

This means that hydrogen gas diffuses or effuses about 3.74 times faster than nitrogen molecules. By observing the methodology in this example, you can see that we can also use Graham's Law to determine an approximate value for the molecular weight of an unknown gas if we know the relative rates of effusion and the molecular weight of one of the gases.

Thomas Graham (1805–1869), Scottish chemist famous for his gas law and his work in colloid chemistry.

CURIOSITY FILE: Graham invented the term "colloid" and created a means of dialysis to separate colloids from crystalloids. • In one of his lesser known papers, Graham describes his work for assessing the purity of commercially available coffee. • For many years, the German Colloid Society offered the Thomas Graham prize for exceptional achievement in colloid science. The prize included a memorial coin. • Graham's Law helped the United States make the atomic bomb that was dropped on Japan.

> In nature, there are no abrupt transitions, and the distinctions of class are never absolute.
>
> —Thomas Graham, describing crystalloid and colloid states of materials, in "Liquid Diffusion Applied to Analysis"

> Thomas Graham spent his life in reading the book of Nature, and giving to mankind knowledge of the truths which he found there. His greatness is to be measured not merely by the amount and importance of the knowledge which he thus gave, but even more by the singleness and strength of purpose with which he devoted his whole life to labors of experimental philosophy.
>
> —A. W. Williamson, "The Late Professor Graham," *Nature*, November 4, 1869

> Thomas Graham was a shy, retiring man, most of whose life was spent in his laboratory. . . . When he came into the lecture theatre [in Glasgow], to deliver his first [chemistry] lecture to a large audience, he looked around in dismay and fled.
>
> —"The Victorian Age, Part Two," in *Cambridge History of English and American Literature (1907–1921)*

Thomas Graham was born in Glasgow, Scotland. His father, a textile manufacturer, had always wanted Graham to become a minister in the Church of Scotland and opposed Graham's growing interest in chemistry. Luckily for Graham, his mother and sister were supportive of his interest in science, which helped Graham achieve his scientific dreams.

In 1814, he went to the high school at Glasgow, and for the next five years he was never absent from school. Graham entered the University of Glasgow in 1819 at the age of 14 and received an M.A. degree in 1826 before continuing his studies for two years at Edinburgh. According to his obituary in *Nature*:

> Young Graham's mother seems to have been his guardian angel, sympathizing with his hopes and his sorrows; and certainly his feelings towards her would have been very inadequately described

> by that frigid word [respect]. While studying at Edinburgh he earned, for the first time in his life, some money by literary work, and the whole sum was expended in presents to his mother and sisters.

In 1830, he was appointed professor of chemistry at the Andersonian University, Glasgow. His mother, who was on her deathbed, lived to hear the good news of his professorship.

In 1834, Graham was elected a fellow of the Royal Society, and a few years later he became professor of chemistry at the University College, London. George Kaufman writes in the *Dictionary of Scientific Biography*:

> His time was then fully occupied in teaching, writing, advising on chemical manufactures, and investigating fiscal and other questions for the government.... With the death of John Dalton in 1844, Graham was left as the acknowledged dean of English chemists, the successor of Joseph Black, Joseph Priestly, Henry Cavendish, William Wollaston, Humphry Davy, and John Dalton.... As a lecturer, Graham was somewhat nervous and hesitant.

Graham received the Royal Medal of the Royal Society in 1837 and in 1853, and the Copley Medal of the Royal Society in 1862. In 1837, he was also appointed professor of chemistry at the London University. He stayed at this university until 1855, at which point he succeeded Sir John Herschel as Master of the Royal Mint.

Diffusion was the area of research on which Graham focused, and his measurements of the relative velocities of particles in gases or liquids are considered to be his most important work. Graham's interest in gas diffusion was stimulated by the work of German chemist Johann Döbereiner (1780–1849), who observed that hydrogen gas diffused out of a thin crack in a glass bottle faster than the surrounding air diffused in to replace it. Graham measured the rate of diffusion of gases through plaster plugs, through very fine tubes, and through small holes. In his most famous experiment, Graham allowed hydrogen to escape through a very small hole in a plate of platinum. He performed the same experiment with oxygen and determined that each hydrogen particle escaped through the plate four times as fast as each oxygen particle.

In 1829, Graham submerged a glass cylinder, with an open top and bottom, in a glass of water in order to study the diffusion of gases and, in particular, the rate at which two gases mix. One of the cylinders was first plugged with plaster that had holes sufficiently large to allow gas flow in and out of the cylinder. Next, he filled the cylinder with hydrogen (H_2) gas, and he found that the water level in the cylinder rose slowly because the H_2 molecules inside the cylinder diffused through the porous plaster more rapidly than the outside air molecules entered the cylinder. By studying the rate at which the water level in this cylinder changed, Graham discovered the rate at which different gases mixed with air. In particular, Graham found that the rates at which gases diffuse is inversely proportional to the square root of their densities. As mentioned above, Graham obtained similar results when he studied gas effusion rates, by using a plate with a pinhole and measuring the rate at which the gas escaped into a vacuum. Here, too, he found that the rate of gas effusion was inversely proportional to the square root of either the density or the molecular weight of the gas.

In the 1850s, Graham studied methods for identifying inappropriate substances that sellers might be mixing with coffee in order to cheat consumers. In 1857, he published "Report on the Mode of Detecting Vegetable Substances Mixed with Coffee for the Purpose of Adulteration" in the *Journal of the Chemical Society of London*. Today, techniques such as infrared spectroscopy and coffee "fingerprints" have been used as tools for classification, authentication, and quality assessment for coffee. For example, several bad coffee mills have been closed for annually using 20,000 kilograms (20 metric tons) of peat for adulteration of coffee.

Graham also studied diffusion of substances in solution and discovered that some apparent solutions were actually suspensions of particles that were too large to pass through a parchment filter. In 1861, he coined the word *colloid* for gluelike materials that diffuse very slowly through porous membrane, and today Graham is considered the father of colloid chemistry. During his research, he divided particles into two classes: crystalloids such as salt, which diffuse quickly, and substances such as starch, gum, and gelatin that diffuse slowly and do not form crystals. Graham wrote in an 1861 *Philosophical Transactions of the Royal Society*:

> As gelatine appears to be its type, it is proposed to designate substances of the class as *colloids* (from Greek κόλλα, meaning glue), and to speak of their peculiar form of aggregation as the *colloidal condition of matter*. Opposed to the colloidal is the crystalline condition. Substance affecting the latter form will be classed as *crystalloids*.... Fluid colloids appear to have always a pectous [curdled] modification; and they often pass under the slightest influences from the first into the second condition.... The colloidal is, in fact, a dynamical state of matter; the crystalloid being the statical condition.

Today, we generally consider colloids as solutions in which the dispersed particles are between 10^{-7} and 10^{-4} centimeters in diameter, and these particles cannot generally be separated by filtration or gravity alone. Graham used the term "dialyzer" for the mechanism that he developed to separate colloids (which dialyzed slowly) from crystalloids (which dialyzed rapidly). Through dialysis, Graham employed various membranes for separating colloids from water and from substances readily dissolved in water, including salts and sugars.

Colloids were an active area of interest even before Graham coined the word. For example, in 1856, British physicist and chemist Michael Faraday (1791–1867) made the first systematic study of colloidal gold and suggested various factors responsible for the stability of these dispersions. Faraday's colloidal gold, which he called *activated gold*, was a suspension of submicrometer-sized particles in water. The liquid became a gorgeous red color if it contained particles smaller than 100 nm in diameter.

Colloidal gold had been known since ancient times, when it was used to stain glass; however, Faraday was one of the first to conduct serious research into the nature of the colloid. He was also the first to determine that the brilliant color resulted from the minute size of the gold particles in the colloid.

Returning to Graham, he explained the differences between three phosphoric acids and established the concept of polybasic compounds. For example, he elucidated the difference between phosphoric acid (modern notation: $3H_2O{\cdot}PO_5$), pyrophosphoric acid ($2H_2O{\cdot}P_2O_5$), and metaphosphoric acid ($H_2O{\cdot}P_2O_5$). Graham also determined that several phosphate salts of sodium exist, including Na_3PO_4, Na_2HPO_4, and NaH_2PO_4.

Graham's Law of Effusion allows scientists to make many practical discoveries. For example, as alluded to above, it is possible to determine the molecular mass of an unknown gas in the following manner. Imagine a device with two chambers. A mystery gas is inserted on one side of the chamber and is allowed to flow through the orifice to a vacuum chamber, via the process of effusion, until the pressure on the two sides is the same, as determined by pressure gauges in each chamber. The time required for the pressure equilibrium to occur is recorded for the mystery gas. The next step is to perform the same experiment for a known gas, such as nitrogen. Given the molecular mass of nitrogen, its rate of effusion, and the rate of effusion of the mystery gas, we may use Graham's Law to determine the molecular mass of the mystery gas.

Graham's Law had particularly practical applications in the 1940s, when it was used in nuclear reactor technology to separate radioactive gases that had different diffusion rates due to the molecular weights of the gases. In particular, a diffusion chamber, which was several hundred yards in length, was used to separate two isotopes of uranium, U-235 and U-238.

These were chemically reacted with fluorine to produce the gas uranium hexafluoride. The less massive uranium hexafluoride molecules containing fissionable U-235 would travel down the chamber slightly faster than the more massive molecules containing the U-238.

During World War II, this separation process was important because the United States sought to develop an atomic bomb, which required the isolation of U-235 for the nuclear fission chain reaction. To separate U-235 and U-238, the government built a gaseous diffusion plant in Clinton, Tennessee, at the cost of $100 million. (This was an especially huge cost in 1940s dollars!) The plant used diffusion through porous barriers and processed the uranium for the Manhattan Project, which allowed the United States to produce the uranium used in the atomic bomb dropped on Japan in 1945.

In order to perform the isotope separation, the gaseous diffusion plant required 4,000 stages in a space that was one-half mile long and six stories high. The porosity of the various barriers had to be high in order to sustain high flow rates and sufficiently durable so as to not be dissolved by the highly corrosive hexafluoride. I wonder what Graham would have thought if he could have known that roughly seventy-five years after his death, his simple law directly contributed to the demise of more than 100,000 people through the use of just two bombs.

At 9 o'clock in the evening of Thursday, September 16, 1869, Graham's spirit effused into the afterlife. He died at his house, No. 4 Gordon Square, in the London Borough of Camden. Today, Graham is honored by a bronze statue unveiled in Glasgow in 1872. The statue sits in George Square, Glasgow, and its arm rests on a book, the cover of which shows experimental equipment. The statue was paid for by wealthy industrial James Young, who had been a student of Graham's.

For those readers who wish to study photos and learn more about Graham's statue, see Ray McKenzie's *Public Sculpture of Glasgow*. For those who want additional background that justifies Graham's title as the "father of dialysis," see the readable introduction in Garabed Eknoyan and colleagues' *History of Nephrology 2*.

As for other applications of Graham's Law, James Trefil in *The Nature of Science* notes that

> A rather surprising application of Graham's law is to the construction of spacecraft on which humans will be spending long periods of time. . . . Given enough time, air will leak through the materials of which the spacecraft's hull is made, just as it leaks out of a birthday balloon. . . . [This will be solved] by providing a means of generating gases on board to replace those lost to the vacuum of space.

FURTHER READING

Eknoyan, Garabed, and the International Association for the History of Nephrology, *History of Nephrology 2* (New York: Karger Publishers, 1997).

Graham, Thomas, "Liquid Diffusion Applied to Analysis," *Philosophical Transactions of the Royal Society* (London), 151: 183–224, 1861.

Kaufman, George, "Thomas Graham," in *Dictionary of Scientific Biography*, Charles Gillispie, editor-in-chief (New York: Charles Scribner's Sons, 1970).

McKenzie, Ray, *Public Sculpture of Glasgow* (Liverpool, U.K.: Liverpool University Press, 2001).

Trefil, James, *The Nature of Science: An A-Z Guide to the Laws and Principles Governing Our Universe* (New York: Houghton Mifflin Company, 2003).

Ward, A. W., and A. R. Waller, editors, *The Cambridge History of English and American Literature* (New York: G. P. Putnam's Sons, 1907–1921); see www.bartleby.com/cambridge/.

Williamson, A., "The Late Professor Graham," *Nature*, 1(1): 20–22, November 4, 1869; see www.nature.com/nature/first/professorgraham.html.

INTERLUDE: CONVERSATION STARTERS

From the intrinsic evidence of his creation, the Great Architect of the Universe now begins to appear as a pure mathematician.

—James Hopwood Jeans, *The Mysterious Universe*, 1930

Physics makes progress because experiment constantly causes new disagreements to break out between laws and facts, and because physicists constantly touch up and modify laws in order that they may more faithfully represent the facts.

—Pierre Duhem, *The Aim and Structure of Physical Theory*, 1962

Modern science is a newcomer, barely four hundred years old. Though indebted in deep ways to Plato, Aristotle, and Greek natural philosophy, the pioneers of the "new philosophy" called for a decisive break with ancient authority. In 1536, Pierre de La Ramee defended the provocative thesis that "everything Aristotle said is wrong."

—Peter Pesic, "Bell and the Buzzer: On the Meaning of Science," *Daedalus*, Fall, 2003

Some of the scientists most closely involved, and some of the most observant philosophers of science, have taken the view that the laws of nature were: invented by man (Einstein, Bohr, Popper); not invented by man (Planck);

expressions of a real underlying order in the world (Einstein); working models justified only by their utility (von Neumann, Feynman);...steps on the road toward complete understanding (Feynman, Deutsch); steps on a road that has no end (Born, Popper, Kuhn)....

—Michael Frayn, *The Human Touch*

Scientists use *theory* in one way, the public another—and opponents of evolution have expertly exploited this disconnect....For truly solid-gold, well-established science, let's stop using the word *theory* entirely. Instead, let's revive much more venerable language and refer to such knowledge as "law."

—Clive Thompson, "A War of Words," *WIRED*

FARADAY'S LAWS OF INDUCTION AND ELECTROLYSIS

π *England, 1831 and 1833.* Induction Law: A changing magnetic field produces an electric field. Electrolysis Law: During electrolysis, the amount of chemical change that a current produces is proportional to the amount of electricity used, and the amounts of chemical change produced by the same quantity of electricity in different substances is proportional to their equivalent weights.

CROSS REFERENCE: JOSEPH HENRY, JAMES CLERK MAXWELL, HUMPHRY DAVY, HEINRICH HERTZ, GUGLIELMO MARCONI, HANS ØRSTED, ANDRÉ-MARIE AMPÈRE, FRANÇOIS ARAGO, CHARLES-AUGUSTIN COULOMB, OTTO VON GUERICKE, HERMANN VON HELMHOLTZ, LENZ'S LAW, AVOGADRO'S NUMBER, AND MAXWELL'S EQUATIONS.

In 1831, Charles Darwin sailed on his famous journey around the world on the *H.M.S. Beagle*. The Scottish mathematical physicist James Clerk Maxwell was born. He would later become famous for developing a set of equations expressing the basic laws of electricity and magnetism. London Bridge was opened. The first horse-drawn buses were used in New York City. Samuel Francis Smith wrote the words to "My Country, 'Tis of Thee."

FARADAY'S LAW OF INDUCTION (1831)

English scientist Michael Faraday's greatest discovery was that of electromagnetic induction. In 1831, he noticed that when he moved a magnet through a stationary coil of wire, he always produced an electric current in the wire. American scientist Joseph Henry (1797–1878) carried out similar experiments at about the same time. Today, this induction phenomenon plays a crucial role in electric power plants.

Faraday also found that if he moved a wire near a stationary permanent magnet, a current flowed in the wire whenever it moved. When Faraday experimented with an electromagnet and caused the magnetic field surrounding the electromagnet to change, he then detected electric current flow in a nearby but separate wire.

Scottish physicist James Clerk Maxwell (1831–1879) later suggested that changing the magnetic flux produced an electric field that not only caused electrons to flow in a nearby wire but also existed in space, even in the absence of electric charges. In other words, according to Maxwell, a conducting loop served only as an instrument to reveal the presence of an electrical field that was always induced around a changing magnetic field, even if a conducting loop was not present. Maxwell expressed the change in magnetic flux and its relation to the induced electromotive force (ε or emf) in what we call Faraday's Law of Induction:

$$\boxed{\varepsilon = -\frac{d\phi_m}{dt}}$$

Here, ϕ_m is flux of the magnetic field through a circuit. As a rough example of flux in our daily lives, one can imagine that flux is the amount of water that flows through a cross section of a hose each second. For our electromagnetic example, Faraday imagined a magnetic field as composed of many lines of induction, along which a small magnetic compass would point. The collection of the lines that intersect a given area indicates the magnetic flux.

According to the induction equation, if the magnetic field is changed in any way, a corresponding emf ε will exist in the nearby electrical circuit that is usually detected by observing a current in the circuit. The emf in a circuit may be thought of as the work done per unit charge by the electric field as the charge moves around the complete circuit.

Let's take a closer look at the induction equation. First, we see that the magnitude of the emf induced in a circuit is proportional to the rate of change of the magnetic flux impinging on the circuit. The induced emf is in units of volts if the rate of change of magnetic flux is expressed in webers per second.

The direction of the emf (as indicated by the negative sign in the formula) is described by Lenz's Law, which states that the emf and induced current are in such a direction as to tend to oppose the change that produced them. The Russian-German physicist Heinrich Lenz (1804–1865) stated his law in 1833. For example, if we attempt to increase the flux through a circuit, a current will be induced that tends to decrease the flux. If the reverse of Lenz's Law were true, the flux would increase the current, which would in turn increase the flux, and this would increase the power in the circuit without limit. Infinite power is not produced because an induced electric current flows in a direction such that the current opposes the change that induced it.

Note that if we have a coil with N turns, an emf appears in every turn, and these emfs are additive. In tightly wound coils, the induced emf ε can be approximated by

$$\boxed{\varepsilon = -N\frac{d\phi_m}{dt}.}$$

In 1941, Donald Kerst at the University of Illinois employed Faraday's Law in an interesting way when Kerst invented the betatron. This device accelerates electrons to high energies by moving them in a circular orbit within an evacuated doughnut-shaped tube. The electrons are accelerated by an electric field that is produced by a magnetic flux. This flux is generated by an electromagnet. Each time an electron circulates around the loop, it falls through a potential difference equal to the induced emf, namely, $\varepsilon = d\phi_m/dt$. The highly energetic electrons produced by the betatron can be used for basic physics research or to produce penetrating X-rays useful for cancer therapy. A 100 mega-electron volt (MeV) betatron can produce an electron that travels at 0.999986 times the speed of light.

Let's solve a practical problem involving a betatron through which an electron completes 2×10^5 trips around the loop of the device before it is ejected against a metal plate to produce X-rays. If, during this time, we find $d\phi_m/dt$ is 400 volts, what is the energy and speed of the ejected electron? To solve this problem, note that each time around the doughnut, the electron falls through a potential difference of $\varepsilon = 400$ volts. After the electron has made 2×10^5 trips, we may think of the electron as having "fallen" through $(2 \times 10^5) \times 400 = 8 \times 10^7$ volts. The energy is therefore 80 MeV. Using formulas for the kinetic energy of this high-speed electron moving at relativistic speeds, we can find that it is moving at a velocity equal to 0.99998 times the speed of light.

FARADAY'S LAW OF ELECTROLYSIS (1833)

Before discussing Faraday's Law of Electrolysis, let us review some fundamental chemistry. Electrolysis is the passage of an electric current through a conducting solution, or a molten salt, that is decomposed in the process. When a direct electric current is passed through an electrolyte—such as the molten salt or an aqueous solution of a salt—chemical reactions take place at the contacts between the electric circuit and the solution. Usually, electrodes are immersed in the electrolyte. The electrode that is attached to the negative pole of the battery, which supplies electrons to the electrolyte, is called the cathode. The electrode that is attached to the positive pole of the battery, which accepts electrons from the electrolyte, is called the anode.

You may recall the electrolysis of water performed in your high school chemistry class. An electrical current is applied between a pair of metal electrodes, which are immersed in the liquid, and hydrogen and oxygen gas is produced:

$$2H_2O_{(liquid)} \rightarrow 2H_{2(gas)} + O_{2(gas)}$$

(In practice, pure water is a poor conductor of electricity, and one may add dilute sulfuric acid in order to establish a significant current flow.) The energy required to separate the ions is provided by an electrical power supply. The electrolysis of water suggests why the process is called electrolysis—the suffix "lysis" comes from the Greek, meaning to split. Some scientists suggest that this kind of chemical reaction could be important in making hydrogen, an energy source for powering future motors and engines. Nuclear submarines can remain submerged for extremely long times because they can generate oxygen for breathing using electrolysis of the water around them.

For the case of molten sodium chloride, an electric current is used to split a compound of NaCl into its elements by the following electrolysis reaction:

$$2NaCl_{(liquid)} \rightarrow 2Na_{(liquid)} + Cl_{2(gas)}$$

Faraday's Law of Electrolysis has two parts. First, the amount of chemical change produced by current during electrolysis is proportional to the quantity of electricity used. In other words, the quantity of elements separated by passing an electrical current through a molten or dissolved salt is proportional to the amount of electric charge passed through the circuit.

Second, the amounts of chemical changes produced by the same quantity of electricity in different substances are proportional to their equivalent weights. Similarly, the mass of the resulting separated elements is proportional to the atomic masses of the elements. This observation provided evidence that atoms contained discrete particles of electricity.

Today, chemists often write Faraday's Law of Electrolysis as

$$\boxed{m = \frac{Q}{qn} \cdot \frac{M}{N_A} = \frac{1}{96{,}485\ \mathrm{C}} \cdot \frac{QM}{n},}$$

where m is the mass of the substance produced at the electrolysis electrode, in units of grams; Q is the total electric charge that passed through a conducting solution or molten salt, in coulombs; and q is the charge of an electron (1.602×10^{-19} coulombs per electron). The variable n is the valence number of the substance when present as an ion in solution. Valence numbers range in value from -4 to $+7$ and describe the combining behavior of the atoms in chemical reactions. For example, iron can have a valence of $+2$ or $+3$. Hydrogen always has valence of $+1$. M is the molar mass of the substance, in grams per mole, and N_A is Avogadro's number (6.022×10^{23} ions per mole).

Note that the value 96,485.3383, usually expressed in units of coulombs per mole, is known as the Faraday constant (F) and may be thought of as the amount of electric charge in 1 mole of electrons. In other words, $F = N_A \cdot q$. The symbol C denotes coulombs.

We can use Faraday's Law of Electrolysis in a practical problem to determine the amount of a substance consumed or produced at an electrolysis electrode, given that this amount should be directly proportional to the amount of electricity that passes through the electrolytic cell. Imagine a circuit in which 1 amp of current flows for 1 second. By definition, this corresponds to 1 coulomb of charge being transferred. Our challenge is to predict the number of grams of sodium metal that forms at the cathode in an electrolysis experiment when a 20-amp current is passed through molten sodium chloride for a period of 8 hours. First, let's calculate the amount of electric charge that flows through the cell:

$$20 \text{ amperes} \times 8 \text{ hours} \times \frac{60 \text{ min}}{1 \text{ hour}} \times \frac{60 \text{ sec}}{1 \text{ min}} \times \frac{1 \text{ C}}{1 \text{ ampere sec}} = 576{,}000 \text{ C}$$

Again, the symbol C denotes coulombs.

We now use Faraday's constant to determine the number of moles of electrons (e^-) that are transferred when 576,000 coulombs of electric charge flow through the cell:

$$576{,}000 \text{ C} \times \frac{1 \text{ mol e}^-}{96{,}485 \text{ C}} = 5.97 \text{ mol e}^-$$

According to the electrolysis equation ($Na^+ + e^- \rightarrow Na$), for the reaction that occurs at the cathode of this cell, we produce 1 mole of sodium for every mole of electrons. Thus, we produce 5.97 moles, or 137.25 grams, of sodium in 8 hours:

$$5.97 \text{ mol Na} \times \frac{22.99 \text{ g Na}}{1 \text{ mol Na}} = 137.25 \text{ g Na}$$

This means that we would have to run this electrolysis experiment for more than a day to prepare a pound of sodium.

Michael Faraday (1791–1867), British physicist and chemist famous for his law that expresses the relationship between a changing magnetic field and an electric field, as well as for his experiments in electrolysis.

CURIOSITY FILE: The Biblical Book of Job was the Bible story of most interest to Faraday—and the section most thoroughly annotated in his own hand. Faraday believed himself to be an instrument to reveal the truth of God's creation. • In the early 1840s, Faraday suffered a nervous breakdown and also became an Elder of the Sandemanian Church. The confluence of these events led to a decrease in the quantity of Faraday's scientific work. • Faraday's handsome face was printed on Great Britain's 20-pound note (1991–1993). • The farad (F) is a unit of electric capacitance. The faraday (Fd) is a unit of electric charge. • Faraday took lessons in drama and elocution to reduce his cockney accent. • Humphry Davy, for whom Faraday worked, often experimented with gases by inhaling them—a practice that was nearly fatal but led to the discovery of nitrous oxide, which came to be known as laughing gas.

> His own relation to [nature] produced in Faraday a kind of spiritual exaltation. His religious feeling and his [science] could not be kept apart; there was a habitual overflow of the one into the other.
>
> —John Tyndall, *Faraday as a Discoverer*

> Michael Faraday was born in the year that Mozart died.... Faraday's achievement is a lot less accessible than Mozart's [but]... Faraday's contributions to modern life and culture are just as great.... His discoveries of electromagnetic rotation and magnetic induction laid the foundations for modern electrical technology... and made a framework for unified field theories of electricity, magnetism, and light.... Faraday argued that the familiar properties of bodies reside not in mater but in forces filling all space.
>
> —David Gooding, "New Light on an Electric Hero," *Times Higher Education Supplement*

> Our physicists led lives in social worlds that covered the full middle-class range, from lower to upper, but rarely found themselves above or below these stations. By far the most prominent exception... is Michael Faraday, born in a London slum.
>
> —William H. Cropper, *Great Physicists*

Michael Faraday was born in Newington Butts in South London, England. His father, a blacksmith, made so little money that young Faraday was sometimes given a single loaf of bread that was expected to feed him for a week. Faraday and his family were Sandemanians, a Christian sect founded in Scotland in 1730. The sect emphasized love and community spirit, and they also believed in the literal truth of the Bible.

According to L. Pearce Williams, writing in the *Dictionary of Scientific Biography*,

> Sandemanism gave him both a sense of the necessary unity of the universe derived from the unity and benevolence of its Creator and a profound sense of the fallibility of man.... The speculations... which led him to the [electricity] experiments and the courage which permitted him to publish physical heresies owe something to his unquestioning belief in the unity and interconnections of all phenomena.

Faraday believed that God sustained the universe and that Faraday was doing God's will to reveal truth through careful experiments and through his colleagues, who tested and built upon his results. He accepted every word of the Bible as literal truth, but meticulous experiments were essential in this world before any other kind of assertion could be accepted. "In my early life," he wrote, as quoted in Bence Jones's *The Life and Letters of Faraday*, "I was a very imaginative person, who could believe in the *Arabian Nights* as easily as in the *Encyclopedia*, but facts were important to me, and saved me. I could trust a fact, and always cross-examined an assertion." Throughout much of his life, he reperformed other scientists' experiments that were described in the literature, before he would accept their assertions and conclusions.

Many of the first Sandemanians were individuals who left the Presbyterian Church of Scotland and the Church of England. Most Sandemanian churches of Faraday's time were led by elders, pastors, or bishops, who were chosen without regard to formal education or occupation and who were treated equally while in their positions. Sandemanians never ate strangled animals or blood. Their faith also told them they should not acquire wealth.

The Sandemanian sect gradually faded from the world scene after Faraday's death, and the last of the Sandemanian churches in America disappeared in 1890. Faraday described the Sandemanians as "a very small and despised sect of Christians." Robert Sandeman, church founder, proclaimed, "That God exists is evident from the intricate contrivances of Nature. Let him who doubts cast up his eyes at the heavens and all doubt must vanish."

Faraday had almost no formal education. He later wrote, "My education was of the most ordinary description, consisting of little more than the rudiments of reading, writing, and arithmetic at a common day school." At the age of 13, when he could barely read or write, he quit school to find a job.

Thomas West, author of *In the Mind's Eye*, suggests that Faraday probably had dyslexia or some related learning disability. For example, Faraday did have poor memory, great difficulty with spelling and punctuation, and a relative inability to do mathematics. But he did have a powerful visual sense. Maxwell once said that Faraday was able to construct mental pictures of lines of force in order to visualize how they might fill space and shape themselves into meaningful patterns.

Faraday delivered newspapers and sold and bound books in order to supplement his family's income. Literacy was on the rise in Europe, partly due to improved printing presses. As a result, books were being sold in record numbers. Faraday enjoyed the books he encountered at his job and gradually improved his reading skills. He carried Dr. Isaac Watt's *The Improvement of the Mind* in his pocket as a means to better himself and improve the way he interacted with others. According to the book, three ways existed to become smarter: attend lectures, take careful notes, and interact with people of like interests.

Faraday's love of science was kindled by his serendipitous encounter with the 127-page entry "Electricity" in the *Encyclopaedia Britannica*, which he happened to be rebinding for a client. He immediately studied some of the simpler observations in the encyclopedia by doing experiments with old bottles and discarded lumber. He also built a hand-cranked device that produces electrical sparks. Faraday learned that although scientists had been aware of electricity for centuries, they still had little understanding of it.

A second chance encounter also changed his life forever. In 1812, the great English chemist Humphry Davy (1778–1829) was temporarily blinded by a chemical explosion, and as a result, Faraday became Davy's assistant. [For more on this explosion and how it relates to the work of French Chemist Pierre Dulong (1785–1838), see "The Dulong-Petit Law of Specific Heats."] Faraday himself experienced several explosions, the most terrible of which occurred when he held a tube of nitrogen trichloride between his thumb and finger. When the chemical exploded, it nearly blew his hand off.

Faraday published his first scientific paper, "Analysis of Caustic Lime of Tuscany," in 1816. In the 1820s, he investigated various oils used for heating and lighting. In 1825, Faraday isolated the chemical compound that we now call benzene. After several experiments, he discovered that the new compound had equal numbers of carbons and hydrogens; thus, he named it "carbureted hydrogen." In 1820, he produced the first known compounds of chlorine and carbon, C_2Cl_6 and C_2Cl_4.

In 1821, Faraday married Sarah Barnard and was happy with his marriage throughout his life, noting that even though Sarah was not an intellectual like he was, her emotional support was all that he needed. Sarah never studied chemistry because, as she said (quoted in L. Pearce Williams's "Michael Faraday"), "Already chemistry is so absorbing, and exciting to him that it often deprives him of his sleep, and I am quite content to be the pillow of his mind." Faraday nearly missed the opportunity of marrying this 23-year-old daughter of a Sandemanian elder in his church because he had initially hurt her feelings by writing a poem that suggested love distracted men from their important work. Luckily for Faraday, he realized his potential loss and then applied his same passion for science to persistence in winning back her heart.

Faraday's discovery of the properties of electromagnetism and his related physics work are described above. His pioneering work on the conversion of electrical to mechanical energy was published as "On Some New Electro-Magnetical Motions, and on the Theory of Magnetism" in the October 1821 issue of the *Quarterly Journal of Science*. For this research, Faraday placed a vertical wire in the center of a pool of mercury and sent an electric current flowing through it, from bottom to top. He then placed a bar magnet in the pool and tethered it to the container bottom so that it stood upright and was loose at the upper end. When the current was on, the magnet began to revolve, as if it were being propelled by an invisible current. In a sense, Faraday had just demonstrated the world's first electric motor!

As Faraday's scientific fame spread, Humphry Davy began to despise him. Davy's jealousy ran wild, and he accused Faraday of stealing ideas from other colleagues. Davy campaigned to prevent Faraday from being elected to the Royal Society. Despite Davy's attempts, Faraday was made a member of the Royal Society in 1824.

In 1831, Faraday used a galvanometer (electric current detector) and an iron ring to show that an electrical current could be induced by another current. In particular, Faraday wrapped a wire around one segment of an iron "doughnut" and wrapped another wire around another segment of the doughnut. His plan was to send a current through the first wire wrapping to produce a kind of swirling magnetic "tornado" that would propagate through the doughnut and induce a current in the second wrapping. What he found was that the moment he turned on the current, electricity was induced for just a moment in the second wrapping. He waited, and nothing more happened, until he shut off the current in the first wrapping and then found a momentary current in the second. What he had discovered was that the electric current in the first wire wrapping produced a magnetic swirl that in turn caused a second electric current to flow in the other wrapping whenever the magnetic swirl either increased or decreased. In 1831, he presented his law, which in simple English can be stated as follows:

1. Whenever a magnetic force changes, it produces electricity.
2. The faster the magnetic force changes, the more electricity it produces.

A few weeks later, Faraday showed that a permanent magnet could be used to generate electrical current and that a magnetic force could be converted to an electrical force. His demonstration made use of a copper disk that produced an electrical current in a circuit when that disk was rotated between poles of a magnet. In the 1860s and 1870s, Maxwell built upon Faraday's foundations to formulate an electromagnetic field theory. In fact, Faraday's descriptive theory of lines of force moving between bodies with electrical and magnetic properties enabled Maxwell to derive the mathematical theory of the propagation of electromagnetic waves. In 1865, Maxwell used a mathematical approach to demonstrate that electromagnetic phenomena propagate at the velocity of light as waves through space. This insight provided the foundation of radio communication, confirmed experimentally in 1888 by the German physicist Heinrich Hertz (1857–1894), with practical applications later demonstrated by Italian-Irish inventor Guglielmo Marconi (1874–1937).

Faraday's accomplishments were not simply theoretical curiosities. His laws led to electric dynamos that created a constantly changing magnetic force by spinning a magnet. The motion of the magnet is powered by falling water, steam, or by various fuels. Once the magnet is rotating, it can be used to produce a steady output of electricity. The faster the magnets spin, the larger the electric output.

Let's review some of Faraday's additional experiments. Similar to the experiment in which he used coils around an iron doughnut, in a related experiment he placed a stationary coil A very close to a stationary coil B. Coil A has a galvanometer placed in series. Coil B had a switch, and a battery to pump electrons through the coil when the switch was closed. Faraday turned the circuit on and noticed that the galvanometer deflected momentarily. When the switch was opened, the galvanometer again deflected momentarily in the opposite direction. Experiments showed that an induced emf in coil A occurred whenever the current in coil B was *changing*.

In Faraday's own words from his "On the Induction of Electric Currents" published in 1832:

> Two hundred and three feet of copper wire in one length were passed round a large block of wood; other two hundred and three feet of similar wire were interposed as a spiral between the turns of the first, and metallic contact everywhere prevented by twine. One of these helices was connected with a galvanometer and the other with a battery of a hundred pairs of plates four inches square, with double coppers and well charged. When the contact was made, there was a sudden and very slight effect at the galvanometer, and there

> was also a similar slight effect when the contact with the battery was broken. But whilst the voltaic current was continuing to pass through the one helix, no galvanometrical appearances of any effect like induction upon the other helix could be perceived, although the active power of the battery was proved to be great....

Before Faraday performed his electromagnetic experiments, the individual who was perhaps closest to clearly elucidating the relationship between electricity and magnetism was Danish physicist Hans Ørsted (1777–1851). In 1820, Ørsted discovered that an electric current caused the needle of a magnetic compass to move, as if the electricity were somehow behaving like a magnet.

Around the same time, French physicists André-Marie Ampère (1775–1836) and François Arago (1786–1853) created an electromagnet when they discovered that an electric current in the shape of corkscrew also behaved like a magnet and attracted iron. French physicist Charles-Augustin Coulomb (1736–1806) had found that magnetism and electricity had similar characteristics, namely, that their forces decreased as the square of the distance. German scientist Otto von Guericke (1602–1686) had showed that both electrical and magnetic phenomena exhibited polarity and could both repel and attract. With the ideas of these great physicists swimming in his head, Faraday thought that electricity and magnetism were essentially interchangeable. If electricity could behave like a magnet—that is, produce a magnetic field as Ampère had shown—Faraday had wondered if magnetism could be used to produce electricity.

Faraday was also interested in electrolysis, the means for producing chemical changes through reactions at electrodes. Using galvanometers and electrolysis equipment, he showed that chemical action is exactly proportional to the quantity of electricity that passes through the solution and that the amounts of substances deposited or dissolved by the same quantity of electricity are proportional to their chemical equivalent weights. After Faraday's death, German physicist Hermann von Helmholtz (1821–1894) used Faraday's papers on electrochemistry to promote theories that electricity must be composed of individual particles.

In 1839, Faraday suffered a nervous breakdown, which some biographers consider to be the result of nearly a decade of constant effort to understand the nature of electricity and magnetism. His productive research days were fewer but certainly not completely over.

In 1844, Faraday was suspended as an elder of the Sandemanian church for missing a single Sunday service—the only time he missed a service in his entire life! When he explained that he had been dining with Queen Victoria, the church elders cared little for his flimsy excuse.

In 1845, Faraday discovered that the plane of polarization of light could be rotated when the light passed through glass in the presence of a magnetic field. Today, we call this phenomenon the "Faraday effect." This discovery has important implications because it was one of the first demonstrations of relationship between light and magnetism. In November of 1845, Faraday wrote to Swiss professor Christian Friedrich Schoenbein (1799–1868):

> At present I have scarcely a moment to spare for any thing but work. I happen to have discovered a direct relation between magnetism & light also electricity & light—and the field it opens is so large & I think rich that I naturally wish to look at it first. I actually have no time to tell you what the thing is—for I now see no one & do no thing but just work. My head became giddy & I have therefore come to this place (Brighton) but still I bring my work with me.

Also in 1845, Faraday discovered the magnetic phenomenon that he named diamagnetism, a form of magnetism that manifests only when materials are placed in an externally applied magnetic field. Although all matter exhibits diamagnetism, only those substances, like gold, in which diamagnetism is particularly strong are referred to as diamagnetic. For more information on diamagnetism, see "Curie's Magnetism Law and the Curie-Weiss Law," below.

Faraday's mind deteriorated after about 1855. Around this time he retreated from all social activities, but he still taught chemistry and physics. His 1860 and 1861 Christmas lectures on physics and chemistry for children were eventually edited and used by other teachers. The Royal Institution Christmas lectures for children, begun by Faraday, continue to this day.

According to one story of Faraday's final hours, when he knew that he was dying, someone asked him, "Mr. Faraday, what are your presumptions, your hypotheses now?" He replied, "I do not entrust my head to presumptions at this moment, but to certainties." And then Faraday quoted from 2 Timothy 1:12, "For I know whom I have believed, and am persuaded that he is able to keep that which I committed unto him against that day." He died sitting up in his favorite chair and, at his prior request, had a small funeral attended only by his relatives.

One of the most personal biographies of Faraday was written in 1872 by John Hall Gladstone, a colleague of Faraday's at the Royal Institution. Gladstone wrote about Faraday's last days:

> When his faculties were fading fast, he would sit long at the western window, watching the glories of the sunset; and one day, when his

wife drew his attention to a beautiful rainbow that spanned the sky, he looked beyond the falling shower and the many-colored arch, and observed, "He hath set his testimony in the heavens." On August 25, 1867, quietly, almost imperceptively, came the release. There was a philosopher less on Earth, and a saint more in heaven.

A lunar crater with a diameter of 69 kilometers was named after Faraday and approved in 1935 by the International Astronomical Union General Assembly.

During the course of his life in science, Faraday had kept a diary that grew to seven volumes. Starting in 1832, Faraday numbered every paragraph in the diary, ending with number 16,041 at the end of his career. Today, the seven volumes are still a useful resource as an example of one man's creative thinking in science. Faraday had been particularly adept at questioning standard scientific wisdom and suspending judgment until he could study the subject further—he was unusually receptive to new ideas.

Michael Guillen writes about Faraday's contribution to our understanding of the world in *Five Equations That Changed the World*:

> Together with Ørsted, Faraday had shown that electricity could beget magnetism, and magnetism could beget electricity, a genetic relationship so incestuous and circular there was none other like it in Nature.... The son of a common laborer had discerned and written down a great secret of the natural world, one that would spell the end of the Industrial Revolution and the beginning of the Electrical Age.

Finally, the Dover editors of the introduction to Faraday's *The Chemical History of a Candle* honor Faraday with a fitting and lasting tribute:

> Faraday was the greatest physicist of the nineteenth century and the greatest of all experimental investigators of physical nature. He is a member of the small class of supreme scientists, which includes Archimedes, Galileo, Newton, Lavoisier and Darwin. Einstein has said that the history of physical science contains two couples of equal magnitude: Galileo and Newton and Faraday and Maxwell.... Faraday must be accounted a greater scientist than Galileo.

We should reemphasize that Faraday presented his ideas in simple English—he was not a great mathematician. In fact, he had always said he liked to couch laws in ways that ordinary people could understand. Maxwell relied heavily on Faraday's work. About three decades after Faraday made his discoveries, Maxwell published his magnificent *A Dynamical Theory of the Electromagnetic Field*, in which he recast Faraday's simple English describing electromagnetic laws into modern mathematics that looked like

$$\nabla \times \mathbf{E} = \frac{-\partial \mathbf{B}}{\partial t}.$$

This formula essentially says that the amount of electricity produced ($\nabla\times$ **E**) is equal to the rate of change in magnetic field ($-\partial B/\partial t$). If there is no change in the magnetic field, then no electricity is produced. Maxwell would write similar kinds of equations for Gauss's Laws of Electricity and Magnetism and Ampère's Circuital Law of Electromagnetism. In general, Maxwell's Equations are the set of four famous formulas that describe the behavior of the electric and magnetic fields. In particular, they express how electric charges produce electric fields and the fact that magnetic charges cannot exist. They also show how currents produce magnetic fields and how changing magnetic fields produce electric fields. Today, one of several ways of writing Maxwell's Equations is as follows:

$\nabla \cdot \mathbf{D} = \rho/\varepsilon_0$	Gauss's Law of Electricity
$\nabla \cdot \mathbf{B} = 0$	Gauss's Law of Magnetism (no magnetic monopoles exist)
$\nabla \times \mathbf{E} = -\frac{\partial \mathbf{B}}{\partial t}$	Faraday's Law of Induction
$\nabla \times \mathbf{H} = \mathbf{J} + \frac{\partial \mathbf{D}}{\partial t}$	Ampère's Law with Maxwell's extension

Bold letters stand for vectors. **E** is the electric field (volts/meter), **H** is the magnetic field (amperes/meter), **D** is the electric flux density (coulomb/meter^2), **B** is the magnetic flux density (tesla or weber/meter^2), ρ is the free electric charge density (coulomb/meter^3), ε_0 is the permittivity of free space, **J** is the free current density (amperes/meter^2), "$\nabla\cdot$" is the divergence operator (per meter), and "$\nabla\times$" is the curl operator (per meter). The divergence operator measures a vector field's tendency to originate from or converge upon a given point. The curl operator measures a vector field's rotation.

Robert P. Crease writes of the beauty and importance of Maxwell's Equations in "The Greatest Equations Ever":

> Although Maxwell's equations are relatively simple, they daringly reorganize our perception of nature, unifying electricity and magnetism and linking geometry, topology and physics. They are essential to understanding the surrounding world. And as the first field equations, they not only showed scientists a new way of approaching

physics but also took them on the first step towards a unification of the fundamental forces of nature.

In 2004, Crease had conducted a survey in which he asked physicists for their candidates for the greatest equations of all time. The greatest set of physics equations, determined by the most votes, was Maxwell's Equations. Tony Watkins, one of the respondents in the Crease survey, wrote:

> I still vividly remember the day I was introduced to Maxwell's equations in vector notation. That these four equations should describe so much was extraordinary.... For the first time I understood what people meant when they talked about elegance and beauty in mathematics or physics. It was spine-tingling and a turning point in my undergraduate career.... My passion was reignited by four lines of symbols.

Richard Feynman writes on Maxwell's Equations in his *Feynman Lectures on Physics*:

> From a long view of the history of mankind—seen from, say, ten thousand years from now—there can be no doubt that the most significant event of the 19th century will be judged as Maxwell's discovery of the laws of electrodynamics. The American Civil War will pale into provincial insignificance in comparison with this important scientific event of the same decade.

FURTHER READING

Bueche, Frederick, *Introduction to Physics for Scientists and Engineers* (New York: McGraw-Hill, 1975); provides information on the betatron.

Cantor, Geoffrey N., David Gooding, and Frank James, *Michael Faraday* (Amherst, N.Y.: Humanity Books, 1996).

Crease, Robert P., "The Greatest Equations Ever," *Physics World*, October 2004; see http://physicsweb.org/articles/world/17/10/2/1.

Faraday, Michael, *The Chemical History of a Candle* (New York: Courier Dover Publications, 2003).

Faraday, Michael, "On the Induction of Electric Currents," *Philosophical Transactions of the Royal Society of London*, 122: 125–162, 1832.

Feynman, Richard, *The Feynman Lectures on Physics*, volume 2 (Boston: Addison Wesley Longman, 1970).

Gladstone, John Hall, *Michael Faraday* (New York: Harper & Brothers, 1872).

Gooding, David, "Envisioning Explanations—the Art in Science," *Interdisciplinary Science Reviews*, 29: 278–294, 2004.

Gooding, David, "From Phenomenology to Field Theory: Faraday's Visual Reasoning," *Perspectives on Science*, 14(1): 40–65, 2006.

Gooding, David, "Michael Faraday, 1791–1867: Artisan of Ideas," University of Bath; see www.bath.ac.uk/~hssdcg/Michael_Faraday.html.

Gooding, David, "New Light on an Electric Hero," *Times Higher Education Supplement*, July 26, 1991, p. 17.

Guillen, Michael, *Five Equations That Changed the World* (New York: Hyperion, 1995).

Hamilton, James, *Life of Discovery: Michael Faraday, Giant of the Scientific Revolution* (New York: Random House, 2004).

Hirshfeld, Alan, *The Electric Life of Michael Faraday* (New York: Walker & Company, 2006).

James, Frank, *The Correspondence of Michael Faraday* (Herts, U.K.: Institute of Electrical Engineers, 1991).

Jones, Bence, *The Life and Letters of Faraday* (Philadelphia: J. B. Lippincott and Co., 1870).

Ludwig, Charles, *Michael Faraday: Father of Electronics* (Scottdale, Pennsylvania: Herald Press, 1978).

Morus, Iwan, *Michael Faraday and the Electrical Century* (Eastbourne, U.K.: Gardners Books, 2004).

O'Connor, John J., and Edmund F. Robertson, "Michael Faraday: 1791–1867," in *MacTutor History of Mathematics Archive*, School of Mathematics and Statistics, University of St. Andrews, Scotland; see www-history.mcs.st-andrews.ac.uk/history/Mathematicians/Faraday.html.

Thompson, Silvanus Phillips, *Michael Faraday His Life and Work* (London: Cassell & Company, 1901).

Tyndall, John, *Faraday as a Discoverer* (New York: D. Appleton and Company: 1868).

Tweney, Ryan, and David Gooding, *Michael Faraday's "Chemical Notes, Hints, Suggestions and Objects of Pursuit" of 1822* (Herts, U.K.: Peter Peregrinus, Ltd., 1991).

West, Thomas, *In the Mind's Eye: Visual Thinkers, Gifted People with Learning Difficulties, Computer Images, and the Ironies of Creativity* (Amherst, New York: Prometheus Books, 1991).

Williams, L. Pearce, "Michael Faraday," in *Dictionary of Scientific Biography*, Charles Gillispie, editor-in-chief (New York: Charles Scribner's Sons, 1970).

INTERLUDE: CONVERSATION STARTERS

> The more we learn of science, the more we see that its wonderful mysteries are all explained by a few simple laws so connected together and so dependent upon each other, that we see the same mind animating them all.
>
> —Olympia Brown (1835–1900), U.S. minister (first woman ordained in U.S.), Sermon, c. January 13, 1895, Mukwonago, Wisconsin.

In science we aim for a picture of nature as it really is, unencumbered by any philosophical or theological prejudice. Some see the search for scientific truth as a search for an unchanging reality behind the ever-changing spectacle we observe with our senses. The ultimate prize in that search would be to grasp a law of nature—a part of a transcendent reality that governs all change, but itself never changes.

—Lee Smolin, "Never Say Always," *New Scientist*, September 23, 2006

As a conservative, I do not agree that a division of physics into separate theories for large and small is unacceptable. I am happy with the situation in which we have lived for the last 80 years, with separate theories for the classical world of stars and planets and the quantum world of atoms and electrons.

—Freeman Dyson, "The World on a String," *New York Review of Books*, May 13, 2004

Physics isn't Christian, though it was invented by Christians. Algebra isn't Muslim, even though it was invented by Muslims. Whenever we get at the truth, we transcend culture, we transcend our upbringing. The discourse of science is a good example of where we should hold out hope for transcending our tribalism.

—Sam Harris, "The God Debate," *Newsweek*, April 9, 2007

GAUSS'S LAWS OF ELECTRICITY AND MAGNETISM

π *Germany, 1835.* The electric flux across any closed surface is proportional to the net electric charge enclosed by the surface. The magnetic flux across any closed surface is zero.

CROSS REFERENCE: FARADAY'S LAW OF INDUCTION AND ELECTROLYSIS, BODE'S LAW OF PLANETARY DISTANCES, AND KIRCHHOFF'S ELECTRICAL CIRCUIT AND THERMAL RADIATION LAWS.

> In 1835, Texas declared its right to secede from Mexico. The first assassination attempt against a U.S. President was made. (The attempt against President Andrew Jackson's life, while he was in the U.S. capitol, was unsuccessful.) American author Mark Twain was born. Copernicus's book on the motion of Earth, *De revolutionibus orbium coelestium*, was removed from the Catholic Church's catalog of prohibited books, the *Index Librorum Prohibitorum*. This index provided a list of publications that the Catholic Church believed were a danger to Catholics.

GAUSS'S LAW OF ELECTRICITY

Before discussing the formula for Gauss's Law of Electricity, it is useful to review some of the law's key variables. For an electric field, *flux*, which is usually denoted by the symbol Φ, is measured by the number of lines of force that cut through a hypothetical surface in the field. The surface may be open, or it may be a closed surface such as a spherical surface. For closed surfaces, the flux is considered positive if the lines of force point outward everywhere and negative if they point inward. The concept of flux is also discussed in "Faraday's Laws of Induction and Electrolysis," above.

If this discussion of flux seems rather theoretical, we can discuss an example that gives another sense of the meaning of flux. At home, I have a large tropical fish tank. To better understand the concept of flux in electromagnetism, imagine a fish net. I immerse the net inside the water of my tank, which has all kinds of interesting flows due to several circulating pumps. As a crude analogy, the amount of water moving through the net at any given instant in time is related to flux. As I move the net close to the pump outlet, the water speed is high, and the flux through the net is large. If the net is twice as large, then the flux would also be larger even if the water speed is held constant, because more water would be flowing through it.

Imagine a closed surface surrounding a positive charge. For visualization purposes, imagine a beautiful pink balloon, enclosing the charge that floats in its interior. The flux is positive for such a surface because the lines of force point outward from the positive charge. Similarly, the flux is negative for a surface enclosing a negative charge, because the lines of force all point inward for such a surface.

Gauss's Law of Electricity provides the relationship between the electric flux Φ, flowing out of a closed surface, and the electric charge enclosed by the surface:

$$\Phi = \oint_S \mathbf{E} \cdot \mathrm{d}\mathbf{A} = \frac{1}{\varepsilon_0} \int_V \rho \cdot \mathrm{d}V = \frac{q_A}{\varepsilon_o},$$

where **E** is the electric field. Each d**A** is a differential area associated with a particular part of the surface. (You can visualize the normal vectors of the d**A**s as a collection of outward-pointing vectors, like quills on an excited porcupine.) q_A is the charge enclosed by the surface, ρ is the charge density at a point in a volume V, and ε_0 is the permittivity of free space and equals $8.8541878176 \times 10^{-12}$ farads/meter. $\oint_S$ is the integral over the surface S enclosing volume V. The circle on the integral sign indicates that the surface of integration is a closed surface. Other than this surface aspect, this integral is no different than other integrals used by physicists.

If this set of equations seems to be confusing, we can note a few important aspects of the behavior of flux, using simple English. First, the total of the electric flux out of a closed surface is equal to the net charge enclosed by the surface divided by the permittivity. Second, if a surface does not enclose a charge (i.e., a situation in which $q = 0$), then we would predict that the electric flux Φ is 0. Third, the q in the equation refers to the net charge, so, for example, if the surface encloses a positive and negative charge of the same magnitude, the flux is 0. Charge outside the surface makes no contribution to the value of q, and the precise locations of the enclosed charges do not affect the flux value. Gauss's Law of Electricity can be a useful tool for the calculation of electric fields when they originate from a symmetrical charge distribution.

Many physicists have marveled at Gauss's Law of Electricity because its simplicity dictates that no matter how distorted the electric field lines are or how oddly shaped the surface, the flux integral through the closed surface is simply proportional to the net enclosed charge. Gauss formulated his famous law in 1835, but it was not published for another 32 years.

GAUSS'S LAW OF MAGNETISM

Gauss's Law of Magnetism is one of the fundamental equations of electromagnetism and is a formal way of stating the conclusion that no isolated magnetic poles exist. The law may be represented in equation form as

$$\Phi_B = \oint_S \mathbf{B} \cdot d\mathbf{A} = 0$$

The law indicates that the net magnetic flux Φ_B across any closed surface is zero. For a magnetic dipole, the magnetic flux directed inward toward the south pole will equal the flux outward from the north pole. The net flux will always be zero for dipole sources.

If there were a magnetic monopole source, this would give a nonzero area integral. Thus, this law states that there are no magnetic monopoles. A similar finding does not hold for charges in electrostatics because isolated charges may exist, and this lack of symmetry between electric and magnetic fields is a puzzle of sorts. Scientists in the 1900s often wondered precisely why it is possible to isolate positive and negative electric charges but not north and south magnetic poles.

In 1931, British theoretical physicist Paul Dirac (1902–1984) was one of the first scientists to theorize about the possible existence of a magnetic monopole, and a number of efforts through the years have been made to detect magnetic monopoles. However, so far, physicists have never discovered an isolated magnetic pole. Note that if you were to cut a traditional magnet (with a north and south pole) in half, the resulting pieces are two magnets each with its own north pole and south pole. In other words, cutting a magnet does not produce a north-only and south-only piece.

Some theories that seek to unify the electroweak and strong interactions in particle physics predict the existence of magnetic monopoles. However, these hypothetical monopoles would be very difficult to produce using particle accelerators because the monopole would have a huge mass and energy (about 10^{16} giga-electron volts).

Carl Friedrich Gauss (1777–1855), German mathematician and scientist often regarded as one of the greatest mathematicians to have ever lived, famous for his contributions to many areas of mathematics, astronomy, and electromagnetism.

CURIOSITY FILE: The gauss is a unit of magnetic induction (flux density) equal to one maxwell per square centimeter, named in Gauss's honor.

• Gauss was extremely secretive about his work. According to mathematical historian Eric Temple Bell, had Gauss published or revealed all of his discoveries when he made them, mathematics would have been advanced by fifty years. As a result, Bell suggests, the mathematics of today would contain wonders that we can barely imagine. • In the 1990s, the German ten-mark banknote featured Gauss's portrait and a normal distribution curve, also called a Gaussian distribution. • Gauss prohibited his students from taking notes as he lectured so they could pay closer attention to his words. His most famous student was mathematician George Riemann (1826–1866). • Gauss was a man of details, even concerning family matters. For example, he kept a notebook in which he recorded the dates when his children cut their teeth.

> Almost everything, which the mathematics of our century has brought forth in the way of original scientific ideas, attaches to the name of Gauss.
>
> —L. Kronecker, *Zahlentheorie*

> It is hard to appreciate fully the isolation to which Gauss was condemned in childhood by thoughts that he could share with no other.... Ideas came so quickly to him that each one inhibited the development of the proceeding.... He published only about half his recorded ideas and in a style so austere that his readers were few.
>
> —Kenneth O. May, "Carl Gauss," in *Dictionary of Scientific Biography*

> The enchanting charms of this sublime science reveal only to those who have the courage to go deeply into it. But when a woman, who because of her sex and our prejudices encounters infinitely more obstacles than men...succeeds nevertheless in surmounting these obstacles and penetrating the most obscure parts of them, without a doubt she must have the noblest courage, quite extraordinary talents, and superior genius.
>
> —Carl Gauss, 1807 letter to Sophie Germain

> I am convinced more and more that the necessary truth of our geometry cannot be demonstrated, at least not by the human intellect to the human understanding. Perhaps in another world, we may gain other insights into the nature of space which at present are unattainable to us. Until then we must consider geometry as of equal rank not with arithmetic, which is purely a priori, but with mechanics.
>
> —Carl Gauss, 1817 letter to Heinrich Olbers

Carl Friedrich Gauss was a German mathematician, astronomer, and scientist who made significant contributions to many fields, including optics, number theory, analysis, astronomy, differential geometry, geodesy, the theory of errors, and electromagnetism. Like other religious mathematicians, after Gauss proved a theorem, he sometimes said that the insight did not come from "painful effort but, so to speak, by the grace of God." He also once wrote, as quoted in James R. Newman's *The World of Mathematics*:

> There are problems to whose solution I would attach an infinitely greater importance than to those of mathematics, for example touching ethics, or our relation to God, or concerning our destiny and our future; but their solution lies wholly beyond us and completely outside the province of science.

Among the most brilliant mathematicians of all of recorded history, Gauss had a strong influence on many fields of mathematics and science and is ranked beside Euler, Newton, and Archimedes in terms of sheer genius, breadth, and innovation. Yet despite this genius, his life was often unhappy—he usually worked in isolation, and his first wife died early. His second wife was always sick, and his emotional relationship with his sons was poor. According to his sons, Gauss discouraged them from pursing scientific careers because he did not want any second-rate work to be associated with his name.

Carl Gauss was born in Brunswick, now part of Lower Saxony in Germany. Gauss's father held many jobs during his life, including work as a gardener and treasurer of a small insurance fund. Gauss called his father uncouth and domineering. His devoted mother died at age 97 after living with Gauss for 22 years.

Gauss, like other lawgivers in this book, was a childhood prodigy and learned to calculate before he could talk. At age 3, he corrected his father's wage calculations when they had errors. At age 8, according to legend, he shocked his schoolteacher by instantly being able to solve her assignment to find the sum of the first 100 integers. [In order to find the answer 5,050 so quickly, it's possible that Gauss used $1 + 2 + \ldots + n = n(n+1)/2$; however, some scholars today suggest that the entire schoolteacher story is apocryphal.] A different version of the summation legend is presented in Eric Temple Bell's 1937 book *Men of Mathematics* in which the problem Gauss solved was a bit more difficult, which makes Gauss seem even more amazing. Bell writes:

As it was the beginning class, none of the boys had ever heard of an arithmetical progression. It was easy then for [the teacher] to give out a long problem in addition whose answer he could find by a formula in a few seconds. The problem was of the following sort, 81297 + 81495 + 81693 + ...+ 100899, where the step from one number of the next is the same all along (here 198), and a given number of terms (here 100) are to be added.

Jeremy Gray, in his introduction to G. Waldo Dunnington's *Carl Frederich Gauss: Titan of Science*, warns readers that they should be skeptical of at least some of these childhood stories regarding Gauss:

It has become inevitable that we doubt the anecdotes about the young Gauss. They were written down only late in life, they derive from fond but perhaps inaccurate memories of Gauss and his mother. They exaggerate, but such were Gauss's prodigious abilities that they came to be believed.

Gauss enrolled in the Brunswick Collegium Carolinum in 1792, a new science-oriented academy at the time. While still a teenager, he frequently made mathematical discoveries and observations and proved theorems, before finding out that they had been solved or discovered in the past. For example, he discovered Bode's Law of Planetary Distances that predicts the mean distances of planets from the Sun, the binomial theorem for rational exponents, and the arithmetic-geometric mean. We can compute the arithmetic-geometric mean of two positive real numbers x and y by first calculating the traditional mean of x and y, namely, $a_1 = (x + y)/2$. Next, we calculate the geometric mean of x and y by $g_1 = \sqrt{xy}$. Finally, we can iterate the following sequences, which converge to the same number $M(x, y)$, the arithmetic-geometric mean of x and y:

$$a_{n+1} = \frac{a_n + g_n}{2}$$
$$g_{n+1} = \sqrt{a_n g_n}$$

If we determine the arithmetic-geometric mean of 1 and $\sqrt{2}$ and then take the reciprocal, we have expressed the constant that is now called Gauss's constant G in his honor:

$$\frac{1}{M(1, \sqrt{2})} = G = 0.83462684167$$

The arithmetic-geometric mean is useful in computing the values of complete elliptic integrals and for finding the inverse tangent. Gauss also related this mean to infinite series expansions.

While at the Collegium, Gauss accurately calculated the square root of 2 to 50 decimal places: 1.41421356237309504880168872420969807856967187537694—quite a feat for a teenager of his day. He also developed the principle of least squares while searching for patterns in the sequence of prime numbers.

Before he entered the University of Göttingen in 1795, Gauss rediscovered the law of quadratic reciprocity, conjectured by Swiss mathematician and physicist Leonhard Euler (1707–1783) and French mathematician Adrien-Marie Legendre (1752–1833). Gauss was the first to satisfactorily prove this law. He was so interested in this topic that he went on to provide more than seven separate proofs over his lifetime. (For the mathematically inclined reader, the law of quadratic reciprocity concerns the solvability of two related quadratic equations in modular arithmetic.)

In 1796, when Gauss was also still at teenager, he found a means by which to construct of a regular 17-gon (i.e., a heptadecagon or seventeen-sided polygon), using just a ruler and compass. The result was very significant because Gauss was the first to perform such a construction even though attempts had been made since the time of Euclid. For more than 1,000 years, mathematicians had known how to construct, with compass and straight-edge, regular n-gons in which n was a multiple of 3, 5, and powers of 2. Gauss was able to add more polygons to this list, namely, those with a prime number of sides of the form $2^{(2^n)} + 1$, where n is an integer. We can make a list of the first few such numbers: $F_0 = 3$, $F_1 = 5$, $F_2 = 17$, $F_3 = 257$, and $F_4 = 65{,}537$. (Numbers of this form are also known as the Fermat numbers, and they are not necessarily prime.) A 257-gon was constructed in 1832.

When he was older, Gauss still regarded his 17-gon finding as one of his greatest achievements, and he asked that a regular 17-gon be placed on his tombstone. According to legend, the stonemason declined, stating that the difficult construction would essentially make the 17-gon look like a circle.

The year 1796 was an auspicious year for Gauss, when ideas poured like a fountain from an open water faucet: Aside from solving the heptadecagon construction (March 30), Gauss invented modular arithmetic and presented his quadratic reciprocity law (April 8) and the prime number theorem (May 31). He proved that every positive integer is represented as a sum of at most three triangular numbers (July 10). He also discovered solutions of polynomials with coefficients in finite fields (October 1).

The prime number theorem states that the number of primes less than n can be approximated by $n/(\ln n)$. This theorem was first conjectured by Gauss and finally proved in 1896 by French mathematician Jacques Hadamard (1865–1963) and Belgian mathematician Charles de la Vallée Poussin (1866–1962), who worked independently. Their proofs relied on complex analysis, and at the time, no one thought a more simplified proof could be constructed. The great mathematical event of 1949 was an elementary proof of the prime number theorem, given by Norwegian mathematician Atle Selberg (born 1917) and Hungarian mathematician Paul Erdös (1913–1996). Incidentally, the theorem can be used to derive a related theorem: For each number greater than 1, there is always at least one prime number between it and its double. Also, one can show from the prime number theorem that the average "gap" between primes less than n is $\ln(n)$. (If you examine the first few primes: 2, 3, 5, 7, 11, and 13, you will notice that the differences between successive primes vary as: 1, 2, 2, 4, 2 . . .).

Regarding Gauss's discovery that every number is expressible as the sum of at most three triangular numbers, note that Gauss kept a diary for most of his adult life. In one of his most famous diary entries, dated July 10, 1796, was the single line "EYPHKA! num = $\Delta + \Delta + \Delta$," which signifies his triangular number discovery. (Triangular numbers can be represented by a growing triangular grid of points. The first few triangular numbers are 1, 3, 6, 10, 15, 21, ...)

Gauss always had so many ideas that he seemed not to have the time to explore them all in great detail. The following lists some highlights of Gauss's many investigations throughout his life. For example, Gauss:

- provided four proofs of the fundamental theorem of algebra, which states that every polynomial equation with complex coefficients has as many solutions as the highest power of the variable
- completely analyzed cyclotomic equations $x^n - 1 = 0$.
- invented the heliotrope, a device that reflects sunlight over long distances for the purpose of surveying lands
- discovered Kirchhoff's Electrical Circuit and Thermal Radiation Laws in 1833 with German physicist Wilhelm Weber (1804–1891), laws that deal with the conservation of charge and energy in electrical circuits
- created the world's first telegraph with Weber
- conducted work in theoretical physics, capillarity, mechanics, optics, crystallography, and acoustics
- expressed the magnetic potential on any location on the surface of Earth, using an infinite series of spherical functions
- proved that any system of lenses is equivalent to an appropriately chosen single lens

- contributed to our knowledge of electromagnetism, surface curvatures, the least-squares method, hypergeometric functions, and differential geometry
- wrote on cartography and the theory of map projections

Gauss also used mathematical methods to precisely predict the location of the asteroid Ceres. The Italian astronomer Giuseppe Piazzi (1746–1826) had originally discovered Ceres in 1800, but the asteroid later disappeared behind the Sun and could not be relocated. Austrian astronomer Franz Xaver von Zach (1754–1832), as quoted in Curtis Wilson's "Carl Friedrich Gauss," noted that "without the intelligent work and calculations of Doctor Gauss we might not have found Ceres again." Interestingly, Gauss kept his methods a secret to maintain an advantage over his contemporaries and to enhance his reputation. Later in his life, he sometimes published scientific results as a cipher, so that he could always prove that he had made various discoveries before others had.

In 1801, Gauss published the first systematic textbook on algebraic number theory, *Disquisitiones arithmeticae*. In 1803, he met Johanna Osthoff, the daughter of a proprietor of a local tannery. Gauss was smitten from the start and wrote the following heartfelt letter to her on July 12, 1804:

> My true friend, receive favorably the fact that I pour out my heart, in writing, before you, about an important matter.... I have a heart for your silent angelic virtues.... You, dear modest soul, are so far removed from all vanity that you yourself do not realize your own value; you don't know how richly and kindly heaven has endowed you. But my heart knows your worth—O! more than it can bear with repose. For a long time it has belonged to you. You won't repel it? Can you give me yours?... Yes, dearest, so warmly do I even love you, that only possession of you can make me happy, if you are of the same feeling. Dearest, I have exposed to you the inner part of my heart: passionately and in suspense am I waiting for your answer.

Gauss and Johanna became engaged on November 22, 1804, and he told a friend, "Life stands like an everlasting spring with new glittering colors before me." In 1805, they married and soon had a son and daughter. Alas, just a few years later, Johanna, her baby, and Gauss's father died. The death of Johanna in particular plunged Gauss into a depression. However, less than a year later, he married Minna Waldeck, his deceased wife's best friend. He had three children with his new wife.

Gauss's philosophy regarding the pursuit of knowledge is revealed in his 1808 letter to Hungarian mathematician Farkas Bolyai (1775–1856):

> It is not knowledge, but the act of learning, not possession but the act of getting there, which grants the greatest enjoyment. When I have clarified and exhausted a subject, then I turn away from it, in order to go into darkness again. The never-satisfied man is so strange; if he has completed a structure, then it is not in order to dwell in it peacefully, but in order to begin another. I imagine the world conqueror must feel thus, who, after one kingdom is scarcely conquered, stretches out his arms for others.

In 1809, he published his second book, *Theoria motus corporum coelestium in sectionibus conicis Solem ambientium* (Theory of the Motion of the Heavenly Bodies Moving about the Sun in Conic Sections). The book was a treatise on the motion of planets; the first volume discussed differential equations, and the second volume dealt with methods for estimating the path of the orbit of a planet.

Throughout his life, Gauss did his best to resist chronic hypochondria and depression. In 1834, he wrote to his former student Christian Gerling that he felt like a stranger in this world. Yet despite his relative isolation and aloofness, his genius established him as the era's foremost mathematician.

Biographer Kenneth O. May writes in the *Dictionary of Scientific Biography* that although Gauss had a million ideas swirling in his brain, he often did not skillfully explain and promote many of his ideas or cause a revolution in scientific thinking. As discussed above, Gauss often withheld publication of his key insights. May writes:

> The contrast between knowledge and impact is now understandable. Gauss arrived at the two most revolutionary mathematical ideas of the nineteenth century: non-Euclidean geometry and non-communative algebra. The first he disliked and suppressed. The second appears as quaternion calculations in a notebook of about 1819 without having stimulated any further activity.

Marcus du Sautoy in *The Music of the Primes* also writes about Gauss's hesitancy to share some of his ideas:

> On the back page of his book of logarithms, Gauss recorded the discovery of his formula for the number of primes up to *N* in terms of the logarithm function. Yet despite the importance of the discovery, Gauss told no one what he had found. The most the world heard of his revelation were the cryptic words, "You have no idea how much poetry there is in a table of logarithms."

David Samuels in "Knit Theory" explains why Gauss did not publish his discovery of hyperbolic geometry:

> Hyperbolic geometry, conceived by mathematician Carl Gauss in 1816, . . . describes a world that is curving away from itself at every point, making it the precise opposite of a sphere. . . . Gauss never published the idea, perhaps because he found it inelegant. In 1825, the Hungarian mathematician Janos Bolyai and the Russian mathematician Nikolai Lobachevsky independently rediscovered hyperbolic geometry.

After Gauss died, scientists discovered many novel theories and findings in his private notes that he had failed to disclose during his lifetime. So many new ideas were uncovered that his influence continued after his death and for the remainder of the 1800s. Gauss's mind was like an explosion in a mathematics factory, and mere mortals were sifting through the wondrous rubble for decades. He expressed his eternal love for mathematics in a letter to his friend and biographer Sartorius von Waltershausen (1809–1876), who preserved Gauss's thoughts in *Gauss zum Gedächtniss* (1856):

> Mathematics is the queen of sciences and arithmetic the queen of mathematics. She often condescends to render service to astronomy and other natural sciences, but in all relations she is entitled to the first rank.

In the same publication, Gauss is quoted as jokingly saying that some problems needed to be considered in the afterlife, when he had access to a new perspective:

> According to his frequently expressed view, Gauss considered the three dimensions of space as specific peculiarities of the human soul. . . . We could imagine ourselves, he said, as beings which are conscious of but two dimensions; higher beings might look at us in a like manner, and continuing jokingly, he said that he had laid aside certain problems which, when in a higher state of being, he hoped to investigate geometrically.

Gauss always had a facility with languages. In 1840, he studied Sanskrit for a short time, and when he was 62, he studied Russian, mastering it in two years and using it in correspondence. One reason for his learning Russian was his desire to read Russian mathematician Nikolai Lobachevsky's (1792–1856) work on non-Euclidean geometry in the original. He also often read English and toward the end of his life completed Edward Gibbon's *The History of the Decline and Fall of the Roman Empire*.

Gauss believed in the immortality of the soul and in the afterlife. He also believed in an omniscient and omnipotent God. Toward the end of his life, heart failure developed, and he died in Göttingen at age 78 on February 23, 1855, with friends and relatives by his side. His funeral was held three days later, with numerous students, friends, and townspeople in attendance. Von Waltershausen, his close friend, delivered a funeral sermon.

A lunar crater with a diameter of 177 kilometers was named after Gauss and approved in 1935 by the International Astronomical Union General Assembly. Gauss also has an asteroid named after him—1001 Gaussia.

Gaussia is also the name of giant copepods (crustaceans related to crabs and lobsters) that produce bright bioluminescent displays. To understand why such creatures received Gauss's name, recall that the famous German South Polar Expedition in 1901–1903 employed a ship named *Gauss* named after Carl Friedrich. The series of reports from the expedition, not published until 1931, revealed many new creatures, such as the large copepods, several species of which were given the name of the ship.

Shortly after Gauss's death, his brain was preserved in alcohol. In contrast to the average of 1,360 grams for males and 1,230 grams for females, the weight of Gauss's brain was 1,492 grams, and it was said to have had "highly developed convolutions." However, in a paper published in the 1999 *Proceedings of the Gauss Society*, researchers at the Max Planck Institute for Biophysical Chemistry and the University of Göttingen carefully studied his brain using magnetic resonance tomography and said they detected nothing grossly unusual about Gauss's brain. For example, the sylvian fissure—one of the main clefts dividing the brain and which was unusual in Einstein's brain—looked normal in Gauss's brain.

In addition to the unit of gauss for magnetic flux density, Gauss is honored by having the Gauss error function named in his honor. This function is useful in the fields of probability, statistics, and partial differential equations, and it may be expressed as

$$\mathrm{erf}(x) = \frac{2}{\sqrt{\pi}} \int_0^x e^{-t^2} \mathrm{d}t.$$

The Gauss hypergeometric function, also named in his honor, is

$${}_pF_q(\alpha_1, \ldots \alpha_p; b_1, \ldots, b_q; x).$$

This can be defined in the form of a hypergeometric series for which the ratio of successive terms is

$$\frac{c_{k+1}}{c_k} = \frac{P(k)}{Q(k)} = \frac{(k+\alpha_1)(k+\alpha_2)\ldots(k+\alpha_p)}{(k+b_1)(k+b_2)\ldots(k+b_q)(k+1)}x.$$

The function ${}_2F_1(\alpha, b; c; x)$, corresponding to $p = 2$ and $q = 1$, is known as Gauss's hypergeometric function and was the first of such functions to be studied.

Another function named after Gauss, the Gaussian function, is of the form

$$f(x) = a^{e^{-(x-b)^2/c^2}}$$

for some real constants $a > 0$, b, and c.

A Gaussian integer is a complex number whose real and imaginary part are both integers. The prime number elements of Gaussian integers are known as Gaussian primes.

A Gaussian distribution in a variable x with mean μ and variance σ^2 is expressed as a probability function:

$$P(x) = \frac{1}{\sigma\sqrt{2\pi}}e^{-(x-\mu)^2/(2\sigma^2)}$$

Statisticians and social scientists often refer to this as a "normal distribution" or "bell curve."

The Gaussian gravitational constant is $k = 0.01720209895(A^{3/2} S^{-1/2} D^{-1})$, where A is the mean radius of the orbit of Earth around the Sun, D is the mean rotation period of the Earth around its axis with respect to the Sun, and S is the mass of the Sun.

Gauss called

$$\Gamma = 9^{9^{9^9}}$$

"a measurable infinity." The number has $10^{369,693,100}$ digits, a number far larger than the number of atoms in the visible universe. If typed on paper, Γ would require $10^{369,693,094}$ miles of paper strip, according to mathematician and author Joseph Madachy. If the ink used in printing Γ was a one-atom thick layer, a million copies of our visible universe would not contain enough matter to print the number. Shockingly, the last 10 digits of Γ have been computed. They are 1,045,865,289.

Before we leave Gauss, I should point out that many famous mathematicians in addition to Gauss—like Srinivasa Ramanujan (1887–1920), James Hopwood Jeans (1877–1946), Georg Cantor (1845–1918), Blaise Pascal (1623–1662), and John Littlewood (1885–1977)—believed that inspiration had a divine aspect. As mentioned, Gauss said he once proved a theorem "not by dint of painful effort but so to speak by the grace of God."

Felix Klein's lectures on the development of mathematics from 1914–1919 include the following tribute to Gauss, as reported in George M. Rassias's *The Mathematical Heritage of C. F. Gauss*:

> He had only two peers, Archimedes and Newton, who were equally gifted. In common with both, Gauss had the unusually long life span which makes possible a full development of personality. Archimedes personifies the scientific achievements of classical antiquity, Newton is the initiator of higher mathematics, while Gauss represents the emergence of a new mathematical era.

FURTHER READING

Bell, E. T., *Men of Mathematics: The Lives and Achievements of the Great Mathematicians from Zeno to Poincaré* (New York: Touchstone; reissue edition, 1986).

Bühler, W. K., *Gauss, a Biographical Study* (New York: Springer, 2005).

Dunnington, G. Waldo, *Carl Frederich Gauss: Titan of Science* (New York: Hafner Publishing, 1955; reprint edition, Washington, D.C.: Mathematical Association of America, 2004).

Dunnington, G. Waldo, "The Sesquicentennial of the Birth of Gauss," *Scientific Monthly*, 24: 402–414, May 1927; see www.mathsong.com/cfgauss/Dunnington/1927/.

du Sautoy, Marcus, *The Music of the Primes: Searching to Solve the Greatest Mystery in Mathematics* (New York: Harper Perennial, 2004).

Hall, Tord, *Carl Friedrich Gauss* (Cambridge, Mass.: MIT Press, 1970).

Hayes, Brian, "Gauss's Day of Reckoning," *American Scientist*, 94(3): 200, May/June 2006; see www.americanscientist.org/template/AssetDetail/assetid/50686.

Madachy, Joseph S. *Madachy's Mathematical Recreations* (New York: Dover, 1979).

May, Kenneth O., "Carl Gauss," in *Dictionary of Scientific Biography*, Charles Gillispie, editor-in-chief (New York: Charles Scribner's Sons, 1970).

Newman, James R. (ed.) *The World of Mathematics* (New York: Simon and Schuster, 1956).

Rassias, George M., *The Mathematical Heritage of C. F. Gauss* (River Edge, N.J.: World Scientific, 1991).

Samuels, David, "Knit Theory," *Discover Magazine*, 27(3): 41–42, March 2006.

Tent, M. B. W., *Prince of Mathematics: Carl Friedrich Gauss* (Wellesley, Mass.: A. K. Peters, Ltd., 2006).

von Waltershausen, Sartorius, *Gauss zum Gedächtniss* (Leipzig, 1856).

Wilson, Curtis, "Carl Friedrich Gauss," in *Landmark Writings in Western Mathematics 1640–1940*, Ivor Grattan-Guinness, editor (Amsterdam: Elsevier, 2005).

INTERLUDE: CONVERSATION STARTERS

> People say to me, "Are you looking for the ultimate laws of physics?" No, I'm not; I'm just looking to find out more about the world and if it turns out there is a simple ultimate law which explains everything, so be it; that would be very nice to discover. If it turns out it's like an onion with millions of layers, and we're just sick and tired of looking at the layers, then that's the way it is. . . .
>
> —Richard Feynman, *The Pleasure of Finding Things Out: The Best Short Works of Richard P. Feynman*

> The idea of eternally true laws of nature is a beautiful vision, but is it really an escape from philosophy and theology? For, as philosophers have argued, we can test the predictions of a law of nature and see if they are verified or contradicted, but we can never prove a law must always be true. So if we believe a law of nature is eternally true, we are believing in something that logic and evidence cannot establish.
>
> —Lee Smolin, "Never Say Always," *New Scientist*, September 23, 2006

> The equations [of physics] are lovely, describing how a baseball arcs parabolically between Earth and sky or how an electron jumps around a nucleus or how a magnet pulls a pin. The ugliness is in the details. Why does the top quark weigh roughly 40 times as much as the bottom quark . . . ?
>
> —George Johnson, "Why Is Fundamental Physics So Messy?" *WIRED* magazine, February, 2007

POISEUILLE'S LAW OF FLUID FLOW

France, 1840. Flow rate in a tube is determined by the viscosity of the fluid, the change in pressure along the tube, and radius of the tube. In particular, the flow rate is (1) directly proportional to the pressure difference between the ends of the tube, (2) directly proportional to the fourth power of its internal radius, and (3) inversely proportional to its length and to the viscosity of the fluid.

CROSS REFERENCE: GOTTHILF HAGEN, LIONEL WILBERFORCE, AND THE HAGEN-POISEUILLE LAW.

> In 1840, American explorer Charles Wilkes circumnavigated Antarctica, claiming what became known as Wilkes Land for the United States. Wilkes's journey was the last "all-sail" naval mission to completely encircle the globe. Russian composer Peter Tchaikovsky was born. The first railroad dining cars were used in the United States.

Poiseuille's Law provides a precise mathematical relationship between the flow rate of a fluid in a pipe and the pipe width, fluid viscosity, and pressure change in the pipe. In particular, the law states that

$$\boxed{Q = \frac{\pi r^4}{8\mu}\frac{\Delta P}{L},}$$

where Q is the fluid flow rate in the pipe, r is the internal radius of the pipe, ΔP is the pressure difference between two ends of the pipe, L is the pipe length, and μ is the viscosity of the fluid. The law, named after French physiologist Jean Louis Marie Poiseuille (1799–1869), is also sometimes called the Hagen-Poiseuille Law after the German physicist and engineer Gotthilf Heinrich Ludwig Hagen (1797–1884) for his similar findings in 1839.

The law assumes that the fluid under study is exhibiting laminar (i.e. smooth, nonturbulent) flow and is incompressible (i.e., the density of the fluid remains constant for isothermal pressure changes). The fluid is also assumed to be in a steady state, which means that the speed at any point inside of the tube remains the same. Additionally, the pipe or tube is assumed to have a constant cross-sectional shape. Poiseuille experimentally derived his law in 1838 and published it in 1840.

This law has practical applications in medical fields and, in particular, in the study of flow in blood vessels. Note that the r^4 term ensures that the radius of a tube plays a major role in determining the flow rate Q of the liquid. If all other parameters are the same, a doubling of the tube width leads to a sixteenfold increase in Q. Practically speaking, this means that we would need sixteen tubes to pass as much water as one tube twice their diameter. From a medical standpoint, Poiseuille's Law can be used to show the dangers of atherosclerosis—if the radius of a coronary artery decreases twofold, the blood flow through it will decrease 16 times. It also explains why it is so much easier to sip a drink from a wide straw as compared to a slightly thinner straw. For the same amount of sipping effort, if you were to suck on a straw that is twice as wide, you would obtain 16 times as much liquid per unit time of sucking.

Similarly, Poiseuille's Law explains how the body can regulate blood flow. For example, in times of stress or strong demand, our bodies must sometimes direct more oxygen to one region of the body by reducing supply to less critical regions. Because blood flow rates depend on an r^4 dependence, the processes of vasodilation and vasoconstriction offer effective flow control mechanisms. We also use the process of vasodilation and vasoconstriction to regulate core body temperature, despite fluctuations in the temperature of the environment. On cold days, the nervous system may constrict blood flow (vasoconstriction) to the arms and legs to reduce the amount of colder blood returning to the main part of the body, which could produce a dangerous drop in body temperature. Vasoconstriction is one mechanism whereby the recreational drug MDMA (ecstasy) can cause hyperthermia, that is, unusually high body temperature.

As with many laws in this book, Poiseuille's Law becomes less accurate when some of the assumptions are violated, for example, in situations in which the fluid flow is not in a steady state or is not laminar. In the human body, the flow is not steady state because of the beating of the heart. Turbulence may be present in the major arteries where the blood moves rapidly. Similarly, when studying the airways leading to the lungs, researchers must keep in mind that Poiseuille's Law was derived using rigid, smooth, nonbranching tubes. The lung airways do not have these characteristics. Nevertheless, Poiseuille's Law is still useful, because it can give an overall understanding on the kinds of trends that scientists expect to see in many practical settings, and various correction factors can be applied for nonideal situations.

Poiseuille's Law can be used to find the viscosity of a fluid by performing a Poiseuille flow experiment. By experimentally determining Q (e.g., in mL/min) for different values of ΔP (e.g., by imposing different pressures on the mouth of a capillary tube), scientists can make a plot of $\Delta P/L$ versus Q. The slope of the straight line through the data points can then be used to determine the viscosity μ using Poiseuille's Law. In 1891, Lionel Robert Wilberforce (1861–1944) extended Poiseuille's Law for use in turbulent fluid flow. Wilberforce taught at the University of Liverpool, where he was professor of physics.

Today, Poiseuille's Law is used in many diverse branches of science. As just one example, my favorite application of the law is in the study of flows in pine tree wounds, and in particular, one study titled "Applicability of Poiseuille's Law to Exudation of Oleoresin from Wounds on Slash Pine," by several researchers working at the Southeastern Forest Experiment Station in Lake City, Florida. The researchers demonstrated how the equations for Poiseuille's Law could be modified and applied to the resin duct system of twelve slash pines and how the law could be used to select particular parent trees that would likely produce progeny having a high capacity for oleoresin production.

Poiseuille's Law has relevance when planning the placement of irrigation pipes. Long pipes are undesirable in certain circumstances because the flow rate is inversely proportional to the length of the pipe and thus a large pump would be required. The law is also useful in planning the number and sizes of pipes needed to fulfill the water requirements of a large city that is supplied by pipes coming from a dam.

As a final practical demonstration of Poiseuille's Law, consider that when an enlarged prostate constricts the urethra and decreases its radius, we can blame Poiseuille's Law for why even a small constriction can have dramatic effects on flow rate.

Jean Poiseuille (1797–1869), French physician and physiologist famous for his work with fluid flows.

CURIOSITY FILE: In addition to the aforementioned applications, the $1/r^4$ dependence in Poiseuille's Law helps us understand why a balloon catheter may be so helpful in angioplasty, a medical procedure used to enlarge a vessel occluded, for example, by atherosclerosis. A small increase in the radius of a blood vessel can cause a dramatic improvement in the flow of blood to a deprived organ such as the heart.

> Poiseuille's name is permanently associated with the physiology of the circulation of blood through arteries.
>
> —Kurt Pedersen, "Jean Poiseuille," in *Dictionary of Scientific Biography*

Jean Poiseuille was born in Paris, France. Various authoritative sources list his year of birth as either 1797 (e.g., several French sources and the *Dictionary of Scientific Biography*) or 1799 (e.g., *Encyclopaedia Britannica*). His father was a carpenter. In 1815, Poiseuille began his studies at the École

Polytechnique in Paris. His education focused on physics and mathematics, and in 1828 he received his doctorate of science.

Poiseuille's dissertation was titled *Recherches sur la force du coeur aortique* ("Research into the Aortic Force of the Heart"), and in this work he showed that blood pressure rises during expiration of the lungs and falls on inspiration. Of Poiseuille's dissertation, Thomas Soderqvist writes in *The Historiography of Contemporary Science and Technology*:

> Using a specially designed "hemodynamometer," Poiseuille performed experiments that had, according to a contemporary reviewer, a "characteristic of precision which commands confidence in their results even if one is a foreigner to mathematics and cannot follow the author through his calculations." Contrary to his own expectations, Poiseuille found that the average force calculated from a long series of measurements was the same in arteries at different distance from the heart.

Poiseuille's other papers include

- "Research Concerning the Origin of Motion of the Blood in the Veins" (1832)
- "Research about the Origin of Motion of the Blood in the Capillary Vessels (1839)
- "Research on the Movement of Liquids in Pipes of Small Diameters" (1840)

Poiseuille's most famous work dealt with the flow of blood through narrow tubes. In 1842, he was elected to the Académie de Médecine in Paris. He was elected inspector of the primary schools in Paris in 1860.

As discussed above, in 1838, Poiseuille had experimentally derived what we today call Poiseuille's Law, and he published his findings in 1840. The law provides a mathematical relationship between the flow rate Q in a tube and the fluid viscosity, inner tube width, and pressure change in the tube. Poiseuille's Law has a wide variety of applications in hydrology and medicine.

Poiseuille also improved existing methods for measuring blood pressure by using a mercury manometer. He decreased the rate of troublesome blood coagulation by placing potassium carbonate into the connection between the manometer and artery.

Because Poiseuille was so fascinated by the forces that affected the flow of blood in the smaller blood vessels of the body and because it was difficult to work with blood because of its tendency to coagulate, most of his experiments that led to his famous law were performed using thin glass tubes filled with water. By using compressed air to exert a pressure,

Poiseuille forced water through the tubes and measured the resulting flows. By varying the amount of air pressure applied and the radius of the tubes, Poiseuille discovered that the rate at which fluid passes through the tube increases proportionately to the pressure applied and also as the fourth power of the tube radius. However, the law that Poiseuille announced in 1840 did not specify the precise constant of $\pi/(8\mu)$ shown in the formula, and Poiseuille just considered this quantity to be a constant. Poiseuille also conducted related experiments and determined a temperature dependence of Q. Louis A. Bloomfield discusses Poiseuille's Law in *How Things Work: The Physics of Everyday Life*:

> It's hardly surprising that the flow rate depends . . . on the pressure difference, pipe length, and viscosity; we've all observed that low water pressure or a long hose lengthens the time needed to fill a bucket with water and that viscous syrup pours slowly from a bottle. But the dependence of the flow rate on the fourth power of pipe diameter may come as a surprise.

Bloomfield goes on to explain that the two most common garden hoses in the United States have diameters of 5/8 inch and 3/4 inch. This seems to be a very minor difference, but the 3/4-inch diameter hose can carry about 1.2^4 times as much water (i.e., roughly twice as much water) per unit time as the 5/8-inch diameter hose.

In 1839, German hydraulics engineer Gotthilf Heinrich Ludwig Hagen found the same law as Poiseuille, but Hagan's work was not appreciated at the time, and Poiseuille had no knowledge of his work.

In 1860, Jacob Hagenbach, professor of mathematics and physics at the University of Basel, named the law after Poiseuille. Both Hagenbach and German physicist Franz Neumann (1798–1895) independently determined the value of $\pi/(8\mu)$ that we use today.

Modern chemists and physicists honor Poiseuille's work by using the unit of "poise" as a unit of viscosity (resistance to flow).

FURTHER READING

Bloomfield, Louis A., *How Things Work: The Physics of Everyday Life* (Hoboken, N.J.: John Wiley & Sons, 2006).

Pedersen, Kurt, "Jean Poiseuille," in *Dictionary of Scientific Biography*, Charles Gillispie, editor-in-chief (New York: Charles Scribner's Sons, 1970).

Schopmeyer, C. S., François Mergen, and Thomas C. Evans, "Applicability of Poiseuille's Law to Exudation of Oleoresin from Wounds on Slash Pine," *Plant Physiology*, 29(1): 82, 1954.

Soderqvist, Thomas, *The Historiography of Contemporary Science and Technology* (Amsterdam, The Netherlands: Harwood Academic Publishers, 1997).

"Viscosity," Transtronics, Inc.; see xtronics.com/reference/viscosity.htm.

INTERLUDE: CONVERSATION STARTERS

When we look at the glory of stars and galaxies in the sky and the glory of forests and flowers in the living world around us, it is evident that God loves diversity. Perhaps the universe is constructed according to the principle of maximum diversity, a principle that says the laws of nature ... are such as to make the universe as interesting as possible. As result, life is possible, but not too easy. Maximum diversity often leads to maximum stress. In the end we survive, but only by the skin of our teeth.

—Freeman Dyson, "New Mercies: The Price and Promise of Human Progress," *Science & Spirit* July/August, 11(3): 17, 2000

But to believe a law is useful and reliable is not the same thing as to believe it is eternally true. We could just as easily believe there is nothing but an infinite succession of approximate laws. Or that laws are generalizations about nature that are not unchanging, but change so slowly that until now we have imagined them as eternal.

—Lee Smolin, "Never Say Always," *New Scientist*, September 23, 2006

Superstring theory turns out to be more complex than the universe it is supposed to simplify. Research suggests that there may be 10^{500} universes ... each ruled by different laws. The truths that Newton, Einstein, and dozens of lesser lights have uncovered would be no more fundamental than the municipal code of Nairobi. ... Physicists would just be geographers of some accidental terrain. ...

—George Johnson, "Why Is Fundamental Physics So Messy?" *WIRED* magazine, February, 2007

JOULE'S LAW OF ELECTRIC HEATING

England, 1840. The amount of heat produced by a steady electric current through a conductor is proportional to the resistance of the conductor, to the square of the current, and to the duration of the current.

CROSS REFERENCE: JOHN DALTON, CLAUSIUS'S LAW OF THERMODYNAMICS, FARADAY'S LAWS OF INDUCTION AND ELECTROLYSIS, FOURIER'S LAW OF HEAT CONDUCTION, THE JOULE-THOMSON EFFECT, AND THE FIRST LAW OF THERMODYNAMICS.

> In 1840, Great Britain issued the world's first official adhesive postage stamp. German physicist Friedrich Kohlrausch was born (see "Kohlrausch's Laws of Conductivity," below). Indoor bowling lanes first opened (in New York). Samuel Morse was granted a U.S. patent for his telegraph.

Joule's Law of Electric Heating states that the amount of heat H generated by a steady electric current flowing in a conductor may be calculated using

$$\boxed{H = K \cdot R \cdot I^2 \cdot t,}$$

where R is the resistance of the conductor, I is the constant current flowing through the conductor, and t is the duration of current flow. If the units for resistance are in ohms, current in amperes, time in seconds, and heat in calories, the constant K has the value 0.2390 calories/joule. If the heat is measured in joules, then K is 1.

When electrons travel through a conductor with some resistance R, the electric potential energy that the electrons lose is transferred to the resistor as heat. A classical explanation of this heat production involves the lattice of atoms in a conductor. The collisions of the electrons with the lattice cause the amplitude of thermal vibration of the lattice to increase, thereby raising the temperature of the conductor. This process is known as Joule heating.

Note that the law may often be applied to nonmetallic conductors. For example, Joule's Law may be observed in semiconductor materials. Also note that Joule heating produced by alternating current may be determined by observing the time average of the parameters in Joule's Law.

Joule's Law and Joule heating play a role in modern electrosurgical techniques in which the heat at an electrical probe is determined by Joule's Law. In such devices, current flows from an "active electrode" through the biological tissue to a neutral electrode, and the ohmic resistance of the tissue is determined by the resistance of the area in contact with the active electrode (e.g., blood, muscle, or fatty tissue) and the resistance in the total path between the active and neutral electrode. In electrosurgery, the duration (t in Joule's Law) is often controlled by a finger switch or foot pedal. The precise shape of the active electrode can be used to concentrate the heat so that it can be used for cutting (e.g., with a point-shaped electrode) or coagulation, which would result from diffuse heat that is produced by an electrode with a large surface area.

Joule heating is evident in a range of disparate phenomena. For example, in the late 1970s, researcher Thibaut Damour of the Observatoire de Paris published "Black-Hole Eddy Currents," in which he describes electromagnetic fields associated with black holes. Damour determined that a form of Joule heating was associated with the currents under study.

James Joule (1818–1889), British physicist famous for his research on the conservation of energy and for his law of heat production in electrical conductors.

CURIOSITY FILE: Joule's gravestone in Sale, Greater Manchester, is inscribed with "772.55," a mechanical equivalent of heat (in units of foot-pounds) that he determined. • Joule performed electrical tests on a woman servant, using a powerful battery. As he increased the voltage, she was told to report her sensations. The experiment continued until she became unconscious. • German physicist Julius Robert von Mayer (1814–1878) actually discovered Joule's mechanical equivalent of heat before Joule; however, von Mayer's manuscript on the subject was poorly written, confusing, and initially unnoticed. After Joule had won so much acclaim for work in this area, von Mayer had a nervous breakdown, attempted suicide, and was institutionalized.

> Joule's insufficient mathematical education did not allow him to keep abreast of rapid developments of the new science of thermodynamics, to the foundation of which he had made a fundamental contribution.... By the middle of the century, the era of the pioneers was closed, and the leadership passed to a new generation of physicists who possessed the solid mathematical training necessary to bring the new ideas to fruition.
>
> —L. Rosenfeld, "James Joule," in *Dictionary of Scientific Biography*

> Believing that the power to destroy belongs to the Creator alone, I affirm... that any theory which, when carried out, demands the annihilation of force, is necessarily erroneous.
>
> —James Joule, "On the Rarefaction and Condensation of Air," *Philosophical Magazine*, 1845

> In response to the tide of Darwinism then sweeping the country... 717 scientists [including James Joule] signed a remarkable manifesto entitled *The Declaration of Students of the Natural and Physical Sciences*, issued in London in 1864. This declaration affirmed their confidence in the scientific integrity of the Holy Scriptures. The list included 86 Fellows of the Royal Society.
>
> —James G. Crowther, *British Scientists of the Nineteenth Century*

> The story of Joule's rapid progress, from dilettante to a position of eminence in British science, can hardly be imagined in today's world of research factories and prolonged scientific apprentices. The theme that dominated Joule's research... was the belief that quantitative equivalences could be found among thermal, chemical, electrical, and mechanical effects.
>
> —William H. Cropper, *Great Physicists*

James Joule was born in Salford, near Manchester, England. He was the son of a wealthy brewer and was educated at home with his brother. From 1834 to 1837, the famous John Dalton (see "Dalton's Law of Partial Pressures," above) educated the two siblings in elementary mathematics, the scientific method, and chemistry. Joule noted that, "It was from his instruction that I first formed a desire to increase my knowledge by original researches." Joule was fascinated by electricity, and he enjoyed experimenting by giving electric shocks to his brother and to the family's servants.

Joule, like his father, was a religious Christian. He married Alice Amelia in 1847. According to some authors, during his honeymoon in the Alps, Joule and a colleague attempted an experiment to measure the temperature difference between the top and bottom of a large waterfall, but they may never have accurately performed the experiment for practical reasons. Armed with a huge thermometer, Joule found that he was unable to make proper measurements because of the large amounts of spray.

Joule's idea was that the water near the base of a waterfall should be about one degree Fahrenheit warmer than the water at the top for every 800 feet of drop. This change of temperature would have been mostly due to the kinetic energy that turned to heat as the water crashed at the bottom of the waterfall.

Joule's wife, Alice, died in 1854, leaving him with two children to raise. Shortly after his wife's death, Joule's family sold the brewery, and thereafter, Joule led a relatively secluded life, immersed in his experiments.

In a letter to *Philosophical Transactions*, Joule wrote,

> Any of your readers who are so fortunate as to reside amid the romantic scenery of Wales or Scotland could, I doubt not, confirm my experiments by trying the temperature of the water at the top and at the bottom of a cascade. If my views be correct, a fall of 817 feet (249m) will of course generate one degree [Fahrenheit] of heat, and the temperature of the river Niagara will be raised about one fifth of a degree by its fall of 160 (48m) feet.

Joule's prediction was fairly accurate. Today, we know that every 768-foot (234-meter) length of the waterfall height should generate a heat change corresponding to 1°F (0.55°K).

Joule's interest in science was motivated to a great degree by his Christian faith. For example, in 1873 he wrote the following in his notes for an address to be given at a meeting as President of the British Association for the Advancement of Science (he did not deliver the address due to poor health):

> After the knowledge of, and obedience to, the will of God, the next aim must be to know something of His attributes of wisdom, power and goodness as evidenced by His handiwork. . . . It is evident that an acquaintance with natural laws means no less than an acquaintance with the mind of God therein expressed.

Joule had begun his independent research at age 19, and throughout much of his life, he performed his key experiments at home in laboratories that he had usually built at his own expense. He is most famous for his careful measurements and experimental work. However, his relative lack of mathematical sophistication made it difficult for him to keep up with new theories in thermodynamics.

Joule was interested in energy, partly because Europe in the 1830s was in the midst of a technological revolution. Industry depended on steam engines that usually burned wood or coal in order to convert water into steam. The high-pressure steam moved the pistons of the engine. As discussed in "Clausius's Law of Thermodynamics" (below), scientists at this time were interested in the flow of energy and the efficiency of engines. It was also a time when English chemist and physicist Michael Faraday

(1791–1867) (see Faraday's Laws of Induction and Electrolysis," above) had discovered electromagnetic induction, and many wondered about the precise connections between work and heat.

Joule devoted himself to conservation laws involving energy and heat in a series of ambitious experiments. Through his research in the 1840s, he discovered that approximately 838 foot-pounds of work resulted in the heating of 1 pound of water through 1 degree Fahrenheit; that is, 1 Btu was roughly equal to 838 foot-pounds. (The currently accepted value is 778 foot-pounds.) In other words, he found a value for the amount of work necessary to produce a unit of heat. This was perhaps the first timc a scientist had asserted that a measured quantity of heat was equivalent to a corresponding amount of mechanical work.

Many years later, the British Association for the Advancement of Science was concerned with standards of electric resistance and asked Joule to determine the mechanical equivalent of heat from the thermal effects of electric currents. This time, Joule's experiments yielded a value of 783, and Joule felt that this value was more accurate than his older value determined by a frictional method. He finally arrived at a value of 772.55 foot-pounds for the amount of work that must be expended at sea level in order to raise the temperature of 1 pound of water from 60°F to 61°F.

In 1840, Joule found that the amount of heat produced by an electrical current is proportional to the square of the current value and with the resistance of the conductor. For this experiment, he varied current intensity and resistance in a circuit and measured slight changes in the temperature of water in which a coiled part of an electric circuit was dipped. Joule's success was due in part to his use of remarkably accurate thermometers.

Joule's numerous experiments throughout his life convinced him that all forms of energy were equivalent. In 1843, at age 24, he wrote in "On the Calorific Effects of Magneto-Electricity, and on the Mechanical Value of Heat":

> I shall lose no time in repeating and extending these experiments, being satisfied that the grand agents of nature are, by the Creator's fiat, *indestructible*; and that whenever mechanical force is expended [work is dissipated], an exact equivalent of heat is *always* obtained.

Joule's experiments, together with the work of other scientists of his time, suggested a general principle of conservation of energy. His work invited speculation that heat is associated with the motion of particles, but the development of his conservation principles did not require a clear theory of the atomic nature of matter.

Joule worked with his friend the British mathematician and physicist William Thomson (1824–1907), performing careful experiments on the behavior of gases in order to determine the extent to which potential energy might be stored or released during expansion of the gas. In a popular lecture presented in Manchester in 1847, Joule seemed to abolish the older caloric theory of matter that incorrectly held that changes in temperature are due to the transfer of an invisible and weightless fluid called caloric (see "Fourier's Law of Heat Conduction," above). At the lecture, he said,

> The most prevalent opinion until of late, has been that [heat] is a *substance* possessing, like all other matter, impenetrability and extension. We have however shown that heat can be converted into living force [kinetic energy] and into attraction through space [potential energy]. It is perfectly clear, therefore, that unless matter can be converted into attraction through space, which is too absurd an idea to be entertained for a moment, the hypothesis of heat being a substance must fall to the ground.

He concludes the same speech on a poetic note in which he muses about energy, the prophet Ezekiel, and the cosmos:

> When we consider our own frames, "fearfully and wonderfully made," we observe in the motion of our limbs a continual conversion of heat into living force, which may be either converted back again into heat or employed in producing an attraction through space, as when a man ascends a mountain. Indeed the phenomena of nature, whether mechanical, chemical, or vital, consist almost entirely in a continual conversion of attraction through space, living force, and heat into one another. Thus it is that order is maintained in the universe—nothing is deranged, nothing ever lost, but the entire machinery, complicated as it is, works smoothly and harmoniously. And though, as in the awful vision of Ezekiel, "wheel may be in middle of wheel," and every thing may appear complicated and involved in the apparent confusion and intricacy of an almost endless variety of causes, effects, conversions, and arrangements, yet is the most perfect regularity preserved.

Another important principle of Joule's is that when a gas under high pressure is passed through a porous plug or small opening, the gas usually undergoes a change in temperature according to the Joule-Thomson effect:

$$\left(\frac{\partial T}{\partial P}\right)_H = \frac{T\left(\frac{\partial V}{\partial T}\right)_P - V}{C_P}$$

The left side of the equation refers to the change in temperature with pressure at constant enthalpy (heat content). The enthalpy is constant because no heat is added to or removed from the system. The expression $(\partial V/\partial T)_P$ refers to the change of volume with temperature at constant pressure. C_P is the molar specific heat at constant pressure. The quantity on the left, $(\partial T/\partial P)_H$, is also sometimes referred to as the Joule-Thomson coefficient, which may be either negative, positive, or zero, depending on the temperature and pressure of the gas.

The Joule-Thomson inversion temperature is the temperature at which the coefficient is zero, and its value varies with the gas under study. At temperatures that are above the inversion temperature, gases are warmed when passed through the small opening. At temperatures below the inversion temperature, gases are cooled. For example, if a tank of carbon dioxide gas is opened in a laboratory room to the atmosphere, one may see a spray of fine dry-ice particles cooled to about −78°C emerging from the aperture. On the other hand, hydrogen and helium gas warm upon expansion under the same conditions.

The effect is named after James Joule and William Thomson, who happened to be another committed Christian. They investigated the effect in 1852 by following earlier work by Joule that concerned the expansion of gases. The Joule-Thomson effect has practical application today in industries in which manufacturers need to liquefy gases.

In 1861, Joule and his children moved to a new house, but his neighbors did not want him to bring his steam engine, which he used for his experiments. He wrote to Thomson:

> As for the olefiant gas experiment, I will try it as soon as I can get rid of the effect of a most villainous attack on me in the neighborhood and convince the people that the report that my experiments will burn up the vegetation is an infamous and malicious falsehood.

In the end, the local authorities prevented Joule from bringing the huge engine.

In 1870, he was awarded the Copley Medal by the Royal Society of London. He also acted as president for the British Association for the Advancement of Science in 1872 and 1887.

Today, Joule is remembered for helping to establish that mechanical, electrical, and heat energy are all related and can be converted to one another. Thus, he provided experimental validations for many elements of the Law of Conservation of Energy, also known as the First Law of Thermodynamics.

His interest in heating had involved ingenious experiments with viscous fluids in cylinders, moving paddle wheels, and microscopes. As one step in showing the conservation law, he had passed water through a perforated cylinder and was able to measure the slight viscous heating of the fluid. Here, he obtained a mechanical equivalent of 770 foot-pounds/Btu. This value was close to a value he had computed before in his electrical experiments, which suggested to Joule that work and heat can convert to one another. He also determined the mechanical equivalent of heat by measuring temperature changes produced by the movement of a paddlewheel submerged in water and turned by a falling mass that was connected to the paddlewheel. Although Joule was not the first person to establish the equivalence of heat energy and mechanical energy (sometimes referred to as "the mechanical equivalent of heat"), his various accurate demonstrations were influential in establishing such equivalences. His research revealed his skill in calibrating and reading thermometers, sometimes with the aid of a microscope, and his initial experience with the use of the thermometers may have come from exposure to instruments in his father's brewery.

Today, Joule is honored by the joule, the unit representing the mechanical equivalent of heat, abbreviated J. The joule is a unit of energy, or work, with units of $kg \cdot m^2/s^2$. A lunar crater with a diameter of 96 kilometers was named after Joule and approved in 1970 by the International Astronomical Union General Assembly.

Until the day he died, Joule strongly affirmed his belief in God, the creator. He wrote, as quoted in James G. Crowther's *British Scientists of the Nineteenth Century*, "After the knowledge of, and obedience to, the will of God, the next aim must be to know something of His attributes of wisdom, power and goodness as evidenced by His handiwork."

In a posthumous tribute to Joule, Thomson said that Joule

> had the genius to plan, the courage to undertake, the marvelous ability to execute and the keen perseverance to carry through to the end the great series of experimental investigations by which Joule discovered and proved the conservation of energy in electric, electromagnetic, and electrochemical actions, and in the friction and impact of solids, and measured accurately, by means of the friction of fluids, the mechanical equivalent of heat, cannot be generally and thoroughly understood at present. Indeed, it is all the scientific world can do just now in this subject to learn gradually the new knowledge gained.

FURTHER READING

Cardwell, Donald, *James Joule: A Biography* (Manchester, U.K.: Manchester University Press, 1991).

Crowther, James G., *British Scientists of the Nineteenth Century* (London: Paul, Trench, Trubner & Co., Ltd., 1935).

Damour, Thibaut, "Black-Hole Eddy Currents," *Physical Review D*, 18(10–15): 3598–3604, November 1978.

Heilbronner, Edgar, and Foil A. Miller, *A Philatelic Ramble Through Chemistry* (Zurich, Switzerland: Helvetica Chimica Acta, 2004).

James, Ioan, *Remarkable Physicists: From Galileo to Yukawa* (New York: Cambridge University Press, 2004).

Joule, James, "On the Calorific Effects of Magneto-Electricity, and on the Mechanical Value of Heat," *Philosophical Magazine*, 23: 263-276, September, 1843.

"James Prescott Joule," in *Notable Names Database*, Soylent Communications; see www.nndb.com/people/275/000049128/.

KLS Martin Group, "Electrosurgery Manual," Gebrüder Martin GmbH & Co.; see www.klsmartin.com/fileadmin/download/Sonderdrucke_PDF/90-604-02-04_09_06_Handbuch_HF.pdf..

Lamont, Ann, "James Joule: The Great Experimenter Who Was Guided by God," in *Answers in Genesis*; see www.answersingenesis.org/creation/v15/i2/joule.asp.

Rosenfeld, L., "James Joule," in *Dictionary of Scientific Biography*, Charles Gillispie, editor-in-chief (New York: Charles Scribner's Sons, 1970).

INTERLUDE: CONVERSATION STARTERS

We are in the age of search culture, in which Google and other search engines are leading us into a future rich with an abundance of correct answers along with an accompanying naive sense of conviction. In the future, we will be able to answer the questions—but will we be bright enough to ask them?

—John Brockman, *What We Believe but Cannot Prove*

God may well have created the universe and the laws of nature, but nature is still a machine, mechanically changing and comprehensible as such.

—Robert Todd Carroll, "Intelligent Design," *The Skeptic's Dictionary*

Our minds arise from the functioning of our physical brains, and the very precise physical laws that underlie that functioning are grounded in the mathematics that requires our brains for its existence.

—Roger Penrose, "What Is Reality?" *New Scientist*

Milkmaids have known for a long time that freshly churned butter is considerably warmer than the cream from which it is made. (We will soon see the elegant scientific use to which Joule put this observation.) ... Count Rumford [Benjamin Thompson] argued, the mechanical work performed by the horses in moving the tool against the resistance to friction was transformed into an equivalent amount of random microscopic motion, or heat. ... The idea which these observations suggest—that heat is a form of motion—dates at least as far as classical Greek times.

—Robert Eisberg and Lawrence Lerner, *Physics: Foundations and Applications*

KIRCHHOFF'S ELECTRICAL CIRCUIT AND THERMAL RADIATION LAWS

Germany, 1845 and 1859. Electrical Circuit Laws: The sum of the currents into a circuit junction equals the sum of the currents out of the junction. The sum of the voltage changes around a circuit loop are zero. Thermal Radiation Law: The ratio of absorptive and emissive power for a radiating body is a function of wavelength and temperature.

CROSS REFERENCE: PLANCK'S LAW OF RADIATION, JOSEPH VON FRAUNHOFER, GEORG OHM, AND JOSIAH GIBBS.

> In 1845, Florida and Texas became U.S. states. Edgar Allan Poe published "The Raven." The rubber band was invented and patented in England. American author Henry David Thoreau began his experiment in simple living at the beautiful Walden Pond in Massachusetts. *Scientific American* began publication.

ELECTRICAL CIRCUIT LAWS (1845)

Kirchhoff's Electrical Circuit Laws focus on the relationships between currents at a circuit junction and the voltages around a circuit loop.

Kirchhoff's Current Law

Kirchhoff's Current Law is a restatement of the principle of conservation of electrical charge in a system. In particular, at any point in an electrical circuit, the sum of currents flowing toward that point is equal to the sum of currents flowing away from that point. This law is often applied to the intersection of several wires to form a junction, like an X-shaped junction or a T-shaped junction, in which current travels toward the junction for some wires and away in other wires. We can also express this law by stating it in terms of the sum of the instantaneous currents entering the junction, or node:

$$\boxed{i_{\text{in}}^1 + i_{\text{in}}^2 + i_{\text{in}}^3 + \ldots + i_{\text{out}}^1 + i_{\text{out}}^2 + i_{\text{out}}^3 + \ldots = 0}$$

Here, i_{in} represents all the currents that flow into a circuit junction, and i_{out} represents all the currents that flow out of a junction. The superscripts designate the different wires in a junction.

Kirchhoff's Voltage Law

Kirchhoff's Voltage Law is a restatement of the conversation of energy in a system: The sums of the electrical potential differences around a circuit must be zero. Imagine we have a circuit with junctions. If we start at any junction and follow a succession of circuit elements that form a closed path back to the starting point, the sum of the changes in potential encountered in the loop is equal to zero. (Elements include conductors, resistors, and batteries.) We may express this relationship as

$$\boxed{v_1 + v_2 + v_3 + v_4 + \ldots = 0,}$$

where the various values for v are the voltage changes in components along a closed path in the circuit. As an example, voltage rises may occur when we follow the circuit across a battery (traversing from the – to + ends of a typical battery symbol in a circuit drawing). As we continue to trace the circuit in the same direction away from the battery, voltage drops may occur, for example, as a result of the presence of resistors in a circuit. Kirchhoff's Voltage Law means that at each instant of time, the sum of any voltage rises is equal to the sum of the voltage drops. The values for these rises and drops are determined by assessing the circuit elements in the same direction around the closed loop.

Note that the aforementioned statements of Kirchhoff's Electrical Circuit Laws make certain assumptions. For example, for the Current Law, we assume that the charge density does not change in time as might occur if there is an accumulation of a net positive or negative charge. If currents and potentials are changing with time, Kirchhoff's laws are close approximations if the capacitance, inductance, and resistance of the wires that connect circuit elements are much smaller than the capacitance, inductance, and resistance of the circuit elements themselves.

Kirchhoff's laws may be generalized to alternating currents if we assume all voltages and currents in a circuit are sinusoids of the same frequency. In this case, the algebraic sums are replaced by vector sums, and the Current Law simply states that the vector sum of the currents at a node at any instant is zero, and the Voltage Law states that the vector sum of the voltages around a loop at any instant is zero.

Finally, we have thus far assumed that the battery is simply a provider of emf (electromotive force). However, in reality, the potential difference across the battery terminals, called the terminal voltage, is not simply the emf of the battery. Terminal voltage actually decreases slightly as the current increases. Thus, a real battery is often represented by a source of emf plus an internal resistor. The internal resistance is usually quite small and may be ignored for many problems.

KIRCHHOFF'S THERMAL RADIATION LAW (1859)

An object radiates a unique spectrum that depends on the object's temperature and emissivity, a term that will be defined shortly. This radiation is called "thermal radiation" because it depends primarily on the temperature of an object.

To understand Kirchhoff's Thermal Radiation Law, note that physicists consider a "blackbody" to be an object that absorbs all electromagnetic energy incident upon it and that does not reflect or transmit any energy. The term was introduced by Kirchhoff, and he noted that such objects also emit the maximum possible amount of radiant energy at any given temperature. However, in nature, no real material exists that can completely absorb all radiation incident on it. All materials in our world actually reflect part of the incident radiation on them and emit less radiant energy than a blackbody at the same temperature. Nevertheless, many materials can approximate the characteristics of a blackbody, thus making Kirchhoff's Thermal Radiation Law very useful.

Consider a blackbody in thermal static equilibrium. Kirchhoff's Thermal Radiation Law states that the ratio of the radiated energy R from an object to the absorbed energy A is a constant C and dependent upon the wavelength and the temperature:

$$\boxed{\frac{R}{A} = C}$$

Let's explore the law further by defining *emissivity* as the ratio of an object's radiant energy to the radiant energy of a blackbody with the same temperature as the object. A true blackbody would have an emissivity of 1, while any real object would have an emissivity of less than 1. The *absorptivity* of the object is the fraction of incident energy that is absorbed by the body. One consequence of this law is that good reflectors are poor emitters. Thus, thermal blankets sometimes have reflective coatings and only slowly lose heat by radiation.

In his 1859 paper *Über den Zusammenhang von Emission und Absorption von Licht und Wärme* ("On the Relation Between Emission and Absorption of Light and Heat"), Kirchhoff concluded "that for rays of the same wavelength at the same temperature the ratio of emissive power and absorptive power is the same for all bodies." This is one way to express his law, but he reformulated the law in the same paper in terms of the ratio of emissive and absorptive powers, which "is a function of the wavelength and the temperature." Modern physicists write this as

$$\frac{e}{a} = f(T, \lambda),$$

where T is temperature, λ is wavelength, e is emissivity, and a is absorptivity. The function f describes the unique emission spectrum for all bodies that are black, that is, that absorb all incident radiation ($a = 1$).

All matter emits radiant energy as "thermal radiation," simply because the matter has a temperature and a constant motion of its atoms or molecules. Additionally, matter partially absorbs, and converts to heat, the radiant energy emitted by the environment of the object.

Kirchhoff also showed that for any body in thermal equilibrium, the emitted power equals the absorbed power. This means that the total emissivity equals the total absorptivity, sometimes expressed as $e = a$. For the special case of a perfect reflector, $a = 0$, and thus $e = 0$, which implies that a perfect reflector does not radiate.

According to Evitherm, the European "Virtual Institute for Thermal Metrology," emissivity is an important concept today in a variety of theoretical and practical settings:

> At high temperatures or in evacuated environments, thermal radiation is the main mode of heat transfer. Total emissivity governs the amount of thermal radiation lost or gained by an object and can therefore either cool or heat it, respectively. The reliable prediction of energy gains and losses to and from such structures as buildings, greenhouses, radomes [domes that house radar antennas], space vehicles, and industrial process plants has become an important part of energy conservation and control.

The concept and formulas associated with emissivity are useful in the field of radiation thermometry in order to deduce the temperature of objects from a measurement of their thermal radiation and the use of Planck's Law of Radiation (see part IV).

In the early 1860s, Kirchhoff made three assertions concerning the characteristics of spectra of objects. (A spectrum shows the variation in the intensity of an object's radiation at different wavelengths.) Today, scientists sometimes refer to these assertions as Kirchhoff's Laws of Spectral Formation, and they describe various scenarios and associated spectra:

1. A hot opaque body, such as a hot solid or dense gas, produces a continuous spectrum. A continuous spectrum is one in which electromagnetic radiation, such as light, is radiated over a continuous range of wavelengths.
2. A hot, transparent (i.e., low-density) gas produces an emission line ("bright-line") spectrum. In other words, the gas emits mostly at specific wavelengths according to the electron configuration of the atoms in the gas.
3. The combination of a cool, transparent (i.e., low-density) gas in front of a source of continuous emission produces an absorption-line ("dark-line") spectrum. In other words, the continuous spectrum, when passed through a low-density gas, produces a continuous spectrum with dark gaps at the same wavelengths as the emission lines described in the second assertion about spectra.

These kinds of rules are approximations but nevertheless are very useful in understanding the spectra of stars. Astronomical objects with different temperatures and compositions emit different spectra, and astronomers can deduce significant information about such objects by studying their spectra.

A NOTE ON SPECTRA

Bright lines in atomic spectra occur when electrons jump from higher energy levels down to lower energy levels. The color of the lines depends on the energy difference between the energy levels. Dark absorption lines in spectra can occur when an atom absorbs light and the electron jumps to a higher energy level. The particular values for the energy levels are identical for atoms of the same type.

By looking at absorption or emission spectra, we can tell what chemical elements produced the spectra. In the 1800s, various scientists noticed that the spectrum of the Sun's electromagnetic radiation was not a smooth curve from one color to the next but had numerous dark lines. This suggested that light was being absorbed at certain wavelengths. These dark lines are called Fraunhofer lines after the Bavarian physicist Joseph von Fraunhofer (1787–1826), who recorded such lines.

Some readers may find it easy to imagine how the Sun can produce a radiation spectrum, but not how it can also produce dark lines. How can the Sun absorb its own light?

You can think of stars as fiery gas balls that contain many different atoms emitting light at a range of colors. Light from the surface of a star, the *photosphere*, has a continuous spectrum of colors, but as the light travels through the outer atmosphere of a star, some of the colors (i.e., light at different wavelengths) are absorbed. This absorption is what produces the dark lines. This phenomenon reminds me a little of turning down the color intensities on a television so that the picture becomes black and white. In the case of the Sun, it is as if many different controls are reducing the intensity of many different narrow ranges of color. On a television of this sort, you would still see a color picture, but certain shades of color might be missing. Imagine watching *Baywatch* if all the red were suddenly absorbed. The red bathing suits would turn black. In stars, the missing colors, or dark absorption lines, tell us exactly what chemical elements are in the outer atmosphere of stars.

Scientists have catalogued numerous missing wavelengths in the spectrum of the Sun. By comparing the dark lines with spectral lines produced by chemical elements on Earth, astronomers have found more than seventy elements in the Sun.

Gustav Kirchhoff (1824–1887), German mathematician famous for his laws concerning electrical circuits and electromagnetic radiation.

CURIOSITY FILE: Kirchhoff discovered that the velocity of electrical signal propagation in an uninsulated wire was approximately equal to the velocity of light in a vacuum. • In 2006, John Mather of NASA's Goddard Space Flight Center in Greenbelt, Maryland, received the Nobel Prize in Physics, in part due to his work that confirmed the fit between the theoretical curve for blackbody radiation at a temperature of 2.7°K and the cosmic background radiation, thus proving that the universe was a near-perfect blackbody at some point in time after the Big Bang.

> Spectrum analysis, which, as we hope we have shown, offers a wonderfully simple means for discovering the smallest traces of certain elements in terrestrial substances, also opens to chemical research a hitherto completely closed region extending far beyond the limits of the earth and even of the solar system. Since in this analytical method it is sufficient to see the glowing gas to be analyzed, it can easily be applied to the atmosphere of the sun and the bright stars.
>
> —Gustav Kirchhoff and Robert Bunson, "Chemical Analysis by Observation of Spectra," *Annalen der Physik und der Chemie*, 1860

Gustav Kirchhoff was born in Königsberg, Prussia. His father was a lawyer. Kirchhoff was a talkative child and small for his age. When he was 18, Kirchhoff attended the Albertus University of Königsberg. While studying under German physicist Franz Neumann (1798–1895), he conducted his famous research of the behavior of electrical currents.

Kirchhoff extended the work of German physicist Georg Ohm (1789–1854) and announced his circuit laws in 1845, which allowed scientists to better understand and calculate currents, voltages, and resistances in electrical circuits with various loops. (Note that Kirchhoff was only a 21-year-old student at the time.) These laws, discussed above, concern the flow of current and the behavior of voltages in circuits with junctions or nodes.

In 1847, Kirchhoff graduated from Königsberg and married Clara Richelot, who was the daughter of his mathematics professor. They moved to Berlin in the same year, and he taught at the University of Berlin from 1848 to 1850. In 1850, he was appointed as extraordinary professor at the University of Breslau. In 1854, he was appointed professor of physics at the University of Heidelberg, where he collaborated with German chemist Robert Bunsen (1811–1899).

In 1859, Kirchhoff proposed that each element had a unique and characteristic spectrum, and he asserted his Law of Thermal Radiation, which suggested, among other things, that for a given atom or molecule, the emission and absorption frequencies are the same. In other words, a material capable of emitting a particular spectral line also absorbs strongly at the same frequency. Kirchhoff and Bunsen examined the spectrum of the Sun in 1861 and identified chemical elements in the atmosphere of the Sun—including two new elements, cesium and rubidium, that they discovered during their investigations. John Emsley explains the discovery of cesium in *Nature's Building Blocks*:

> Kirchhoff and Bunsen took about 30,000 liters of mineral water which they boiled down, and from which they removed the lithium, sodium, potassium, magnesium, calcium and strontium salts. The remaining liquor was sprayed into a Bunsen flame and the light produced was analyzed using a spectroscope, showing two blue lines very close together. This has never seen before, so Bunsen and Kirchhoff immediately realized that they had stumbled on a hitherto unknown element.

Kirchhoff is famous for being the first to explain the dark lines in the spectrum of the Sun as caused by absorption of light at particular wavelengths as the light passes through gases in the atmosphere of the Sun. More generally, Kirchhoff and Bunsen founded the theory of spectral analysis in which a researcher could perform chemical analyses merely by studying the light emitted by heated materials. They also demonstrated that every element radiates an identifying fingerprint of colored light when heated. Kirchhoff and Bunsen write in their 1860 paper "Chemical Analysis by Observation of Spectra":

> Spectrum analysis should become important for the discovery of hitherto unknown elements. If there should be substances that are so sparingly distributed in nature that our present means of analysis fail for their recognition and separation, then we might hope to recognize and to determine many such substances in quantities not reached by our usual means, by the simple observation of their flame spectra. We have had occasion already to convince ourselves that there are such now unknown elements.

Kirchhoff also showed that if an object at a particular temperature is in "radiative equilibrium," the ratio of the absorptive and emissive powers for each wavelength depends only on the wavelength and temperature of the object. He coined the term "blackbody radiation" in 1862.

In 1869, his wife Clara died, leaving Kirchhoff to raise his four children alone. This task would have been difficult for any man, but it was made especially challenging by his disability, which due to an earlier accident, forced him to spend his life on crutches or in a wheelchair. In 1872, he married Luise Brömmel, a woman who supervised a local eye clinic.

Perhaps his best-known work is the four-volume *Vorlesungen über mathematische Physik* (Lectures on Mathematical Physics, 1876–94). L. Rosenfeld in the *Dictionary of Scientific Biography* gives Kirchhoff a fitting eulogy:

> In all his work, Kirchhoff strove for clarity and rigor in the quantitative statement of experience, using a direct and straightforward approach and simple ideas. His mode of thinking is as conspicuous in his contributions of immediate practical value (the laws of electrical networks) as in those with wide implications (the method of spectral analysis).... The excellence of Kirchhoff as a teacher can be inferred from the printed text of his lectures... [that] set a standard for the teaching of classical theoretical physics in German universities....

After Kirchhoff died, German physicist Max Planck (1858–1947) succeeded him as the chair of theoretical physics at the famous University of Berlin. A lunar crater with a diameter of 24 kilometers was named after

Kirchhoff and approved in 1935 by the International Astronomical Union General Assembly.

FURTHER READING

Emsley, John, *Nature's Building Blocks: An A–Z Guide to the Elements* (New York: Oxford University Press, 2003).

Evitherm, the Virtual Institute for Thermal Metrology, "What Physical Processes Determine Emissivity?"; see www.evitherm.org/default.asp?ID=221.

Pickover, Clifford, *The Stars of Heaven* (New York: Oxford University Press, 2001).

Rosenfeld, L., "Gustav Kirchhoff," in *Dictionary of Scientific Biography*, Charles Gillispie, editor-in-chief (New York: Charles Scribner's Sons, 1970).

Schirrmacher, Arne, "Experimenting Theory: The Proofs of Kirchhoff's Radiation Law before and after Planck," Münchner Zentrum für Wissenschafts- und Technikgeschichte (Munich Center for the History of Science and Technology), March 2001; see www.mzwtg.mwn.de/arbeitspapiere/Schirrmacher_2001_1.pdf.

INTERLUDE: CONVERSATION STARTERS

> You find that many laws of nature are themselves connected to others by still deeper laws, that those deeper laws have even deeper connections, and so on. Eventually, at the very center of the web, you find a relatively small number of laws that cement the whole framework together.... These are sometimes referred to as "laws of nature." ... To paraphrase *Animal Farm*, all laws of nature are equal, but some laws are more equal than others.... Of course, there is no universal agreement among scientists as to what the overarching principles are ..., but you would be hard pressed to find a scientist who doesn't agree that they exist.
>
> —James S. Trefil, *The Nature of Science: An A–Z Guide to the Laws and Principles Governing Our Universe*, 2003

> A law of nature is a law and hence a conceptual and linguistic entity, and a law of nature refers to nature, i.e. to the real word. At first glance, it is not quite clear how these two aspects fit together.... The world (universe) is ordered and structured by laws. Popper called this assumption the "law of lawfulness."
>
> —Peter Mittelstaedt and Paul A. Weingartner, *Laws of Nature*, 2005

How often might a man, after he had jumbled a set of letters in a bag, fling them out upon the ground before they would fall into an exact poem, yea, or so much as make a good discourse in prose? And may not a little book be as easily made by chance, as this great volume of the world? How long might a man be in sprinkling colours upon a canvas with a careless hand, before they could happen to make the exact picture of a man? And is a man easier made by chance than his picture? How long might twenty thousand blind men, which should be sent out from the several remote parts of England, wander up and down before they would all meet upon Salisbury Plains, and fall into rank and file in the exact order of an army? And yet this is much more easy to be imagined than how the innumerable blind pans of matter should rendezvous themselves into a world.

—John Tillotson, *Maxims and Discourses, Moral and Divine*, 1719

CLAUSIUS'S LAW OF THERMODYNAMICS

⚛ π ⚗ *Germany, 1850.* The entropy change of the universe, for any given process, must be greater than or equal to zero. Heat flows spontaneously from a hotter to a colder object but not vice versa.

CROSS REFERENCE: FIRST LAW OF THERMODYNAMICS, THE SECOND LAW OF THERMODYNAMICS, THE THIRD LAW OF THERMODYNAMICS, HESS'S LAW OF CONSTANT HEAT SUMMATION, GIBBS FREE ENERGY, SADI CARNOT, LUDWIG BOLTZMANN, WALTER NERNST, JAMES JOULE, AND JAMES CLERK MAXWELL.

> In 1850, California was admitted as the 31st U.S. state. American Express was founded by Henry Wells & William Fargo. African-American abolitionist Harriet Tubman started to help slaves escape to the northern free states and Canada using the Underground Railroad.

Clausius's Law, which is the Second Law of Thermodynamics in one of its early formulations, states that the total entropy, or disorder, of an isolated system tends to increase over time as it approaches a maximum value. For a closed thermodynamic system, entropy can be thought of as measure of the amount of thermal energy unavailable to do work. In 1865, German physicist and mathematician Rudolph Clausius (1822–1888) stated the First and Second Laws of Thermodynamics in the following form: "*Die Energie der Welt ist konstant; die Entropie der Welt strebt einen Maximum zu.*" In other words,

1. The energy of the universe is constant.
2. The entropy of the universe tends to a maximum.

Thermodynamics is the study of heat and, more generally, the study of transformations of energy in all its forms. The implications of the Second Law of Thermodynamics can be stated many ways. For example, the Second Law implies that all matter and energy in the universe tends to evolve toward a state of uniformity. We also indirectly invoke the Second Law when we consider that a house, body, or car—without maintenance—deteriorates over time. Or, as William Somerset Maugham wrote in *Of Human Bondage*, "It's no use crying over spilt milk, because all of the forces of the universe were bent on spilling it." Woody Allen in his 1992 movie *Husbands and Wives* states the law and the tendency toward disorder in a modern way:

It's the Second Law of Thermodynamics—sooner or later everything turns to shit. That's my phrasing, not the *Encyclopaedia Britannica*'s.

The second law can be described mathematically in several ways. For its simplest expression, we may write

$$\frac{dS}{dt} \geq 0,$$

where S is the entropy, and the change in entropy with respect to time (dS/dt) is equal to zero only when entropy is at its maximum equilibrium value.

When thinking about Clausius's Law, it is useful to make a distinction between spontaneous processes (e.g., the cooling of hot water at room temperature, which occurs without needing to be driven by some external source) and nonspontaneous processes (which do not happen on their own). Early in his career, Clausius stated the Second Law as "heat does not transfer spontaneously from a cool body to a hotter body." Today, we say that the entropy of an isolated system generally increases when a spontaneous change occurs. Although the total amount of energy must be conserved in any process, the *distribution* of that energy changes in an irreversible manner. Hot objects cool, and cool objects do not spontaneously become hot.

The Second Law is also sometimes written as

$$dS = \frac{dq_{\text{reversible}}}{T},$$

where dS is the change in entropy of a system, dq is the energy transferred to the system as heat, T is the temperature in degrees kelvin, and "reversible" indicates that the heat transfer must be done in a reversible fashion, that is, without producing entropy other than in the system. When the temperature is low, a given amount of heat transferred to the system yields a greater change in entropy than when the temperature is high. These equations are useful because they are not limited to applications involving engines and mechanical devices that use heat in their operations but can also be applied to understand countless natural processes. These formulations are often referenced in discussions ranging from philosophy to astronomy.

Austrian physicist Ludwig Boltzmann (1844–1906) expanded upon these compact definitions for entropy and Clausius's Law when he interpreted entropy as a measure of the disorder of a system due to thermal motion of molecules of a system. Thus, if the temperature is low, adding a quantity of heat to the system may elicit a relatively large additional disorder in the thermal motion of molecules in the system.

The Second Law of Thermodynamics places constraints upon the direction of heat transfer as well as the efficiencies that may be achieved by engines that are powered by heat. The Second Law says that, in a closed system, we cannot finish any real physical process with as much useful energy as we had to start with, because some energy is always wasted. Thus, the Second Law can be used to show the impossibility of most classes of proposed perpetual motion machines.

From another perspective, the Second Law says that two adjacent systems in contact with each other tend to equalize their temperatures, pressures, and densities. For example, when a hot piece of metal is lowered into a tank of cool water, the metal cools and the water warms until each is at the same temperature. An isolated system that is finally at equilibrium can do no useful work without energy applied from outside the system, which is yet another way to understand how the Second Law prevents us from building many classes of perpetual motion machines.

Creationists sometimes argue that because the Second Law of Thermodynamics drives nature to disorder and chaos, the intricate and ordered patterns of life could not have arisen spontaneously without an intelligent designer. But nothing could be further from the truth. For example, we do not need to invoke an intelligent designer whenever we see an ordered snowflake crystal. The Second Law of Thermodynamics does not preclude pockets of order from arising naturally, especially when we consider that life-giving energy is continually pumped into our world in the form of sunlight. The Sun drives photosynthesis, which has enabled complex life to flourish. We should note the Second Law expresses a tendency, which means that it is highly unlikely that entropy will decrease in a closed system at a particular instant, but entropy decrease is possible. Additionally, while it is true that the overall entropy of a complete, or closed, system generally increases when spontaneous change occurs, in cases of spontaneously interacting *subsystems* of a closed system, some may gain entropy, while other subsystems may lose entropy.

As with other laws in this book, the Second Law of Thermodynamics was due to more than one person. French physicist Nicolas Léonard Sadi Carnot (1796–1832) in 1824 realized that the efficiency of converting heat to mechanical work depended on the difference of temperature between hot and cold objects. As discussed above, Boltzmann derived the Second Law of Thermodynamics from probability arguments involving the motion of particles. Other scientists, such as American mathematician Claude E. Shannon (1916–2001) and German-American physicist Rolf Landauer (1927–1999), have shown how the Second Law and the notion of entropy also apply to communications and information theory.

Another thought experiment to consider is one in which the air molecules in your living room are very likely to be distributed evenly, because of the overall result of the random molecular motions. However, an infinitesimally small chance exists that the air molecules could travel to one corner of the room, thus suffocating you. In fact, the so-called Fluctuation Theorem states that the chances of the Second Law being violated increase as the system under study gets smaller. While we cannot predict the movement of individual molecules, the tendency toward homogeneity on the macroscopic scale is predictable. Similarly, sandcastles are very unlikely to arise spontaneously from the beach—the improbable rarely happens.

An amoeba builds its "improbable structure" from the disordered materials around it, but it does this at the expense of increasing the entropy around it. A crystal or a planet forms because it represents a lower potential energy than alternative configurations and is therefore more probable. As discussed above, it is only the overall entropy of closed systems that increases; the entropy of individual components of a closed system may and often do decrease.

Every day, most of us work toward "counteracting" this entropic "urge" to turn our bodies to dust and to cause sandcastles to fade. We burn fuel to create skyscrapers from concrete; we metabolize food to build proteins from amino acids; and, as physical chemist Peter W. Atkins once said, "The random fluctuations of electrical currents in neuronal circuits within the brain can be changed into coherent thoughts and works of art." A local region of space can become beautifully structured if is coupled to nearby pockets of disorder, as when a molecule of ATP "disintegrates" to ADP within a cell. Of course, for most of these examples, we are fighting an eventual losing battle, and finally we die, disintegrate, and become at equilibrium with our environment.

Corey S. Powell likens entropy to the shuffling of a card deck in his article "Welcome to the Machine":

> Entropy is one of those words that almost everyone has heard and almost nobody can really explain.... [It's the] amount of disorder and information in a system. [Consider] a fresh, unshuffled deck of cards. In that state it has low entropy and contains little information. Just two pieces of data (the hierarchy of suits and the relative ranks of the cards) tell you where to find every card in the deck without looking. [After shuffling,] the deck has a lot of entropy and a lot of information. If you want to locate a particular card, you have to hunt through the entire deck. There is only one perfectly ordered state but about 10^{68} disordered ones, which is why you will never, ever accidentally shuffle the deck back into its original order.

In 2002, chemical physicists in Australia showed that the Second Law could be temporarily suspended in microscopic systems, and today physicists realize that the Second Law does not apply at atomic scales over very short times in which situations that "violate" the Second Law become more likely. The research in 2002 was notable because it showed that the Second Law can be routinely broken at the micrometer-size scale for up to 2 seconds.

According to Matthew Chalmers's "Second Law of Thermodynamics 'Broken'," this team at the Australian National University in Canberra, led by Denis Evans, measured changes in the entropy of latex beads, each a few micrometers across and suspended in water. The researchers found entropy changes that were negative over time intervals of a few tenths of a second, which may be thought of as a portion of nature that runs in reverse. The beads gained energy from the random motion of the water molecules, which the researchers likened to a "small-scale equivalent of the cup of tea getting hotter." However, over time intervals of more than 2 seconds, an overall positive entropy change was measured and normality was restored. The team's experiment provided the first evidence that the Second Law of Thermodynamics can be violated at appreciable time and length scales. The Australian team is currently reviewing how their results may affect the field of nanotechnology and our understanding of molecular interactions inside living creatures. Evans's work was important because it experimentally verified results that showed what other scientists would have expected: that the Second Law doesn't always apply at small size scales.

Before proceeding to the biography of Clausius, I should note that a Third Law of Thermodynamics exists. From a classical physics perspective, the Third Law states that as a system approaches absolute zero temperature (0°K, −273.15°C, or −459.67°F), all processes cease and the entropy of the system approaches a minimum value. The Third Law was developed by German chemist Walther Nernst (1864–1941) around 1905. The modern version of the law can be stated as follows: As temperature goes to absolute 0, the entropy S approaches a constant S_0. Classically speaking, the entropy of a pure and perfectly crystalline substance is 0 if the temperature could actually be reduced to absolute zero.

According to the creative folks at everything2.com, it is amusing to ponder the classical ramifications of what it means to be at absolute zero:

> Since heat is a measure of average molecular motion, zero thermal energy means that the average atom does not move at all. Since no atom can have less than zero motion, the motion of every individual atom must be zero if the average molecular motion (heat) is zero.

> Thus, if the entire universe had an average heat of zero, the universe . . . would be over as far as we're concerned: no motion, no reactions, no observers. If the universe becomes an unobservable motionless system, does it really exist?

As discussed above, using a classical analysis, all motion stops at absolute zero. However, I should note that the ramifications of absolute zero are more complicated because quantum mechanical "zero-point motion" allows systems in their lowest possible energy state (i.e., "ground state") to have a finite probability of being found over extended regions of space. Thus, two atoms bonded together are not separated at some single distance from each other, but can be thought of as undergoing rapid vibration with respect to one another, even at absolute zero.

The phrase "zero-point motion" is used by scientists to describe the fact that atoms in a solid—even a super-cold solid—do not remain at exact geometric lattice points, but rather, a probability distribution exists for both their positions and momenta. Scientists have achieved temperatures of less than a ten-thousandth of a degree centigrade above absolute zero.

It is impossible to cool a body to absolute zero by any finite process. If scientists could cool objects to absolute zero, all such bodies would still have a definite energy, called the zero-point energy.

I conclude this section with some quotations relating to the Second Law. Michio Kaku in *Hyperspace* reminds us that British scientist and author C. P. Snow had a useful way of remembering the three laws of thermodynamics:

1. *You cannot win* (i.e., you cannot get something for nothing, because matter and energy are conserved).
2. *You cannot break even* (you cannot return to the same energy state, because there is always an increase in disorder; entropy always increases).
3. *You cannot get out of the game* (because absolute zero is unattainable).

Astrophysicist Sir Arthur Stanley Eddington writes in *The Nature of the Physical World*:

> The law that entropy increases—the Second Law of Thermodynamics—holds, I think, the supreme position among the laws of Nature. If someone points out to you that your pet theory of the Universe is in disagreement with Maxwell's equations—then so much the worse for Maxwell's equations. If it is found to be contradicted by observation—well, these experimentalists do

bungle things sometimes. But if your theory is found to be against the Second Law of Thermodynamics I can give you no hope; there is nothing for it but to collapse in deepest humiliation.

Rudolf Clausius (1822–1888), German mathematical physicist who formulated the Second Law of Thermodynamics and who generally advanced the field of thermodynamics.

CURIOSITY FILE: Clausius's work was largely theoretical, and he never published any experimental research.

> The history of thermodynamics is a story of people and concepts. The cast of characters is large. At least ten scientists played major roles in creating thermodynamics, and their work spanned more than a century. The list of concepts, on the other hand, is surprisingly small: there are just three leading concepts in thermodynamics: energy, entropy, and absolute temperature.
>
> —William H. Cropper, *Great Physicists*

> No part of science has contributed as much to the liberation of the human spirit as the Second Law of thermodynamics. Yet, at the same time, few other parts of science are held to be so recondite. Mention of the Second Law raises visions of lumbering steam engines, intricate mathematics, and infinitely incomprehensible entropy. Not many would pass C. P. Snow's test of general literacy, in which not knowing the Second Law is equivalent to not having read a work of Shakespeare.
>
> —Peter W. Atkins, *The Second Law*

> All flesh shall perish together, and man shall turn again to dust.
>
> —Job 34:15

> Amazingly, this process of generating entropy is universal. It is what happens when a candle burns, when the sun shines and when your stomach digests your lunch. In every instance, there is an inexorable, irreversible trend toward disorder and an increase in the total amount of information in the world.
>
> —Corey S. Powell, "Welcome to the Machine," *New York Times Book Review*

Rudolf Clausius was born in 1822 in Köslin, Prussia—his parent's fourteenth child. The year 1822 is important in the history of thermodynamics because it is the year in which French engineer Sadi Carnot was in the process of finishing his grand work "Reflections on the Motive Power of Fire," which would one day inspire Clausius to investigate the properties of heat and energy flow as well as their profound implications for machines and men.

As a boy, Clausius attended a small school that his father had established. Clausius was curious about nature and loved to collect seashells. In 1840, he entered the University of Berlin. Only two years earlier, steamships first crossed the Atlantic Ocean.

In 1843, his mother died while giving birth to her eighteenth child, and Clausius began to help raise the younger children.

His paper on the theory of heat, published in 1850, helped him secure a teaching position at the Royal Artillery and Engineering School in Berlin, and in 1855 he became professor of mathematical physics at the École Polytechnicum, a prestigious new university in Zurich. While in Zurich, he met an attractive young woman, Adelheid Rimpau, and they were wed in 1859. In 1870, he was wounded while leading a student ambulance corps in the Franco-Prussian War. In 1875, his wife died while giving birth to their sixth child. Michael Guillen, in *Five Equations That Changed the World*, eloquently writes of Clausius feelings at the time:

> How ironic, how cruel and painful, was the timeless struggle between life and death, Clausius lamented bitterly, holding his wife's cooling hand. He had devoted his career to the scientific understanding of heat. . . . The universe as a whole was dying. . . . Indeed, even now at this moment of his most profound grief, the grim imbalance had been maintained: He had lost a wife and gained a daughter, but in his heart and mind, Clausius understood how and why the great equation of life had taken more than it had given.

Clausius's continued pain from his war injury, coupled to his responsibilities for the sole care of his family, may have hampered his scientific progress in these later years of his career. Despite these hardships, Clausius was a caring father. His brother Robert wrote in Clausius's obituary notice, "He was the best and most affectionate of fathers, fully entering into the joys of his children. He himself supervised the schoolwork of his children."

After his wife's death, Clausius longed to return to Germany, and in 1877, he accepted a professorship at the University of Würzburg in Germany. In 1879, he received the Copley Medal of the Royal Society.

Let me briefly review the field of thermodynamics around the time of Clausius before describing his work in more detail. Much of the initial work in this area centered on the operation of engines and how fuel, such as coal, could be efficiently converted to useful work by the engine. Carnot is probably most often considered the "father" of thermodynamics due to his 1824 work *Réflexions sur la puissance motrice du feu* ("Reflections on the Motive Power of Fire"). Sadly, in 1832 Carnot contracted cholera, and by order of the health office, nearly all his books, papers, and other personal belongings had to be burned! This made it difficult for Clausius to initially obtain additional information about Carnot and his ideas.

Carnot worked tirelessly to understand heat flow in machines partly because he was disturbed that British steam engines seemed to be more efficient than French engines. During his day, steam engines usually burned wood or coal in order to convert water into steam. The high-pressure steam moved the pistons of the engine. When the steam was released through an exhaust port, the pistons returned to their original positions. A cool radiator converted the exhaust steam to water, which could be heated again to steam in order to drive the pistons.

Carnot imagined an ideal engine, which we refer to today as a Carnot engine, that would theoretically have a work output equal to that of its heat input and not lose even a small amount of energy during the conversion. This perfectly efficient energy would not violate the First Law of Thermodynamics that states, in essence, that energy is conserved and that the increase in the internal energy of a system is equal to the amount of energy added to the system by heating, minus the amount lost in the form of work done by the system on its surroundings. The First Law of Thermodynamics is actually an extension of results obtained by English physicist James Joule (1818–1889).

After various experiments, Carnot came to realize that no device could perform in this ideal matter—some energy had to be lost to the environment. Energy in the form of heat could not be converted completely into mechanical energy. Carnot also wrote in *Réflexions sur la puissance motrice du feu* (*Reflections on the motive force of heat*), "The production of heat is not sufficient to give birth to the impelling power. It is necessary that there should be cold; without it, the heat would be useless."

During the operation of one of Carnot's heat engines, heat is absorbed from a reservoir at high temperature, and part of this heat is converted into useful work. However, much of the heat is expelled into a low-temperature reservoir and thus "wasted." In a steam engine, the boiler is the high-temperature reservoir, and the condenser is the low-temperature one. The greater the temperature difference between the two reservoirs, the greater the fraction of absorbed heat that is converted into useful work.

Carnot's formula for the thermodynamic efficiency of a machine powered by a heat difference can be expressed as $e = 1 - T_L/T_H$, where e is the efficiency of the machine, T_L the low operating temperature of the machine, and T_H is the high operating temperature—both in degrees kelvin. Most engines require fuel to be burned to supply heat at high temperature to produce an expanding gas. But we must supply heat at a temperature higher than the surroundings. Or, to put in another way, T_L is the absolute temperature of a reservoir to which a substance (e.g., a gas) is exhausted, and T_H is the temperature of the reservoir that supplies the gas (the source). For a machine to attain perfect efficiency, T_L would have to be at absolute zero. Notice that if the temperature of the heat source is the same as the exhaust, the efficiency is zero, and the engine can do no work. The engine must have the opportunity to expel heat energy to a lower temperature object. As discussed above, some of the initial heat supply must always be discarded to the surroundings by the engine exhaust.

For a steam engine, which cannot easily operate at temperatures much higher than 100°C, we would have T_H approximately equal to 373°K. During the cold portion of the engine cycle, the gas cannot be at a temperature lower than that of the surrounding air, for example, around 300°K. Therefore, a good steam engine of Carnot's time will have an efficiency of only about $e = 1 - 300/373$ or 20 percent. Modern power plants for electricity generation operate at around 1,000°F (800°K) with cold sinks at around 212°F (373°K). Theoretically speaking, they can run at a maximum efficiency of 54%. A car engine has a theoretical maximum of about 56%, but in practice, car engines are actually designed with other goals in mind, such as the goal of being lightweight and responsive, and thus only attain about 25% efficiency.

Although Carnot effectively abolished all hopes that a perpetual motion machine could be built, he was able to help engine designers improve their engines so that the engines could work close to their peak efficiencies.

These foregoing discussions have been concerned with devices that work as "cyclical devices" in which, at various parts of their cycles, the device absorbs or rejects heat, and it does work as a result. In summary, although in modern times we have greatly increased the efficiency of such cyclical engines that rely on heat differences, it is impossible to make such an engine that is 100% efficient. This impossibility is yet another way of stating the Second Law of Thermodynamics.

As an additional review of the Second Law, note that three common ways exist for stating the Second Law of Thermodynamics:

1. The Second Law prohibits the complete conversion of heat into work without loss of energy. (Lord Kelvin)
2. The Second Law prohibits heat flowing spontaneously from a colder object to a hotter object. (Clausius)
3. Natural processes are accompanied by an increase in the entropy of the universe. (This entropy principle captures much of what is said in the Clausius and Kelvin statements.)

The actual word "thermodynamics" was coined in 1849 by British mathematician and physicist William Thomson (later Lord Kelvin, 1824–1907) in his paper on the efficiency of steam engines. In 1850, Clausius went further and originated and defined the term *entropy* to denote the heat lost or unavailable for useful work. Entropy comes from the Greek word *entrepein*, meaning to turn. Clausius wrote in *Über verschiedene für die Anwendung bequeme Formen der Hauptgleichungen der mechanischen Wärmetheorie* ("On Several Convenient Forms of the Fundamental Equations of the Mechanical Theory of Heat"), "I have intentionally formed the word *entropy* so as to be as similar as possible to the word *energy*, for the two magnitudes . . . are so nearly allied in their physical meanings that a certain similarity in designation appears to be desirable."

Peter W. Atkins, author of *The Second Law*, writes about the work of Clausius and Kelvin:

> Clausius's first contribution cut closer to the bone than had Kelvin's. In dealing with the theme inspired by Carnot, carried on by Joule, and extended by Kelvin—in a monograph that was titled *Über die bewegende Kraft der Wärme* when it was published in 1850—Clausius sharply circumscribed the problems then facing thermodynamics, and in doing so made them more open to analysis. His was the focusing mind, the microscope to Kelvin's cosmic telescope.

As Atkins notes, the most significant research of Clausius's career on the mechanical theory of heat was *Über die bewegende Kraft der Wärme* ("On the Motive Power of Heat"). According to Yung Sik Kim, author of "Clausius's Endeavor to Generalize the Second Law of Thermodynamics,"

> Clausius first stated the basic idea of the second law of thermodynamics [in his 1850 *Wärme* paper]. He used it in showing that for a "Carnot cycle," which transmits heat between two heat reservoirs at different temperatures and at the same time converts heat into work, the maximum work obtained from a given amount of heat depends solely upon the temperatures of the heat reservoirs and not upon the nature of the working substance.

The 1850 paper was only the start of Clausius's involvement in the study of the Second Law. In the following decade, he published eight more memoirs in which he tried to place the Second Law into a simpler and mathematical form.

Clausius's second most important paper, *Über eine veränderte Form des zweiten Hauptsatzes der mechanischen Wärmetheorie* ("About a Modified Form of the Second Theorem on Mechanical Heat Theory"), contained his further developments of the concept of entropy. At this point in time, Clausius did not actually use the term "entropy" and instead referred to the principle of the equivalence of transformations, which gave rise to his statement that heat cannot of itself go from a cold body to a hot body. In 1865, he wrote his maxim, *Die Entropie der Welt strebt einen Maximum zu* ("The Entropy of the Universe Tends to a Maximum").

Guillen further discusses Clausius's formulation that states why everything in the universe ages and eventually dies ($\Delta S_{\text{universe}} \geq 0$). Guillen likens the universe to a casino, and entropy to money. Engines are like gamblers in the casino:

> Clausius's Law of Entropy Nonconservation was like saying that a casino's positive money changes always exceeded its negative money changes. In other words, a casino's winnings always exceeded it losses; it always made a profit, which was how it stayed in business. A casino existed at the expense of its players, which meant it could keep winning only so long as its players could keep losing. When they had lost everything, when positive money changes ceased to exist, the casino would shut down forever.

Clausius also was interested in possible models of the molecular motions in gases, and he demonstrated that rotational motions were needed to account for the heat present in a gas in addition to translational motions. He established one of the first significant connections between thermodynamics and the kinetic theory of gases. After 1875, Clausius worked on developing various theories of electrodynamics.

In 1871, Clausius worked with Scottish mathematician and physicist James Clerk Maxwell (1831–1879) in an area that came to be known as statistical thermodynamics, which focuses on the mathematical properties of large numbers of particles in a system. In 1875, Austrian physicist Ludwig Boltzmann (1844–1906) formulated a mathematical relationship between entropy S and molecular motion. This relationship is expressed as $S = k\ln W$, where W is the number of possible states of the system, and k is Boltzmann's constant that gives S in useful units. Interestingly, this equation for S is engraved on Boltzmann's tombstone in Vienna! Richard Feynman helps us understand the Boltzmann model of entropy in "Order and Entropy" in his *Lectures on Physics*. In particular, he explains the concepts of order and disorder from a mathematical standpoint:

> Suppose we divide the space into little volume elements. If we have black and white molecules, how many ways could we distribute them among the volume elements so that white is on one side and black is on the other? On the other hand, how many ways could we distribute them with no restriction on which goes where? Clearly, there are many more ways to arrange them in the latter case. We measure "disorder" by the number of ways that the insides can be arranged, so that from the outside it looks the same. *The logarithm of that number of ways is the entropy.* The number of ways in the separated case is less, so the entropy is less, or the "disorder" is less.

John Hutchinson, in "Equilibrium and the Second Law of Thermodynamics," provides a thought experiment to help us visualize the drive of a system to disorder. Imagine placing drop of red ink in a glass of water. At first, the ink is highly concentrated. Therefore, the molecules of the ink are close together. As we watch, the ink disperses. Scientists have shown that dispersal can occur without a change in temperature; thus, no energy enters or is released during the mixing. Even though there is no energetic advantage for the ink molecules to disperse, they do so spontaneously.

Hutchinson asks us to consider a model for dye molecules placed in the water. Imagine a row of squares, each one of which represents a possible location for a molecule, either a water or dye molecule. Here, we can represent the dye molecules with smiley faces ☺. At the start of the experiment, we model a "drop" as three consecutive ☺ symbols:

				☺	☺	☺			

It turns out that only eight drop configurations exist in our ten cells, assuming that the three dye "molecules" are indistinguishable from one another. In contrast, many more ways exist for the dye molecules to be arranged so that they do not form a drop, for example,

	☺			☺		☺			

In fact, 112 different ways exist for arranging the molecules in an unmixed state, again assuming that dye molecules are indistinguishable from one another. The total number of nonidentical arrangements of the molecules is 120. Thus, if we randomly place the three ☺ symbols in the array of 10 boxes, the chances are only 8 out of 120 of creating a drop of ink.

According to kinetic theory, the molecules are in constant random motion and always rearranging themselves. All possible arrangements are equally probable. Because most of the arrangements do not correspond to a drop of ink, this means that most of the time we will not observe a drop. Mixing occurs spontaneously simply because so many more arrangements exist that are mixed than that are not. A spontaneous process occurs because it produces the most probable final state.

Using the terminology of thermodynamics, we can say that 112 ways W (or microstates) exist to create a particular macrostate, in our case, a mixed state. Using the formula $S(W) = k \ln W$, we can calculate the entropy and can get a feel for why the more microstates that exist, the greater the entropy. A macrostate with a high probability (e.g., a mixed state) has a large value for the entropy, and a spontaneous process produces the final state of greatest entropy, which is one way of stating the Second Law of Thermodynamics.

Today, we remember Boltzmann for his work in thermodynamics, heat, and disorder. As discussed above, he used the concept of the atom to explain how heat was a statistical property of the motions of many atoms. However, several of his contemporaries, such as Austrian physicist Ernst Mach (1838–1916) and German chemist Wilhelm Ostwald (1853–1932), argued so forcefully against Boltzmann's position that Boltzmann's depression worsened, and he killed himself in 1906. Boltzmann appeared to have bipolar disorder, and his low emotional periods were only exacerbated by his failing eyesight and arguments with colleagues. All we know for certain is that when he was on holiday with his wife and daughter, he hanged himself while they were swimming. Although Boltzmann's idea of deriving thermodynamics by visualizing gases as made of atoms seems obvious to us today, many physicists of his time criticized the concept of atoms.

Leon Cohen in "A History of Noise" fondly recalls Boltzmann:

> Boltzmann was a grand man in every way—in size, personality, appetite, travel, excitement, charm and, of course, accomplishments. Also, he suffered from depression, got mad as hell at times, attempted suicide, and succeeded. But, basically, he was a nice guy who felt he was revolutionizing science and did not understand why some were opposing him so viciously.... Years later, Lisa Meitner, the discoverer of nuclear fission, would remember Boltzmann's lectures as "the most beautiful and stimulating that I have ever heard.... He himself was so enthusiastic about everything he taught us that one left every lecture with the feeling that a completely new and wonderful world had been revealed."

Let me end this section on Clausius's Law of Thermodynamics with a puzzle that I enjoy presenting to students. As discussed above, it is possible, although extremely unlikely, for you to suffocate right now as all the air molecules in your room suddenly go to one corner—based simply on random movements of the molecules. To better understand the chance of this occurring, consider the following smaller problem. How unlikely is it for just 10 molecules in your room to jump into a corner volume that is 10% the room's volume?

Here's a solution: Consider one air molecule in the room. For simplicity, let's pretend a molecule can jump from its current position to any position in the room. By chance alone, it has a 10% or 1/10 probability of being in a predefined corner room volume that is 1/10 the volume of the room. If you consider two molecules moving in the room, the chance of both being in the corner is much smaller, only (1/10) × (1/10) or 1%. Ten molecules have a 1 in 10,000,000,000 chance of being in the corner. One hundred molecules have a 1 in 1 googol chance of being in the corner, where a googol is 1 with 100 zeros (10^{100}). We would have to wait roughly 10^{80} times the age of the universe for all 100 molecules to migrate to the corner of the room by chance alone. Robert Ehrlich in *What If You Could Unscramble an Egg?* notes that there are 10^{27} air molecules in a typical room. The probability of all of them going to the corner at one moment due to random motions is roughly

$$\frac{1}{10^{10^{27}}}.$$

This number is so tiny that it is about equal to the odds of the Statue of Liberty jumping into the air. Of course, these kinds of simple models are just approximations, because in real gases, molecules diffuse by randomly walking through space, and future positions depend on current positions.

As one final concept relating to the energy and entropy, note that in thermodynamics, the Gibbs free energy is defined as the energy portion of a thermodynamic system available to do work. The definition of Gibbs free energy G can be derived from Clausius's Law and is defined as $G \equiv U + PV - TS$, where U is the internal energy, P is the pressure, V is the volume, T is the temperature, and S is the entropy. This equation, along with the identity $\Delta G = \Delta H - T\Delta S$, where ΔH is the change of enthalpy (heat content), is frequently used in physical chemistry when scientists are interested in determining whether a reaction will occur at some constant temperature and pressure. Any natural process occurs spontaneously if and only if the associated change in G for the system is negative ($\Delta G < 0$). Gibbs free energy is named after American physicist and chemist Josiah Willard Gibbs (1839–1903), who is considered to be one of the greatest nineteenth-century American scientists. According to J. G. Cowther, writing in the *Encyclopaedia Britannica*, Gibbs

> remained a bachelor, living in his surviving sister's household. In his later years he was a tall, dignified gentleman, ... approachable and kind (if unintelligible) to students.
>
> Gibbs was highly esteemed by his friends, but U.S. science was too preoccupied with practical questions to make much use of his profound theoretical work during his lifetime. He lived out his quiet life at Yale, deeply admired by a few able students but making no immediate impress on U.S. science commensurate with his genius.

Gibbs was quoted in the December 1944 issue of the *Scientific Monthly* as saying, "A mathematician may say anything he pleases, but a physicist must be at least partially sane."

William H. Cropper writes in *Great Physicists*:

> Thermodynamics had its Newton: Willard Gibbs. Where Clausius hesitated, Gibbs did not. Gibbs recognized the energy-entropy partnership, and added to it a concept of great utility in the study of chemical change, the "chemical potential." ... Gibbs treatise opened theoretical vistas far beyond the theory of heat sought by Clausius....

A related law, Hess's Law of Constant Heat Summation, was discovered by Swiss-Russian chemist Germain Henri Hess (1802–1850), and it states that the amount of heat released or absorbed in a chemical change depends on the initial and final states, not on the intermediate reactions involved. This law is discussed in somewhat greater detail in the "Great Contenders" section at the end of this book.

Clausius died in Bonn in 1888, and Gibbs wrote the obituary that was published in the Boston-based *Proceedings of the American Academy of Arts and Sciences*. The obituary made it clear that the field of thermodynamics had begun in the year 1850 with the publication of Clausius's famous first paper on the Second Law. Gibbs also mentioned that a statement of the Second Law by Lord Kelvin in 1851 was based on Clausius's 1850 work. Debashish Chowdhury and Dietrich Stauffer write in *Principles of Equilibrium Statistical Mechanics*:

> Clausius' contribution to thermostatics is comparable to those of Newton and Maxwell to mechanics and electromagnetism, respectively. In the obituary J. W. Gibbs remarked that Clausius's first memoir "marks an epoch in the history of physics...." Very little is

known about his personal life. Gibbs remarked that Clausius's "true monument lies not in the shelves of libraries, but in the thoughts of men, and in the history of more than one science."

Today, we honor Clausius by naming a lunar crater with a diameter of 24 kilometers after him. The naming was approved in 1935 by the International Astronomical Union General Assembly. Elizabeth Garber writes about Clausius's uniqueness in *The Language of Physics*:

> Clausius's research was never the familiar nineteenth-century mix of experiment and mathematics. He was a theoretical physicist and never published any experimental research although he was always well aware of it.... He was also one of the first German physicists to be fully competent in contemporary mathematics, and to manipulate it for his own needs.

FURTHER READING

Atkins, Peter, W., *The Second Law* (New York: Scientific American Books, 1984).

Cardwell, Donald, *From Watt to Clausius: The Rise of Thermodynamics in the Early Industrial Age* (Ames: Iowa State Press; reprint edition, 1989).

Carnot, Sadi, *Réflexions sur la puissance motrice du feu* (Paris: Bachelier, 1824).

Chalmers, Matthew, "Second Law of Thermodynamics 'Broken,'" *NewScientist.com*, July 19, 2002; see www.newscientist.com/article.ns?id=dn2572.

Chowdhury, Debashish, and Dietrich Stauffer, *Principles of Equilibrium Statistical Mechanics* (Hoboken, N.J.: Wiley, 2000).

Clausius, Robert, "Obituary notice," in *Proceedings of the Royal Society* (London: Harrison and Sons, 1891)

Clausius, Rudolf, *Über verschiedene für die Anwendung bequeme Formen der Hauptgleichungen der mechanischen Wärmetheorie. Annalen der Physik und Chemie*, 125: 353–400 (1865).

Cohen, Leon, "The History of Noise," *IEEE Signal Processing Magazine*, 22(6): 20–45, November 2005.

Cowther, J. G., "J. Willard Gibbs," *Britannica Concise Encyclopedia*; see concise.britannica.com/ebc/article-9365569.

Cropper, William H., *Great Physicists: The Life and Times of Leading Physicists from Galileo to Hawking* (New York: Oxford University Press, 2001).

Daub, Edward, "Rudolf Clausius," in *Dictionary of Scientific Biography*, Charles Gillispie, editor-in-chief (New York: Charles Scribner's Sons, 1970).

Eddington, Arthur Stanley, *The Nature of the Physical World* (New York: Macmillan, 1928).

Ehrlich, Robert, *What If You Could Unscramble an Egg?* (New Brunswick, N.J.: Rutgers University Press, 1996).

Feynman, Richard, "Order and Entropy," in *Lectures on Physics* (Boston: Addison Wesley Longman, 1963).

Garber, Elizabeth, *The Language of Physics: The Calculus and the Development of Theoretical Physics in Europe, 1750–1914* (Boston: Birkhäuser, 1998).

Guillen, Michael, *Five Equations That Changed the World* (New York: Hyperion, 1995).

Hutchinson, John, "Equilibrium and the Second Law of Thermodynamics," Connexions Web site, March 30, 2005; see cnx.rice.edu/content/m12593/1.2/.

Kaku, Michio, *Hyperspace: A Scientific Odyssey Through Parallel Universes, Time Warps, and the 10th Dimension* (New York: Anchor, 1995).

Kim, Yung Sik, "Clausius's Endeavor to Generalize the Second Law of Thermodynamics, 1850–1865," *Archives Internationales d'Histoire des Science*, 33(111): 256–273, 1983.

Klyce, Brig, "The Second Law of Thermodynamics: Quick Guide to Cosmic Ancestry," in *Cosmic Ancestry*; see www.panspermia.org/seconlaw.htm.

Laidler, Keith, *Energy and the Unexpected* (New York: Oxford University Press, 2003).

Powell, Corey S., "Welcome to the Machine," *New York Times* Book Review, Section 7, p. 19, April 2, 2006.

Rajasekar, S., and N. Athavan, "Ludwig Edward Boltzmann," arXiv.org; see arxiv.org/PS_cache/physics/pdf/0609/0609047v1.pdf.

"Third Law of Thermodynamics," Everything$_2$; see www.everything2.com/index.pl?node=third%20law%20of%20thermodynamics.

Wang, G. M., E. M. Sevick, Emil Mittag, Debra J. Searles, and Denis J. Evans, "Experimental Demonstration of Violations of the Second Law of Thermodynamics for Small Systems and Short Time Scales," *Physical Review Letters*, 89(5): 050601/1–4, July 2002; see link.aps.org/doi/10.1103/PhysRevLett.89.050601.

INTERLUDE: CONVERSATION STARTERS

> Clausius introduced the entropy concept, and supplied the name, but he was ambivalent about recognizing its fundamental importance. He showed in a second simple differential equation how entropy is connected with heat and temperature, and stated formally the law now known as the second law of thermodynamics: that in an isolated system, entropy increases to maximum value. But he hesitated to go further.
>
> —William H. Cropper, *Great Physicists*

> You must remember this: Like atoms, heat is so intangible that it was one of the last concepts in classical physics to be sorted out. In the process, the science of thermodynamics was created. Pollyannas who believe anything is possible should be subjected to a course in thermodynamics.
>
> —Tony Rothman, *Instant Physics: From Aristotle to Einstein, and Beyond*

The origin of nature or the laws of nature... practically mean the same thing. For not one man in a thousand has either time or ability to investigate such a problem for himself. Therefore, all the other 999 ought to believe no statement of anyone else about it, since no man can possibly say that he "knows" anything about the origin of the laws of nature: the most that the cleverest man can do is to select the most probable of the only possible theories.

—Edmund Beckett Grimthorpe, *On the Origin of the Laws of Nature*

Science is about figuring out how the world works, and there are really two kinds of science. One is where you know the rules but have to figure out how they apply in specific situations. The other is where you try to figure out the rules themselves. In this second category there have been revolutions—such as thermodynamics, quantum mechanics, relativity, the genetic code—that change the whole game.... The great questions lie in figuring out the rules.

—Steven Koonin, "What Are the Grand Questions in Science?" in Robert Kuhn's *Closer to Truth*

STOKES'S LAW OF VISCOSITY

Ireland, 1851. The frictional force exerted on a sphere moving in a fluid is proportional to the fluid viscosity and the radius and speed of the sphere.

CROSS REFERENCE: STOKES-EINSTEIN RELATION, NAVIER-STOKES EQUATIONS, STOKES'S LAW OF FLUORESCENCE, CLAUDE NAVIER, SIMÉON POISSON, AND ADHÉMAR BARRÉ DE SAINT-VENANT.

In 1851, Herman Melville published *Moby Dick*. The *New York Times* and Reuters news service were founded. British astronomer William Lassell discovered Ariel and Umbriel, moons of Uranus. The first YMCA opened in the United States. Rowland Hussey Macy founded Macy's department store.

Consider a solid sphere of radius r moving with a velocity v through a fluid of viscosity μ. George Stokes determined that the frictional force F that resists the motion of the sphere can be stated as

$$\boxed{F = 6\pi r \mu v.}$$

Note that this drag force F is directly proportional to the sphere radius. This was not intuitively obvious because some researchers supposed that the frictional force would be proportional to the cross-section area, which erroneously suggest an r^2 dependence. When dealing with tumbling molecules, the radius r is usually considered to be the Stokes radius, that is, the radius of a sphere that diffuses at the same rate as the molecule. The behavior of this imaginary sphere accounts for hydration and shape effects. Stokes's Law tends to be most accurate for small, slowly moving particles in viscous liquids.

Table 8 should be useful in order to get a feel for typical viscosity values μ that are used in Stokes Law. Note that 1 pascal equals 1 $kg/(m{\cdot}s^2)$, and the viscosity of water is, conveniently, about 1 mPa·s or 1 g/(m·s). [In actuality, the viscosity of water depends on temperature. At 293°K (20°C), the viscosity of water is 1.002 cP, where cP stands for centipoise. A centipoise is 1 millipascal second (mPa·s).]

Stokes's Law has many practical applications. For example, the law is considered in industry when studying sedimentation that occurs during the separation of a suspension of solid particles in a liquid. In these applications, scientists are often interested in the resistance exerted by the liquid to the motion of the descending particle.

Consider a scenario in which a particle in a fluid is subject to the forces of gravity. For example, some older readers may recall the popular Prell

TABLE 8 Viscosities of Common Fluids

Fluid	Viscosity μ (Pa·s)
Air (20°C)	1.8×10^{-5}
Water (20°C)	1.0×10^{-3}
Blood (37°C)	4.0×10^{-3}
Canola oil (20°C)	1.0×10^{-2}
Motor oil (20°C)	1
Corn syrup (20°C)	8
Molten lava	1,000

shampoo TV commercial that shows a pearl dropping through a container of the green shampoo. (I believe that this commercial was intended to demonstrate how thick and luxurious the shampoo was, although I am not certain that the more viscous the shampoo, the better it is for your hair!)

The particle (a pearl in this case) starts off with zero speed and then initially accelerates, but the motion of the particle quickly generates a frictional resistance counter to the acceleration. Thus, the particle rapidly reaches a condition of zero acceleration when the force of gravity is balanced by the force of friction. At this point in the sedimentation process, the particle drifts downward at a constant velocity called the terminal velocity. In the absence of turbulence, viscous friction is always in the direction opposite to velocity. Stokes's Law is most accurate in the ideal case in which the particle is thought of as smooth, spherical, and unaffected by the presence of neighboring particles and the walls of the container.

Sedimentation is used in industry to separate particulate material from fluid streams. For example, a sedimentation process, which can be understood and characterized using Stokes's Law, is sometimes used in the food industry for separating dirt and debris from useful materials, separating crystals from the liquid in which they are suspended, or separating dust from air streams. Stokes's Law is used to research and treat urban storm water in order to remove certain kinds of pollutants from wet weather runoff. The law is also important in determining the nature of emissions from volcanic explosions by allowing researchers to study the settling of dust. Additionally, the law is useful for studying the sedimentation of small particle pollutants in a river. Medical researchers use it to study the aerodynamic properties of aerosol particles in order to optimize drug delivery to the lungs.

Stokes's Law also applies to very small raindrops plummeting to Earth in the absence of wind and turbulence. In both the pearl and the raindrop example, v is the terminal velocity of the particle. Stokes's Law would not be accurate for skydivers jumping from planes because of the turbulence the skydiver encounters when plummeting to Earth.

Let us try some practical problems relating to Stokes's Law to get a better feel for its application. Consider a raindrop of radius 0.2 mm descending through air that has a viscosity μ of 1.8×10^{-5} N·s/m^2. What is the terminal velocity v of the raindrop?

From Stokes's Law, we have $F = 6\pi r \mu v$. When the raindrop is traveling at constant velocity, the force of gravity downward (which equals the raindrop mass m times the gravitational constant $g = 9.8$ m/s^2) is exactly balanced by the Stokes's force upward. Thus, we have $mg = 6\pi r \mu v$.

We can solve for the terminal velocity v, giving us:

$$v = \frac{mg}{6\pi r \mu}$$

The mass m of the raindrop may be estimated by assuming that the drop has a roughly spherical volume ($4\pi r^3/3$) and a density ρ of 1,000 kg/m^3. Consolidating all of this information, we have

$$v = \frac{(4\rho\pi r^3/3)g}{6\pi r \mu} = \frac{2\rho g r^2}{9\mu}$$

or

$$v = \frac{2 \cdot (1000\,\text{kg/m}^3)(9.8\,\text{m/s}^2)(0.00002\,\text{m})^2}{9 \times 1.8 \times 10^{-5}\,\text{N·s/m}^2} = 4.8\,\text{m/s}.$$

Notice that in our raindrop problem, the larger the viscosity, the slower the terminal velocity, and the larger the drop radius, the higher the velocity.

For another practical problem, consider a plastic spherical pearl descending in a jar of shampoo. The sphere has a density ρ_p and a radius r. Its velocity downward is v. The density of the shampoo is ρ_s. What is the viscosity of the shampoo? To solve this problem, recall that the forces of the constant-velocity particle are balanced so that the buoyant force B and the Stokes force upward balance the weight W that pulls the sphere downward:

$$B + 6\pi r \mu v - W = 0$$

The buoyant force can be computed from $B = (4\pi r^3/3)\rho_p g$. Solving for viscosity, we have

$$\mu = \frac{W - B}{6\pi r v}.$$

Thus, we can now compute the viscosity of the shampoo.

In our analysis of the falling raindrop above, we determined the formula for computing the terminal velocity of the raindrop. This formula involved the density only of the falling particle of water because the density of air was so small in comparison. However, in instances in which an object is falling through a medium of significant density (e.g., shampoo), the formula for terminal velocity of the falling particle is more generally written as

$$v = \frac{2(\rho_p - \rho_s)gr^2}{9\mu}.$$

If we define a Reynolds number to be $dv\rho/\mu$, where ρ is the fluid density, and d is the diameter of the particle, Stokes's Law is generally accurate for Reynolds numbers less than about 0.3—in other words, it is most accurate for very viscous liquids and small particles. In practice, Stokes's Law is not very accurate for rain droplets, but it is accurate for the smaller falling droplets called "cloud droplets," which typically have diameters of the order of 0.01–0.02 millimeters. Cloud droplets are spherical particles of liquid water, formed by condensation of water vapor. Such drops form a visible cloud. Thus, Stokes's Law can be applied more accurately to fog droplets and should be used with caution for applications involving raindrops of normal size.

In addition to sedimentation problems, Stokes's Law also has application to the study of aerosols, that is, gaseous suspensions of fine solid or liquid particles. In the late 1990s, Stokes's Law was used to provide an accurate and convincing scientific explanation of how micrometer-size, distributed uranium particles can remain airborne for many hours and traverse great distances—and thus possibly have contaminated Persian Gulf War soldiers. Because depleted uranium has a very high density and hardness and is pyrophoric (spontaneously igniting), U.S. cannon rounds often contained depleted-uranium penetrators, and this uranium becomes aerosolized when the rounds impact hard targets such as tanks.

In 1920, Stokes's Law was refined to account for effects that may be generated by the container walls that enclose the viscous fluid. These wall, or edge, effects tend to result in a slower observed velocity because the medium is not continuous and the liquid is slightly compressed against the sides of the container as the sphere descends. The refinement of the law requires researchers to determine the ratio between the particle radius and the inner radius of the container if the container is cylindrical. The smaller the radius of the container, the more important this correction factor becomes.

The following is one expression of a modified Stokes's Law, which uses a corrected value for the viscosity μ_c that is based on the diameter d of the falling particle and the internal diameter d_c of the cylinder:

$$\mu_c = \mu\left[1 - 2.104\frac{d}{d_c} + 2.09\left(\frac{d}{d_c}\right)^3 - 0.95\left(\frac{d}{d_c}\right)^5\right]$$

In 1905, Albert Einstein discussed a relationship he had discovered between the mobility of particles in a liquid and the diffusion constant D, Boltzmann's constant k, and the absolute temperature T. (The mobility is the ratio of the terminal drift velocity of the particle to an applied force.) In particular, Einstein was researching Brownian motion of particles, which refers to the random motion of tiny particles immersed in a fluid. Combining Einstein's findings with Stokes's Law leads to a relationship known as the Stokes-Einstein relation:

$$D = \frac{kT}{6\pi\mu r}$$

D, the diffusion constant or coefficient, gives an indication of the amount of a substance diffusing across a unit area through a unit concentration gradient in a unit time. Thus, the Stokes-Einstein relation can be used to estimate the diffusion coefficient of particles under study. We may think of the Stokes-Einstein formula as a convenient way to express the relationship of the diffusion coefficient, of a spherical Brownian particle in a viscous fluid, to the frictional drag force.

George Stokes (1819–1903), Anglo-Irish physicist famous for his law of friction and for his wide variety of work in chemistry, physics, and mathematics.

CURIOSITY FILE: Stokes wrote very long and formal letters to his fiancé, expressing his love in mathematical terms, and his unusual approach almost caused her to reject the idea of marriage. • Palynologists use Stokes's Law to differentially sort different kinds of palynomorphs (i.e., spores or the remains of microscopic creatures). • Stokes coined the word "fluorescence" after the mineral fluorite, which exhibits a colorful fluorescence. • The Campbell-Stokes recorder, named after Celtic scholar John Francis Campbell (1822–1885) and George Stokes, records the amount of sunshine falling on a particular location on Earth. The device, still sometimes used today, includes a spherical lens that burns an image of the Sun upon a specially prepared card. • In 2004, researchers studying the physics of viscosity at the University of Minnesota filled a swimming pool with a mucuslike guar gum and found that swimmers swam just as fast in the gum as they did in water.

A devoutly religious man, Stokes was deeply interested in the relationship of science to religion. This was especially true toward the end of his life....

—E. M. Parkinson, "George Stokes," in *Dictionary of Scientific Biography*

The well-known theorem in vector calculus which bears Stokes's name is sadly not due to Stokes, but was communicated to him in a letter by Lord Kelvin. The confusion appears to have arisen because Stokes set this theorem as a problem in the Smith's Prize Examination a few years later!

—Alastair Wood, "George Gabriel Stokes 1819–1903"

Barring infection, beer clarity is compromised only by yeast cells and Non-Microbiological Particles (NMP)....All fining agents work by sticking small particles together to form larger aggregates which settle faster according to Stokes's Law. Since the speed of settlement is proportional to the square of the radius, a modest increase in particle size can yield a profound decrease in settlement time.

—Ian L. Ward, "Clear Beer Through Finings Technology"

We refer to scientific measurements that have been made of the atmospheric wind-borne transport of uranium aerosols over distances up to 26 miles (42 km) from their sources. Stokes's well-known physical law helps to explain how airborne transport of depleted uranium particles can occur over large distances.

—Leonard A. Dietz, "Contamination of Persian Gulf War Veterans and Others by Depleted Uranium," July 19, 1996

His work is distinguished by a certain definiteness and finality, and even of problems, which when he attacked them were scarcely thought amenable to mathematical analysis, he has in many cases given solutions which once

> and for all settle the main principles. This result must be ascribed to his extraordinary combination of mathematical power with experimental skill....
> —1911 *Encyclopaedia Britannica*

Stokes was born in Skreen, Ireland, and he was surrounded by religion since birth. His father was the rector of the Skreen parish and taught his son Latin grammar. His mother was the daughter of a rector. His four brothers all became important leaders in the Church of Ireland. After attending school in Dublin, at the age of 16 he went to Bristol College in Bristol, England. According to theoretical physicist and mathematician Joseph Larmor's (1857–1942) *Memoir and Scientific Correspondence of the Late Sir George Gabriel Stokes*, Stokes's mathematics teacher in Bristol College noted:

> His habit...of answering with a plain yes or no, when something more elaborate was expected, is supposed to date from his transference from an Irish to an English school, when his brothers chaffed him and warned him that if he gave long Irish answers he would be laughed at by his school fellows.

In 1837, Stokes entered Pembroke College, Cambridge. He published important papers on the motion of incompressible fluids, such as his 1842 paper, "On the Steady Motion of Incompressible Fluids." Three years later, he published "On the Theories of the Internal Friction of Fluids in Motion," and in 1849, he published "On the Variation of Gravity at the Surface of the Earth," in which his research on the motion of pendulums in fluids provided information on the strength of gravity at different locations on Earth.

He became Lucasian Professor at Cambridge in 1849. [Note that this prestigious position was held by Isaac Newton (1642–1727) and is currently held by astrophysicist and popular author Stephen Hawking (born 1942).] In 1851, Stokes published his paper describing the velocity of spheres moving through viscous liquids, which led to the famous Stokes's Law discussed above.

In 1857, he married Mary Susanna, the daughter of Thomas Romney Robinson (1792–1882), a famous Irish astronomer and physicist. During the process of winning Mary's heart, Stokes had written numerous letters to Mary, some as long as 55 pages! In one letter, he told Mary that he worried that she would not be able to be with a man who sometimes worked on mathematical problems late into the night. According to Larmor's memoir of Stokes, Stokes's love letters to Marry often mixed mathematics and romance. For example, Stokes had written:

> I too feel that I have been thinking too much of late, but in a different way, my head running on divergent series, the discontinuity of arbitrary constants. . . . I often thought that you would do me good by keeping me from being too engrossed by those things.

The oddness of his correspondence initially made her nervous about marrying him, and she told him so. He responded that he was worried that he "should go to the grave a thinking machine unenlivened and uncheered and unwarmed by the happiness of domestic affection."

Eventually, Stokes and Mary put their differences aside, and they married. The cottage in which the Stokes lived was in a large garden area, and he conducted his experiments there "in a narrow passage behind the pantry, with simple and homely apparatus." Stokes's first two daughters died in infancy. One of his sons died in 1893 of an accidental overdose of morphine while training to become a medical doctor.

Stokes was president of the Cambridge Philosophical Society from 1859–1861 and president of the Royal Society of London in 1885. The Royal Society awarded him the Copley Medal in 1893. From 1886 to 1903, he was president of the Victoria Institute of London, the purpose of which was to study relationships between science and religion. He was knighted in 1899. E. M. Parkinson writes of Stokes in the *Dictionary of Scientific Biography*:

> His . . . investigations covered the entire realm of natural philosophy. Stokes systematically explored areas of hydrodynamics, the elasticity of solids, and the behavior of waves in elastic solids, including the diffraction of light, always concentrating on physically important problems and making his mathematical analyses subservient to physical requirements. His few excursions into pure mathematics were prompted either by a need to develop methods to solve specific physical problems or by a desire to establish the validity of mathematics he was already employing.

Like many of the other versatile polymaths in this book, Stokes conducted significant research in a variety of areas, such as those concerning the nature of light, gravity, chemistry, sound, heat, meteorology, and solar physics.

In the early 1840s, Stokes became particularly interested in hydrodynamics and the study of fluid flow. In 1845, he performed various studies of friction and viscosity in fluids. Interestingly, French scientists Claude Navier (1785–1836), Siméon Poisson (1781–1840), and Adhémar Barré de Saint-Venant (1797–1886) had independently derived equations for fluid flow with friction, but Stokes was not familiar with their work when he derived his own equations and obtained results using a different approach and theoretical justification.

Additionally, Stokes was able to extend his equations and theories in order to deduce formulas for the motions of elastic solids. In particular, Stokes developed a series of equations, known today as the Navier-Stokes Equations, which are fundamental partial differential equations that describe the flow of incompressible fluids. The equations also relate pressure and external forces acting on a fluid to the response of the fluid flow.

Over the years, these equations became useful in designing ships and to model the weather and the flow of waters in pipes and around airplane wings. The equations are named after Stokes (who in 1845 published his derivation of the equations in a manner that is currently well appreciated) and its other principle developer, Claude Navier (who published his work in 1822). Stokes reported on his various hydrodynamic theories in 1846 at a meeting of the British Association for the Advancement of Science.

Some other areas of Stokes's research include:

- Oscillations of waves in water (1847)
- Frictional effects of air on the behavior of raindrops and the formation of clouds (1847)
- Periodic series in mathematics and their application to the study of heat, hydrodynamics, and electricity (1847)
- The relationship between the strength of gravity and the shape of Earth's surface (1849)
- A proof that shows why the strength of gravity is less on a continent than on an island (1849)
- The implementation of a new approach for determining the value of the following integral, used in optical studies:

$$\int_0^\infty \cos\left[\frac{\pi}{2}(x^3 - mx)\right] \mathrm{d}x, \text{ for large, real values of } m \text{ (1850)}$$

- Fluid viscosity and its effect on pendulum motion (1850)
- His law that mathematically describes the fall of an object through a liquid (1851)
- The effect of wind on the intensity of sound (1857)
- The effect of clanging bells, modeled as spheres, on a surrounding gas (1868)
- Various studies on light, the polarization of light, and diffraction (1848–1849)
- Measurements of astigmatism in the eye (1849)

- Solutions to differential equations representing the motions of iron railroad bridges (1849)
- The design of instruments for analyzing elliptically polarized light (1851)
- The conduction of heat in crystals (1851)
- Methods for determining the constants for asymptotic solutions of the Bessel equation

$$\frac{d^2y}{dx^2} + \frac{1}{x}\frac{dy}{dx} - \frac{n^2}{x^2}y = y \text{ for a real constant } n \text{ (1868)}$$

- An explanation of fluorescence and spectra (1852–1854)

With respect to this last point, fluorescence usually refers to the glow of an object caused by visible light that is emitted when the object is stimulated via electromagnetic radiation. In 1852, Stokes observed phenomena that behave according to Stokes Law of Fluorescence, namely, that the wavelength of emitted fluorescent light is always greater than the wavelength of the exciting radiation. Stokes published his finding in his 1852 memoir "On the Change of Refrangibility of Light."

Today, we sometimes refer to the process as "Stokes fluorescence"—the reemission of longer wavelength (lower frequency) photons by a molecule that has absorbed photons of shorter wavelengths (higher frequency). The precise details of the process depend on the characteristics of a particular molecule involved in the fluorescence process. Light is generally absorbed by molecules in about 10^{-15} seconds, and this absorption causes electrons to become excited and jump to a higher energy state. The electrons remain in the excited state for about 10^{-8} seconds, and then the electron may emit energy as it returns to the ground state. The phrase "Stokes shift" usually refers to the difference in wavelength or frequency between absorbed and emitted quanta.

Stokes named "fluorescence" after fluorite, a strongly fluorescent mineral. He was the first to adequately explain the phenomenon in which fluorescence can be induced in some materials through stimulation with ultraviolet light. Today, we know that these kinds of materials can be made to fluoresce by stimulation by all kinds of electromagnetic radiation, such as ultraviolet light, visible light, infrared radiation, X-rays, and radio waves.

Stokes experimented for most of his life, his interests ranging from physics to botany. He helped to establish the composition of chlorophyll in his paper "On the Supposed Identity of Biliverdin with Chlorophyll, with Remarks on the Constitution of Chlorophyll," published in the 1864 *Proceedings of the Royal Society*. Some other key works include *Dynamical*

Theory of Diffraction (1849), *Light* (1884), and *Natural Theology* (1891). He did not publish many of his discoveries and only mentioned them during his lectures.

Throughout his life, Stokes was rather humble and generous in giving credit to others. If he discovered that some of his work was done by earlier researchers, he acknowledged the previous work without argument. He often shared his unpublished ideas with other scientists of his time.

According to Larmor, in the 1850s, a fellow member of the Royal Society had written,

> One of the distinguishing characteristic qualities of Sir George was the generous way in which he was always ready to lay aside at once, for the moment, his own scientific work, and give his whole attention and full sympathy to any point of scientific theory or experiment about which his correspondent had sought his counsel.

As David Wilson notes in the introduction to *The Correspondence Between Sir George Gabriel Stokes and Sir William Thomson, Baron Kelvin of Largs*:

> At the century's end, grateful British scientists widely recognized that, unlike Kelvin's, Stokes's mind had for decades been exercised not primarily in his own research programme, but in response to the research of others. As [chemist] Arthur Smithells wrote, "What Stokes did for his generation can hardly be estimated." Kelvin [agreed] that Stokes "gave generously and freely of his treasures to all who were fortunate enough to have the opportunity of receiving from him."

The famous Stokes Theorem in calculus and geometry was actually first stated in 1850, without proof, by Lord Kelvin (William Thomson) in his letter to Stokes. We refer to it today as the Stokes Theorem partly because, starting in 1854, Stokes assigned the proof of this theorem as part of his examinations. James Clerk Maxwell presented the name when he began to refer to it as Stokes Theorem.

In 1891, Stokes wrote in his book *Natural Theology*:

> Admit the existence of God, of a personal God, and the possibility of miracles follows at once. If the laws of nature are carried out in accordance with His will, He who willed them may will their suspension. And if any difficulty should be felt as to their suspension, we are not even obliged to suppose that they have been suspended.

One of Stokes's surviving daughters, along with her husband, lived with Stokes at his cottage and helped care for him after the death of his wife in 1899. Two days after his death in 1903, the *Times* wrote in his obituary that

> Sir G. Stokes was remarkable . . . for his freedom from all personal ambitions and petty jealousies. . . . It is sometimes supposed . . . that minds conversant with the higher mathematics are unfit to deal with the ordinary affairs of life. Sir George Stokes was a living proof that if the mathematician is only big enough, his intellect will handle practical questions so easily and as well as mathematical formulas.

Stokes's obituary also praised his deep Christian faith and his ability to pursue both science and religion with equal fervor:

> No account of his life would be complete without a reference to its religious side. To many, he was one of the prominent instances of the possibility of combining scientific research with the maintenance of Christian convictions. . . . He was often present at the discussion of questions in which science and faith might seem to clash, and he maintained a conservative position to the last.

A lunar crater with a diameter of 51 kilometers was named after Stokes and approved in 1964 by the International Astronomical Union General Assembly. The Stokes crater on Mars was also named after him and is notable for its dark-toned sand dunes.

FURTHER READING

Dake, H. C., and Jack DeMent, *Fluorescent Light and Its Applications* (Brooklyn, NY: Chemical Publishing Company, Inc., 1941), 51–52.

Dietz, Leonard A., "Contamination of Persian Gulf War Veterans and Others by Depleted Uranium," July 19, 1996; see www.wise-uranium.org/dgvd.html.

Hopkin, Michael, "Swimming in Syrup Is as Easy as Water," Nature Publishing Group; see www.nature.com/news/2004/040920/full/news040920-2.html.

Larmor, Joseph, *Memoir and Scientific Correspondence of the Late Sir George Gabriel Stokes* (Cambridge, U.K.: Cambridge University Press, 1907).

Parkinson, E. M., "George Stokes," in *Dictionary of Scientific Biography*, Charles Gillispie, editor-in-chief (New York: Charles Scribner's Sons, 1970).

Wilson, David, *The Correspondence Between Sir George Gabriel Stokes and Sir William Thomson, Baron Kelvin of Largs* (New York: Cambridge University Press, 1990).

Wood, Alastair, "George Gabriel Stokes 1819–1903: An Irish Mathematical Physicist," School of Mathematical Sciences, Dublin City University, Ireland, 1998; see www.cmde.dcu.ie/Stokes/GGStokes.html.

INTERLUDE: CONVERSATION STARTERS

Physics tries to discover the pattern of events which controls the phenomena we observe. But we can never know what this pattern means or how it originates; and even if some superior intelligence were to tell us, we should find the explanation unintelligible.

—James Hopwood Jeans, *Physics and Philosophy*, 1942

The laws of nature are but the transcript of the thoughts of God, immutable and unchangeable. There is no such thing as Chance. Chance has no existence under the constant laws of nature or under any laws. What is called Chance is only the uncalculated result of some known or unknown law of nature.

—Henry Augustus Mott, *The Laws of Nature and Man's Power to Make Them Subservient to His Wishes*, 1882

Whenever a theory appears to you as the only possible one, take this as a sign that you have neither understood the theory nor the problem which it was intended to solve.

—Karl Popper, *Objective Knowledge: An Evolutionary Approach*

The earliest use of the term [law of nature] in English ... dates only from the seventeen century, when systematic science began to take off. The first two examples traced by the *Oxford English Dictionary* are dated 1665—one from the *Transactions of the Royal Society* and one from Boyle—and they relate to a universe set and maintained in motion by the command of God ... Descartes presents in the *Principia Philosophiae* (1644) ... certain rules or laws of nature.

—Michael Frayn, *The Human Touch*

BEER'S LAW OF ABSORPTION

Germany, 1852. The absorbance of a solution is proportional to the concentration of dissolved solute.

CROSS REFERENCE: PIERRE BOUGUER, JOHANN LAMBERT, THE LAMBERT-BEER LAW, AND THE BOUGUER-BEER LAW.

> In 1852, Harriet Beecher Stowe published *Uncle Tom's Cabin.* The first edition of Peter Roget's Thesaurus was also published. Emma Snodgrass was arrested in Boston for wearing pants. The judge released her after giving her some "wholesome advice about her eccentricities."

Imagine shining a light of a particular wavelength λ and intensity I_0 through a cylinder containing a solution. Some of the light will be transmitted through the sample with intensity I, and some of the light will be absorbed. Beer's Law states that the amount of light absorbed by a solution is directly proportional to the concentration of the solute (i.e., dissolved substance) and the path length of the light through the solution. One common way of expressing Beer's Law is

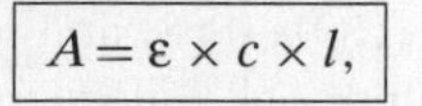

$$A = \varepsilon \times c \times l,$$

where A is the amount of absorbance, c is the concentration of the solution (in moles per liter), l is the path length in centimeters, and ε is a constant of proportionality known as the molar extinction coefficient or molar absorptivity. If $A = 0$, then no photons are absorbed. This law is most accurate when solutions are dilute because interactions between solute molecules can occur at higher concentrations. For example, Beer's Law begins to break down at high concentrations due to electrostatic interactions between molecules in close proximity. The law can also not be relied upon if the sample fluoresces or phosphoresces.

The absorbance A may also be defined as $A_\lambda = -\log_{10}(I/I_0)$, where I is the intensity of light passing through the liquid. From a practical standpoint, absorbance can be used to determine the concentration of the solution and varies with the wavelength of light used in the experiment. In other words, using Beer's Law in the field of spectroscopy, if l and ε are known, then the concentration of a substance in a solution can be deduced from the amount of light it absorbs. The value of the absorption coefficient ε also depends on the substance dissolved and on the wavelength of light.

August Beer (1825–1863), German mathematician, chemist, and physicist famous for his studies of light absorption in solutions.

CURIOSITY FILE: Beer's Law is used in countless interesting applications. For example, many investigators have studied the radiation distribution in plant canopies (the uppermost layer in a forest, formed by the crowns of the trees) and used Beer's Law and modifications of Beer's Law for determining light at a particular height in the canopy. The extinction coefficient varies with the orientation of the leaves.

> "The flux of global radiation at the forest floor can be related to that above the canopy by a form of Beer's law."
>
> —Richard Lee, *Forest Microclimatology*

August Beer was born in Trier, Germany's oldest city, which is situated on the western bank of the Moselle River and near German's border with Luxembourg. Beer studied the natural sciences and mathematics. He worked for several years under the tutelage of German mathematician physicist Julius Plücker (1801–1868) and received his Ph.D. at age 23. He discusses Beer's Law in his book *Einleitung in die höhere Optik* (Introduction to Advanced Optics) and eventually became professor of mathematics at Bonn.

When Beer was in his 30s, he desired to summarize the entire field of mathematical physics. However, due to his death at age 38, many of his papers were published posthumously. These papers dealt with elasticity, magnetism, electrodynamics, and capillarity.

Years before Beer made his discovery of what we now call Beer's Law, both scientists and laypeople had noticed that the intensity of light decreased as it passed through solutions. For example, in 1729, French mathematician Pierre Bouguer (1698–1758) quantified this by stating an absorption law: The fraction of light absorbed by a particular material is directly proportional to the thickness of the material. In Bouguer's 1729 paper *Essai d'optique sur la gradation de la lumière* ("Optical Experiment on the Gradation of Light"), he defined the quantity of light lost by passing through a given extent of the atmosphere. Some consider Bouguer the true discoverer of "Beer's Law." Alas for poor Bouguer, his discovery never made him as famous as it did Beer.

Swiss-German mathematician, physicist, and astronomer Johann Heinrich Lambert (1728–1777)—discussed in "Lambert's Law of Emission" in part II—was more prominent than Bouguer, and he rediscovered and published Bouguer's Law of Absorption. As additional careful experiments were made, scientists noticed that the amount of light absorbed by solutions depended on factors other than the thickness of the material.

In 1852, Beer announced a more complete law of absorption that is known variously as Beer's Law, the Lambert-Beer Law, and the Bouguer-Beer Law. Beer had observed that, in addition to the effect of the sample thickness, the amount of radiation absorbed by a solution was proportional to the concentration of the dissolved substance that is absorbing the radiation. This law established the bedrock on which the field of quantitative spectroscopy is founded, because it provides a simple means for determining concentrations of solutions without having to destroy a portion of the sample.

Today, Beer's Law has many applications. Consider, for example, the need to use ultraviolet radiation to kill microorganisms in drinks. UV light can be used to pasteurize fruit juices, but the antimicrobial effectiveness depends on the UV absorbance of the juice as understood by Beer's Law.

Although Beer did not formulate the exponential absorption law

$$I = I_0 e^{-acx},$$

this relationship has often been called Beer's Law. I is the intensity of light passing through a sample of thickness x, c is the concentration of solute, and a is an absorption coefficient. The first time this formula was referred to as "Beer's Law" occurred in an 1889 paper by B. Walter in *Annalen der Physik*.

FURTHER READING

Ihde, Aaron John, *The Development of Modern Chemistry* (New York: Dover, 1984).

Mavi, Harpal S., *Agrometeorology* (Binghamton, NY.: Haworth Press, 2004); discusses applications of Beer's Law to forest canopies.

INTERLUDE: CONVERSATION STARTERS

> Every atom, being self-existent, had the power in the beginning to adopt what laws of motion it pleased; so they all—by some mysterious universal suffrage, conveyed through the infinity of space...—mutually agreed on the law and intensity of gravity and have steadily kept

to their agreement ever since. If this proposition looks absurd, atheists can blame no one but themselves, for the doctrine of inherent forces cannot be translated in plain English in any other way.

—Henry Augustus Mott, *The Laws of Nature and Man's Power to Make Them Subservient to His Wishes*

How did the universe know, at that moment of beginning, what laws to follow?

—Lee Smolin, "Never Say Always," *New Scientist*, September 23, 2006

You have only to stretch out your hand, close it quickly and you feel that you have caught mathematical air and that a few formulae are stuck to your palm.... Even the sun rays must remember, when passing through the windows, the law to which they are subject according to the will of God, Newton, Einstein, and Heisenberg.

—Leopold Infeld, *Quest: An Autobiography*

There is no mention of laws in Copernicus, or even in Galileo, while Kepler makes no use of the term himself in introducing what are often described as the first truly scientific laws—his three laws of planetary motion.

—Michael Frayn, *The Human Touch*

THE WIEDEMANN-FRANZ LAW OF CONDUCTIVITY

Germany, 1853. In a metal, the ratio of the thermal conductivity to the electrical conductivity is proportional to the temperature.

CROSS REFERENCE: WIEDEMANN-FRANZ-LORENZ LAW AND FOURIER'S LAW OF HEAT CONDUCTION.

In 1853, William Shanks of England completed his calculation of the value of π to 607 digits. Twenty years later, he published a result that extended this precision to 707 decimal places. (Alas, in 1940s, mathematicians discovered that Shanks had made an error in the 528th place, which meant that all the successive digits were incorrect.) Also in 1853, the United States agreed to pay Mexico $10 million for 29,640 square miles (76,770 square kilometers) of land that is today located in southern Arizona and New Mexico. This was known as the "Gadsden Purchase" after James Gadsden, the U.S. minister to Mexico.

The Wiedemann-Franz Law states that the ratio of the thermal conductivity K of a metal to its electrical conductivity σ is proportional to the absolute temperature T:

$$\boxed{\frac{K}{\sigma} = LT}$$

The proportionality constant L is known as the Lorenz number, after Danish mathematician and physicist Ludwig Lorenz (1829–1891), who conducted research in this area. The Lorenz number can be determined from

$$L = \frac{\pi^2}{3}\left(\frac{k}{e}\right)^2,$$

which is equal to 2.45×10^{-8} W·Ω/K^2. Here, e is the elementary electronic charge, and k is the Boltzmann constant.

This empirical law is valid for a limited range of temperatures and is named after German physicist Gustav Wiedemann and his collaborator Rudolf Franz, who reported that the quantity K/σ has the same value for different metals at the same temperature. In 1872, Lorenz discussed in detail how K/σ changed as function of temperature, and sometimes the full law is referred to as the Wiedemann-Franz-Lorenz Law.

TABLE 9 Thermal Conductivities

Material	Thermal Conductivity [W/(m·K)]
Silver	406.0
Copper	385.0
Aluminum	205.0
Concrete	0.8
Styrofoam	0.01

We encounter different thermal conductivities in our daily lives. For example, if a spoon is left in a cup of tea, the spoon handle will become warm because heat from the tea is conducted along the spoon handle. If you were to place a stick of wood in the tea, you would not feel the stick get warm because wood is a much poorer conductor of heat.

If you touch a piece of metal at room temperature, it feels colder than the piece of wood at room temperature because the metal is better able to conduct the heat away from your finger. Different metals, such as silver and aluminum, have different values of thermal conductivity. Some typical thermal conductivity values are given in table 9.

Both heat and electrical conductivity involve the motion of the free electrons in metals. As one heats the metal, average particle velocity increases, which also increases thermal conductivity. On the other hand, electrical conductivity decreases with an increase in temperature because the collisions interfere with the electron motion that corresponds to the transport of charge. At a given temperature, the ratio of the thermal and electrical conductivities is a constant. Those metals that are the best electrical conductors are also the best thermal conductors. "Fourier's Law of Heat Conduction," above, discusses a general expression for heat transfer by conduction.

Until 2001, the Wiedemann-Franz Law was universally acknowledged as applicable to all metals. However, in 2001, University of Toronto researchers showed that the law does not hold for a new class of copper oxide materials at very low temperatures. Although the law is valid for common working temperatures, the quest for additional deviations from the Wiedemann-Franz Law is an ongoing theoretical and experimental effort. Additionally, in the past few years physicists have been exploring the limitations of the law with exotic, new kinds of "wires" that Wiedemann and Franz could have hardly imagined. For example, consider the following excerpt from the paper "Nonlinear Peltier Effect and Thermoconductance in Nanowires," which discusses quantum nanowires (wires with cross sections so small that quantum effects have to be taken into account):

> We find that . . . the Wiedemann-Franz (WF) law, relating the thermal and electric conductances, which holds in classical point contacts, is violated in 3D quantum wires due to the strong energy dependence of the transmission probabilities of the conducting electrons through the wire; in this context we also demonstrate here restoration of the WF law when the energy level quantization effects are less effective, that is for short quantum wires where tunneling contributions to the transmission probabilities become significant, and/or at sufficiently low temperatures when the Fermi distribution is sharpened.

Gustav Wiedemann (1826–1899), German physicist famous for his studies of thermal and electrical conductivity in metals and for his investigations of electromagnetism.

CURIOSITY FILE: Wiedemann's two genius sons came from an extremely intellectual and distinguished pedigree! Their father was famous for the Wiedemann-Franz Law and was professor of physical chemistry at Leipzig. Their maternal grandfather was Eilhard Mitscherlich (1794–1863), famous for his work on chemical isomorphism and similarity of crystal structures. Their mother Clara helped translate into German the Irish natural philosopher John Tyndall's (1820–1893) *Heat as a Mode of Motion*.

> His data for the thermal conductivity of various metals were for long the most trustworthy at the disposal of physicists, and his determination of the ohm in terms of the specific resistance of mercury showed remarkable skill in quantitative research. He carried out a number of magnetic investigations which resulted in the discovery of many interesting phenomena, some of which have been rediscovered by others; they related among other things to the effect of mechanical strain on the magnetic properties of the magnetic metals, to the relation between

> the chemical composition of compound bodies and their magnetic properties, and to a curious parallelism between the laws of torsion and of magnetism.
>
> —1911 *Encyclopaedia Britannica*

Gustav Wiedemann was born in Berlin. His father was a merchant and died when Wiedemann was only 2 years old. His mother died when he was 15. According to the *Proceedings of the Royal Society of London*,

> Wiedemann was thus from an early age thrown much upon his own resources, but the care of a friend secured for him a careful, classical, and scientific education. His inclination to the special study of physics, seems to have been largely due to the influences of [physicist Thomas] Seebeck, who, for several years, was one of his teachers at the Cologne Gymnasium.

In 1844, Wiedemann began his studies of the natural sciences at the University of Berlin. In 1847, he received his doctorate for a dissertation on biuret, a white, crystalline, nitrogenous substance, $C_2O_2N_3H_5$, formed by heating urea. He lectured at the university on many different topics in physics and conducted research on the polarization of light.

In 1851, Wiedemann married the daughter of German chemist Eilhard Mitscherlich. His eldest son, Eilhard, became a physicist and historian of science and author of the 1880 paper "On a Means to Determine the Pressure at the Surface of the Sun and Stars, and Some Spectroscopic Remarks." According to E. Newton Harvey's *A History of Luminescence*, in 1888 Eilhard was the first individual to use the term "luminescence." The younger son, Alfred, became an Egyptologist and author of a popular book that offered readers an accessible overview of the predominant role of religion in ancient Egyptian life.

In 1864, Wiedemann was professor of physics at the University of Basel, and in 1871 he held the first professorship of physical chemistry in Germany—at the University of Leipzig. In 1877, he became editor of the prestigious *Annalen der Physik und Chemie*.

Wiedemann's primary contributions revolved around his studies of electrical conductivity in metals, the rotation of the plane of light polarization that is caused by electric currents, and the thermal conductivity of metals. In 1853, he and his collaborator Rudolph Franz discovered the law discussed above that states that at a constant temperature, the electrical conductivity of metals is approximately proportional to the thermal conductivity of the metal. Wiedemann later worked on a variety of subjects that included:

- The effect of current intensity on osmotic pressure
- The effect of temperature on the magnetization of steel and iron
- Magnetism in chemical compounds
- Vapor pressures of salts containing water
- The construction of new galvanometers, that is, instruments used to detect and measure the direction of small electric currents

His greatest work is often cited as *Die Lehre vom Galvanismus* (1861–1863), which summarized everything known about galvanism, the study of direct electrical current and its effects. The full title of the book, *Die Lehre vom Galvanismus und Elektromagnetismus nebst technischen Anwedungen*, was changed to *Die Lehre von Elektricität* in a later updated edition.

Very little appears to be known about Wiedemann's collaborator Rudolph Franz. Few encyclopedias provide information, and even his date of birth is unknown. I welcome information from readers on the elusive Franz.

Wiedemann's 1899 obituary, appearing in *Physical Review*, commented on his skills as a lecturer:

> As a lecturer upon the elements of physics and of chemistry, Wiedemann was noted for the clearness and simplicity of his exposition and for a delightful fluency and ease of diction. He had personal acquaintanceship with many of the famous experimenters of the first half of the century, and the inexhaustive fund of anecdotes concerning them and their work was delightful and, historically speaking, a valuable element of his lectures. He treated physics from the point of view of the chemist rather than of the mathematician, and chemistry very largely from the standpoint of experimental physics.

FURTHER READING

Bogachek, E. N., A. G. Scherbakov, and Uzi Landman, "Nonlinear Peltier Effect and Thermoconductance in Nanowires," *Physical Review* B, 60(15): 678–682, October 15, 1999.

Harvey, E. Newton, *A History of Luminescence from the Earliest Times until 1900* (Philadelphia: American Philosophical Society, 1957).

Körber, Hans-Günther, "Gustave Wiedemann," in *Dictionary of Scientific Biography*, Charles Gillispie, editor-in-chief (New York: Charles Scribner's Sons, 1970).

"Wiedemann Obituary," *Physical Review* (Series I), 9: 57–58, July 1899.

"Wiedemann Obituary," *Proceedings of the Royal Society of London*, 75: 41–42 1905.

INTERLUDE: CONVERSATION STARTERS

Our brains have evolved to get us out of the rain, find where the berries are, and keep us from getting killed. Our brains did not evolve to help us grasp really large numbers or to look at things in a hundred thousand dimensions.

—Ronald Graham, quoted in Paul Hoffman's "The Man Who Loves Only Numbers," *Atlantic Monthly*, 1987

There may be further laws to discover, to do with the unification of gravity with quantum theory and with the other forces of nature. But in a certain sense, we have for the first time in history a set of laws sufficient to explain the result of every experiment that has ever been done.

—Lee Smolin, "Never Say Always," *New Scientist*, September 23, 2006

The fact that anything at all in the Universe is comprehensible means either that we are very intelligent or that the basics of nature are very simple. Given that we are some sort of chimpanzee, carrying a mere kilogram of glop between our ears, I opt for the latter alternative.

—Vincent Icke, *The Force of Symmetry*

The lesson for the truth of fundamental laws is clear: fundamental laws do not govern objects in reality; they govern only objects in models.

—Nancy Cartwright, *How the Laws of Physics Lie*

FICK'S LAWS OF DIFFUSION

Germany, 1855. The steeper the concentration gradient, the greater the net flux of material by diffusion.

CROSS REFERENCE: FICK'S PRINCIPLE IN THE FIELD OF CARDIAC PHYSIOLOGY.

> In 1855, French chemist George Audemars obtained the first patent for production of artificial silk, which he derived from mulberry bark. The safety match was invented in Sweden. These matches ignited only when struck on a chemically impregnated surface. The Panama Railway became the first railroad to connect the Atlantic Ocean and Pacific Ocean. German mathematician and physicist Carl Friedrich Gauss died.

Fick's laws concern the net transport of material in a liquid through the process of diffusion. Diffusion refers to the physical process whereby particles spread spontaneously through a medium, and in particular, it refers to the movement of particles from higher concentration to lower concentration. One common example of diffusion is the spreading of a drop of ink in water. As time progresses, the ink particles spread evenly throughout the water.

Imagine a 1-square-centimeter area in the *y*, *z*-plane viewed on edge and symbolized by this vertical line, |. This boundary might be thought of as an invisible partition in a glass of water. On one side of | is a drop of ink placed into water. We define the flux J_x as the net amount of substance that diffuses through this unit area per unit time in the *x* direction. Common units of flux are mol/(cm^2·s). If the ink were evenly distributed in the glass of water, then no concentration gradient would exist in the *x* direction, and $J_x = 0$. In this situation, we expect equal numbers of ink particles to cross our boundary | from the left and from the right.

Next, imagine that there is a concentration gradient of the ink in the *x* direction and that more ink exists on the right of | than on the left. If *c* is concentration, then $dc/dx > 0$. Because more particles per unit volume exist to the right of our boundary than to the left, we expect that more molecules per unit time will diffuse through | from the right than from the left. Indeed, a net transport of ink particles will take place in the direction opposite to the concentration gradient.

FICK'S FIRST LAW OF DIFFUSION

Fick's First Law of Diffusion states that the steeper the concentration gradient $\mathrm{d}c/\mathrm{d}x$, the larger the net flux J_x:

$$\boxed{J_x = -D\frac{\mathrm{d}c}{\mathrm{d}x}}$$

Here, D is a proportionality constant called the diffusion coefficient (e.g., the diffusion coefficient for myoglobin protein in water at 20°C is 11.3×10^{-7} cm^2/s). The negative sign indicates that the net transport by diffusion is in a direction opposite to the concentration gradient. Example units for D are cm^2/s, and for the concentration gradient $\mathrm{d}c/\mathrm{d}x$, we have units of mol/cm^4. Note that because a net transport of material, such as ink particles, exists, $\mathrm{d}c/\mathrm{d}x$ is itself a function of time. If we let the ink diffuse in the glass of water for a long time, the system approaches homogeneity, and $\mathrm{d}c/\mathrm{d}x = 0$. Thus, the equation is given in terms of the instantaneous flux at any time t. Numerous laboratory experiments have been used to verify Fick's First Law.

A simple example illustrates Fick's First Law. Consider a liquid-filled cylinder with a diameter of 10 centimeters. Assume that a material is dissolved in the liquid so that its concentration profile decreases linearly along the axis of the cylinder, and the diffusion coefficient of the material is 4×10^{-5} cm^2/s. The concentration of the material is 1 mol/dm^3 at one end of the cylinder, for example, at location x + 10 centimeters ("dm" is the unit for decimeter, equal to 10 cm). The concentration of the material at the other end is 0.5 mol/dm^3 (at position x). What is an approximate value for the flux through the material, assuming that there is no change in the concentration profile?

The change in concentration is $c(x_2) - c(x_1) = -0.5$ mol/dm^3 or -500 mol/m^3. The change in x is $x_2 - x_1 = 10$ cm or 0.1 m. Thus, an approximate value for $\mathrm{d}c/\mathrm{d}x$ is $-(500\ \mathrm{mol/m^3})/(0.1\ \mathrm{m}) = -5,000\ \mathrm{mol/m^4}$. Given the aforementioned value of D and converting to units of m^2/s, we have $J_x = -(4 \times 10^{-9}\mathrm{m^2/s})(-5 \times 10^{-3}\ \mathrm{mol/m^4}) = 2.0 \times 10^{-5}\ \mathrm{mol/(m^2{\cdot}s)}$.

FICK'S SECOND LAW OF DIFFUSION

Fick's Second Law of Diffusion describes how the concentration of material in a gradient changes with time. For example, in a uniform concentration gradient of ink particles, where $\mathrm{d}c/\mathrm{d}x$ is a constant for all values of x, Fick's First Law demands that the flux J_x is the same at all locations, and thus, c will not change with time. One may visualize the flux into every volume element from one side is equal to that going out the other side for a steady-state flow of ink particles by diffusion. However, if the concentration gradient is not the same everywhere, we have Fick's Second Law:

$$\left(\frac{\partial c}{\partial t}\right)_x = D\left(\frac{\partial^2 c}{\partial x^2}\right)_t,$$

where $(\partial c/\partial t)_x$ is the change of concentration with time at position x. For Fick's Second Law, we assume the diffusion coefficient D is a constant and independent of concentration. Experimentalists can actually determine D by using either Fick's First or Second Law. However, because measurement of J can be challenging, the Second Law is more often used to determine the value of D.

Fick's Laws of Diffusion are so general that they have had wide applications in many fields. Edward L. Cussler, in his talks titled "Fick's Second Law or Diffusion for Dummies," describes the diversity of applications. He notes that this law can predict physiological and sensory reactions to the relative sweetness of syrups and sauces if we assume that sweetness depends on the transport of sugar in a solution to the surface of the tongue. As another Cussler example, when a fire ant becomes alarmed, the insect releases a pheromone to warn other ants of potential danger. Fick's Second Law can predict the effect of this pheromone on the other ants, using a very large value for the diffusion coefficient.

Fick's Second Law has also been used to predict the spread of muskrats in Europe following their accidental release in 1905. The law has been used to model the concentration of smokestack contaminants and to simulate the displacement of hunter-gatherers by farmers in Neolithic times.

Note that Fick's laws can apply to solids, gases, and liquids. In a number of research studies, the law has been useful for the study of photosynthesis and the concentration gradient of carbon dioxide inside leaves. Researchers have also used the laws to study diffusion of radon in the open air or diffusion in soils contaminated with petroleum hydrocarbons.

Adolf Fick (1829–1901), German physiologist famous for his laws of diffusion.

CURIOSITY FILE: Fick is credited with the invention of glass contact lenses in the late 1880s, which he tested first on rabbits and then on himself. However, the use of practical and comfortable lenses started in 1948 when California optician Kevin Tuohy invented the soft plastic lens for contacts. • In 1865, Fick and German chemist Johannes Wislicenus climbed a mountain in the Swiss Alps in order to study the relationship between the food they ate and the urine produced. They published their findings in 1866 in "On the Origin of Muscular Power," where they concluded that fats and carbohydrates (and not protein) provided the necessary fuel for their muscles.

> It is astonishing that no one has arrived at the following obvious method by which [the amount of blood ejected by the ventricle of the heart with each systole] may be determined directly, at least in animals. One measures how much oxygen an animal absorbs from the air in a given time, and how much carbon dioxide it gives off. During the experiment one obtains a sample of arterial and venous blood; in both the oxygen and carbon dioxide content are measured. The difference in oxygen content tells how much oxygen each cubic centimeter of blood takes up in its passage through the lungs. As one knows the total quantity of oxygen absorbed in a given time one can calculate how many cubic centimeters of blood passed through the lungs in this time. Or if one divides by the number of heart beats during this time one can calculate how many cubic centimeters of blood are ejected with each beat of the heart. The corresponding calculation with the quantities of carbon dioxide gives a determination of the same value, which controls the first.
>
> —Adolf Fick, "On the Measurement of the Blood Volume in the Cardiac Ventricle," 1870

> A model that predicts Fickian diffusion can be constructed from assuming a simple random-walk model. Such a model (also called a drunkard's walk) assumes that a single molecule follows a series of steps, each of which moves in a direction taken at random. Fick's law applies equally well to molecular diffusion through a static fluid and diffusion across a laminar flow in a channel or duct. In both cases, the diffusion occurs via the random paths of individual fluid molecules.
>
> —Gerard V. Middleton and Peter R. Wilcock, *Mechanics in the Earth and Environmental Sciences*

Adolf Fick was born in Kassel, Germany, the ninth child of Friedrich Fick, a senior municipal architect. As with other lawgivers in this book, Fick had family members who were destined for intellectual greatness. One brother, for example, became a professor of anatomy at Marburg University, and another became a professor of law. Fick attended Marburg and was very interested in mathematics and physics but decided that a career in medicine would most suit him.

Fick was especially interested in the mathematical and physical study of anatomical forms. His first paper, published in 1849, was a study of torque exerted by leg muscles. He obtained his doctorate in 1851, with a thesis on visual problems arising from astigmatism, an eye defect in which the unequal curvature of one or more refractive surfaces of the eye prevents light rays from focusing clearly on one point on the retina. His love of mathematics led him to formulate some of his key thoughts about diffusion in the mid1850s, when he concluded that "the distribution of a compound dissolved in a solvent takes place in the same way as warmth is distributed within a conductive material." He began to believe that diffusion could be described using a mathematical basis reminiscent of Fourier's Law of Heat Conduction. E. L. Cussler writes in *Diffusion: Mass Transfer in Fluid Systems*:

> Fick seemed initially nervous about his hypothesis. He buttressed it with a variety of arguments based on kinetic theory. Although these arguments are now dated, they show physical insights that would be exceptional in medicine today. For example, Fick recognized that diffusion is a dynamic molecular process. He understood the difference between a true equilibrium and steady state.

In order to test his equations, Fick had to determine a way to create a steady-state concentration gradient. He finally achieved this by using a glass cylinder containing crystalline sodium chloride at the bottom and a large volume of water at the top. By frequently changing the water at the top, he produced the desired linear gradient.

In additional experiments, Fick also noted that the volume of gas flow moving across a tissue sheet per unit time is directly proportional to the area of the sheet and the difference in partial pressures between the two sides, but inversely proportional to tissue thickness:

$$V_{\text{gas}} \propto \frac{A \cdot D(P_1 - P_2)}{T}, \ D \propto \frac{S}{\sqrt{m_{\text{w}}}}$$

Here V_{gas} is the volume of gas flow per unit time, A is the area, D is a diffusion constant, P_1 is the pressure on one side of the membrane or tissue, and P_1 the partial pressure on the other side of the membrane of thickness T. The diffusion constant D is directly proportional to the gas solubility S but inversely proportional to the square root of its molecular weight, m_{w}.

Fick married in 1862 and had five children. One son eventually became an anatomist and another became a jurist. In 1868, Fick became full professor of physiology in the faculty of medicine at the University of Würzburg, from which he did not retire until age 70. He died in 1901 after suffering from a cerebral hemorrhage.

Fick is perhaps most respected for the sheer diversity of the subjects he studied, many at the boundaries between physiology and physics. In 1854, he devoted himself to the study of the saddle joint of the thumb and the muscles of the eye and how they operated. In 1855, he developed a differential equation that modeled diffusion. Later, he analyzed the eye's blind spot, retina, color vision, and intraocular pressure. He also developed improved devices for measuring blood pressure and for recording the movements of the torso caused by breathing. In 1856, when only 26, he published *Medizinische Physik* (Medical Physics), in which his fertile mind discussed topics that ranged from gas diffusion to the dynamics of muscles, hydrodynamics, the elasticity of blood vessels, the physics of the eye, and the origin of heat in the body.

In 1870, Fick developed a principle that helped researchers calculate cardiac output by monitoring oxygen levels of the blood. (Cardiac output is often measured in terms of the volume of blood that the heart pumps in a minute.) In particular, Fick showed why the cardiac output can be calculated from the oxygen consumption of breathing divided by the difference in oxygen content between the left and right heart chambers; stated more mathematically, cardiac output (liters/minute) is equal to oxygen consumption (milliliters/minute) divided by the arteriovenous oxygen difference in mixed blood (milliliters/liter). This equation is known as Fick's Principle. Verification of this principle in humans was initially accomplished years after Fick's death, in 1930, by researchers who were able to obtain samples of mixed venous blood by inserting a spinal tap needle just to the right of the sternum. The needle entered the right ventricular chamber by puncturing its wall.

Later studies focused on the speed of blood flow, blood pressure in capillaries, protein metabolism, heat generation in muscles, and the effect of nerve stimulation on muscles. According to William Coleman's *Biology in the Nineteenth Century*, in 1874, Fick stated his reductionist credo in which all life should be viewed in terms of purely mechanical events and that so-called "vital phenomena" are

> caused by the forces inherent in the material bases of the living organism. Since customarily these forces are divided into chemical and physical, we may designate this as the "chemicophysical." In so far, however, as all forces are in final analysis nothing other than motive forces determined by the interaction of material atoms, and in so far as the general science of motion and its causal forces is

called mechanics, we must designate the direction of physiological research as truly "mechanical."

In addition to numerous scientific papers and his aforementioned classic textbook *Medizinische Physik*, Fick wrote *A Compendium of Physiology* and *Handbook of Anatomy and Physiology of the Sense Organs; and Circulation of Blood*. In 1929, two of his sons founded the Adolf Fick Fund, which awards a prize every five years for an exceptional contribution to physiology.

FURTHER READING

Bentley, David J., Jr., "Polymers/Laminations/Adhesives/Coatings/Extrusions," *Paper, Film & Foil Converter* magazine, July 1, 2001; see pffc-online.com/mag/paper_polymerslaminationsadhesivescoatingsextrusions_3/; includes a description of Edward L. Cussler's talk titled "Fick's Second Law or Diffusion for Dummies."

Coleman, William, *Biology in the Nineteenth Century* (New York: Cambridge University Press, 1978).

Cussler, Edward L., *Diffusion: Mass Transfer in Fluid Systems* (New York: Cambridge University Press, 1997).

Middleton, Gerard V., and Peter R. Wilcock, *Mechanics in the Earth and Environmental Sciences* (New York: Cambridge University Press, 1994).

Rothschuh, K. E., "Adolf Fick," in *Dictionary of Scientific Biography*, Charles Gillispie, editor-in-chief (New York: Charles Scribner's Sons, 1970).

Sten-Knudsen, Ove, *Biological Membranes* (New York: Cambridge University Press, 2002).

Tinoco, Ignacio, Kenneth Sauer, and James Wang, *Physical Chemistry* (Englewood Cliffs, N.J.: Prentice-Hall, 1978).

Vandam, Leroy D., and John Fox, "Adolf Fick (1829–1901), Physiologist: A Heritage for Anesthesiology and Critical Care Medicine," *Anesthesiology*, 88(2): 514–518, February 1998.

INTERLUDE: CONVERSATION STARTERS

> Maybe the brilliance of the brilliant can be understood only by the nearly brilliant.
>
> —Anthony Smith, *The Mind*

> Since Galileo's time, science has become steadily more mathematical.... It is virtually an article of faith for most theoreticians... that there exists a fundamental equation to describe the phenomenon they are studying.... Yet... it may eventually turn out that fundamental

laws of nature do not need to be stated mathematically and that they are better expressed in other ways, like the rules governing the game of chess.

—Graham Farmelo, Foreword to *It Must Be Beautiful: Great Equations of Modern Science*

When the scientist attempts to understand a group of natural phenomena, be begins with the assumption that these phenomena obey certain laws which, being intelligible to our reason, can be comprehended. This is not, let us hasten to note, a self-evident postulate which leaves no room for qualifications. In effect, what it does is to reiterate the rationality of the physical world, to recognize that the structure of the material universe has something in common with the laws that govern the behavior of the human mind.

—Arthur March and Ira M. Freeman, *The New World of Physics*

Our knowledge of nature . . . is highly compartmentalized. Why think nature itself is unified?

—Nancy Cartwright, *How the Laws of Physics Lie*

BUYS-BALLOT'S WIND AND PRESSURE LAW

 *Netherlands, 1857.* The wind blows at right angles to the atmospheric pressure gradient.

CROSS REFERENCE: RUDOLF CLAUSIUS AND THE DOPPLER EFFECT.

> In 1857, Elisha Otis's first safety-equipped passenger elevator was installed—in New York City. (This elevator with safety brakes decreased public fears of elevators, which was crucial to the rise of skyscrapers.) Divorce without parliamentary approval becomes legal in Britain. *Atlantic Monthly* magazine was first published. James Clerk Maxwell mathematically proved that the rings of Saturn are composed of many small bodies orbiting the planet.

Buys-Ballot's Law, named after Christoph Hendrik Diederik Buys Ballot, asserts that in the Northern Hemisphere, if a person stands with his back to the wind, the low pressure area will be to his left. This means that wind travels counterclockwise around low pressure zones in the Northern Hemisphere. (This direction is reversed in the Southern Hemisphere.) The law also states that the wind and the pressure gradient are at right angles if measured sufficiently far above the surface of Earth in order to avoid frictional effects between the air and surface of Earth.

The weather patterns of Earth are affected by several planetary features such as Earth's roughly spherical shape, rotation, and the Coriolis effect, which is the tendency for any moving body on or above the surface of Earth, such as an ocean current, to drift sideways from its course due to the rotation of Earth.

We can understand the reason for the swirling air patterns around regions of low pressure by considering air flow into the lowpressure area from the north and south. Air that is closer to the equator is generally traveling faster than air farther away because equatorial air is farther from Earth's axis of rotation. To help visualize this, consider that air farther from the axis must travel faster in a day than air at higher latitudes that is closer to the axis of Earth. Thus, if a low pressure system in the north exists, it will draw air from the south, and this air can move faster than the ground below it because the more northerly part of Earth's surface has a slower eastward motion than the southerly surface. This means that the air from the south will move east as a result of its higher speed.

We can see that the opposite kinds of motion apply for air moving from north to south into a low-pressure region. The northerly air will move south and also west. The net effect of air movement from north and south is a counterclockwise swirl around a low pressure area in the Northern Hemisphere, and a clockwise swirl in the Southern Hemisphere. The Coriolis effect is weak near the equatorial regions; thus, the Buys-Ballot's Law is not applicable at equatorial latitudes.

Christoph Hendrik Diederik Buys Ballot (1817–1890), Dutch meteorologist and physical chemist who explained movements of the wind.

CURIOSITY FILE: In order to better understand the behavior of sound, Buys Ballot conducted experiments involving trumpet players blowing a G note while on moving trains. • The Dutch named an island after Buys Ballot while he was still alive.

> When you place yourself in the direction of the wind, . . . you will have at your left the least atmospheric pressure.
> —C. H. D. Buys Ballot, "On the System of Forecasting the Weather Pursued in Holland," 1863

Buys Ballot was born in Kloetinge, the Netherlands. The son of a Dutch reformed minister, Buys Ballot received his doctorate in 1844. He lectured on mineralogy and geology at the University of Utrecht in 1845. In 1847, he was appointed professor of mathematics. L. C. Palm and colleagues in *The History of Science in the Netherlands* describe his early years as a faculty member:

> Buys Ballot worked on a mathematical theory of matter, in which atoms attracted each other but the surrounding ether particles repelled each other. He published his theory in 1849 under the title "Sketch of a Physiology of the Inorganic Realm of Nature." The book was met with little interest. . . . Disappointed, Buys Ballot abandoned physics and chemistry and concentrated on meteorology, which could hardly be called a science at the time.

In 1854, he founded the Royal Netherlands Meteorological Institute. He married twice and was elected to the Royal Academy of Sciences of Amsterdam in 1855. He became professor of physics in 1867.

Buys Ballot was quite religious and an active member of the Walloon church. One of his chief passions was meteorology, and he continually lobbied for establishing wide networks of meteorological observations that could provide simultaneous data through the use of the telegraph.

In 1845, Buys Ballot had performed one of the first experiments verifying Doppler's idea for sound, which stated that a change in the observed frequency of a sound should occur when the source and observer are in motion relative to each other. The frequency should increase when the source and observer approach each other and decrease when they move apart.

To perform his experiment, Buys Ballot employed a train that carried trumpeters who played a constant note while other musicians listened on the side of the track! Because no accurate devices existed for the measuring musical pitch of the trumpet note while the trumpets moved, Buys Ballot had to rely on the human ear. In particular, Buys Ballot enlisted the help of Utrechters who were known to have a talent for sensitivity to pitch. Some rode on the train. Others stood along the tracks. One trumpeter sounded a G note in the moving train. Three trumpeters were used on the ground as the train passed. The 14 observers, distributed on the train and ground, were asked to quantify the pitch change of the G note. As a result of two days of experimentation, conducted at several speeds, Buys Ballot proved the existence of the Doppler effect, which he then reduced to a formula.

Although today we know that the Doppler effect applies to light as well as sound, Buys Ballot remained skeptical about generalizing his findings. Dev Maulik and Ivica Zalud write in *Doppler Ultrasound in Obstetrics and Gynecology*:

> Buys Ballot proved not only the existence of the Doppler Effect in relation to sound transmission but its angle dependency as well. Incredibly, Buys Ballot still refused to accept the validity of the theory for the propagation of light, and most of the scientific community of the nineteenth century did not acknowledge the validity of Doppler's theory [when used to explain the behavior of light].

Because of the telegraph, it became possible to establish meteorological observation stations that could communicate with one another in order to make better weather predictions. Buys Ballot became a leader in this field of meteorology, and for several years he published weather observations that were provided by a large network of observers.

In 1857, he noticed that in the Netherlands, the wind blew at right angles to the pressure gradient, and he published this observation in the *Comptes Rendus*. In 1863, he stated the law again in *Transactions*, a journal of the British Association for the Advancement of Science. Unknown to Buys Ballot, this rule of wind motion was first actually deduced by the American meteorologists James Henry Coffin (1806–1873) and William Ferrel (1817–1891). In fact, Ferrel was also first to realize that the Buys-Ballot Law results from the deflecting force of Earth's rotation. Although Ferrel first provided the theory for the law, Buys Ballot was the first to provide an empirical validation through his extensive observations. Buys Ballot later acknowledged that Ferrel first discovered the law.

Today, people sometimes use Buys-Ballot's Law to locate hurricanes. For example, if you are in North America, the center and direction of travel of a hurricane can be estimated by Buys-Ballot's Law as follows: Face the wind and extend your right hand out at 90° from the direction you are facing. Your arm is now pointing approximately to the center of the storm. Periodic determinations like this will indicate the relative movement of the storm and on which side of the hurricane's line of motion you are standing.

In 1857, German physicist Rudolf Clausius (1822–1888) published what he felt to be the average speed of oxygen, nitrogen, and hydrogen molecules at the temperature of melting ice, namely, 461 m/s, 492 m/s, and 1,844 m/s, respectively. Buys Ballot examined these numbers and understood a consequence that somehow escaped Clausius and other scientists. If gas molecules really moved so quickly, why didn't one instantly smell ammonia or hydrogen sulfide the moment a vial was opened at the other end of a room? Buys Ballot thought he had refuted the new molecular theory by pointing out a startling contradiction between its predictions and the real world. As a result, Clausius had to make an important modification to his theory and had to assume that molecules were sufficiently large that they collided with one another and could not move very far without such collisions. Thus, a molecule must change its direction many times every second, and may require a long time to escape from a given macroscopic region of space. In this way, the slowness of ordinary gas diffusion, compared with molecular speeds, could be explained.

Buys Ballot's obituary appearing in *Symons's Meteorological Magazine* notes

> On Sunday night, February 2nd, from his well-loved home at Utrecht, passed away the spirit which gave to the world the useful "Buys-Ballot's Law," by which the author will be remembered long after his many personal friends have themselves been removed. Dr. Ballot was director, indeed almost creator, of the Royal Meteorological Institute of the Netherlands. In 1883, a new island, discovered by the Dutch Meteorological Expedition, in 70° 25′ 28′ N., was

named after him as Buys-Ballot's Island.... Dr. Ballot's earliest scientific papers were upon chemistry and physics, but for forty years nearly all his time and thought has been devoted to meteorology.

A lunar crater with a diameter of 55 kilometers was named after Buys-Ballot and approved in 1970 by the International Astronomical Union General Assembly.

FURTHER READING

Burstyn, Harold, "Christoph Buys Ballot," in *Dictionary of Scientific Biography*, Charles Gillispie, editor-in-chief (New York: Charles Scribner's Sons, 1970).

"Buys Ballot Obituary," in *Symons's Meteorological Magazine*, volume 25 (London: Edward Stanford: 1890), p. 8.

Maulik, Dev, and Ivica Zalud, *Doppler Ultrasound in Obstetrics and Gynecology* (New York: Springer, 2005).

Palm, L. C., Albert Van Helden, and Klaas Van Berke, *The History of Science in the Netherlands* (Leiden, The Netherlands: Brill Academic Publishers, 1999).

INTERLUDE: CONVERSATION STARTERS

> Armies of thinkers have been defeated by the enigma of why most fundamental laws of nature can be written down so conveniently as equations. Why is it that so many laws can be expressed as an absolute imperative, that two apparently unrelated quantities (the equations right and left sides) are exactly equal? Nor is it clear why fundamental laws exist at all.
>
> —Graham Farmelo, *It Must Be Beautiful: Great Equations of Modern Science*

> We are in the habit of talking as if [the laws of Nature] caused events to happen; but they have never caused any event at all. The laws of motion do not set billiard balls moving; they analyze the motions after something else... has provided it. They produce no events: they state the pattern to which every event... must conform.... It is therefore inaccurate to define a miracle as something that breaks the laws of Nature.... If I knock out my pipe, I alter the position of a great many atoms, in the long run, and to an infinitesimal degree, of all the atoms there are. It is one more bit of raw material for the laws to apply

to. . . . If God creates a miraculous spermatozoon in the body of a virgin, it does not proceed to break any laws. Nature is ready. Pregnancy follows according to all the normal laws.

—C. S. Lewis, "Miracles," in *The Complete C.S. Lewis Signature Classics*, 2002

Models in the mathematical, physical and mechanical sciences are of the greatest importance. Long ago philosophy perceived the essence of our process of thought to lie in the fact that we attach to the various real objects around us particular physical attributes—our concepts—and by means of these try to represent the objects to our minds. . . . On this view our thoughts stand to things in the same relation as models to the objects they represent.

—Ludwig Boltzmann, 1902 *Encyclopaedia Britannica*

EÖTVÖS'S LAW OF CAPILLARITY

Hungary, 1866. The surface tension of a liquid depends on the temperature and density of the liquid.

CROSS REFERENCE: CHARLES-AUGUSTIN COULOMB, GUSTAV KIRCHHOFF, NEWTON'S LAWS, AND EINSTEIN'S GENERAL THEORY OF RELATIVITY.

> In 1866, Dostoevsky published *Crime and Punishment* in twelve monthly installments. Alfred Nobel, creator of the Nobel Prize, invented dynamite. The Atlantic Cable was successfully completed, allowing reliable transatlantic telegraph communication for the first time. (An earlier version failed soon after it was installed due to deterioration of the insulation, and some members of the press were so skeptical that they hinted that the first line was a mere hoax.) Andrew Rankin patented the urinal. Tennessee became the first Confederate state readmitted to Union. Manuelito, the last Navaho chief, surrendered at Fort Wingate. Ernst Haeckel introduced the word "ecology" (spelled *Oecologie* in German).

Eötvös's Law of Capillarity states the relationship between the surface tension of a liquid and the temperature of a liquid. In particular, we have

$$\boxed{\gamma = k(T_0 - T)/\rho^{3/2},}$$

where the surface tension γ (also called the capillarity constant) of a liquid is related to its temperature T, the critical temperature of the liquid (T_0), and its density ρ. The constant k is approximately the same for many common liquids such as water. Note that T_0 is the temperature at which the surface tension disappears or becomes zero.

The term "surface tension" usually refers to a property of liquids that arises from unbalanced molecular forces at or near the surface of a liquid. As a result of these forces, the surface tends to contract and exhibits properties similar to those of a stretched elastic membrane. Eötvös showed that this surface tension is related to the molecular volume of a fluid, that is, the volume occupied by one mole of the material that makes up the liquid. This volume is numerically equal to the molecular weight divided by the density. Let γ_1 and γ_2 be the surface tensions at temperatures t_1 and t_2, and v_1 and v_2 the corresponding molecular volumes. According to Eötvös's Law, we have

$$(\gamma_1 v_1 - \gamma_2 v_2)^{2/3}/(t_1 - t_2) = k$$

Hence, the surface tension, which may be considered as a *molecular* surface energy, changes in response to temperature in a manner that is independent of the nature of the liquid. Eötvös's Law has been important for chemists because they could use it to determine the molecular weight of a material. The molecular volume v and the molecular weight μ are related by $v = \mu/\rho$, where ρ is the density.

During his experiments, Eötvös had to take special care that the surface of his fluids had no contamination of any kind, and thus he worked with glass vessels that had been closed by melting, and he used optical methods for determining the surface tension. These sensitive methods were based on optical reflection in order to characterize the local geometry of the liquid surface. Because the properties of the surface change in time through heating, he also achieved a higher degree of temperature stability by enclosing the fluid in a closed glass tube. He showed that the molecular surface energy of fluids only depends on the temperature.

Surface tension plays a role in numerous aspects of nature. It allows insects to walk on water. At the surface of the liquid, the molecules are pulled inward by intermolecular forces. Surface tension is often measured in newtons per meter (N/m), a unit that is equivalent to joules per square meter (J/m^2). This second set of units reminds us that surface tension may be considered as surface energy.

Loránd Eötvös (1848–1919), Hungarian physicist famous for his studies of surface tension as well as the gravitational field of Earth.

CURIOSITY FILE: Did you know that water is sprayed on the spot where a diver will enter the water because some divers feel that this turbulence reduces the surface tension of the water, thus resulting in a less painful impact? • Loránd Eötvös was a famous mountain climber, and a peak in the Dolomites in northeast Italy is named after him. • The *eotvos*, which usually is denoted by the symbol E, is a unit of acceleration divided by distance. The gravitational gradient of Earth, which may be thought of as the change in gravitational acceleration from one point on the surface on Earth to another, is sometimes measured in units of eotvos. Gradient anomalies in mountainous areas can be as large as 1,000 eotvos.

> Poets can penetrate deeper into the realm of secrets than scientists.
>
> —Loránd Eötvös, quoted in P. Király, "Eötvös and STEP"

A scientist can soar high like a poet, but also knows how high he flies.

—Loránd Eötvös, quoted in P. Király, "Eötvös and STEP

My unknown country spread out far below the frozen surface of Lake Balaton. I have never seen it and shall never see it, only my instrument sensed it, still how hard it was to part with it when the ice started to melt.

—Loránd Eötvös, describing his gravitational measurements performed with the torsion balance on the ice sheet of Hungary's Lake Balaton in 1901 and 1903, quoted in P. Király, "Eötvös and STEP

Vásárosnaményi Báró Eötvös Loránd was born in Budapest, Hungary. He is also sometimes referred to as Roland, Baron von Eötvös (the German version of the Hungarian name), Loránd Eötvös, or Roland Eotvos (Eötvös is pronounced ut' vush). His father was one of Hungary's foremost writers and political philosophers of the 1800s.

Eötvös entered the University of Budapest in 1865 as a law student but was so fascinated by mathematics and physics that he took private lessons in these subjects in addition to his law courses. When he finally decided that his passion was for science, Eötvös left the law school and entered the University of Heidelberg in 1867, where he obtained his doctorate summa cum laude in 1870. Some of his early papers dealt with the intensity of moving light sources, a theoretical body of work that eventually led to the theory of relativity.

Eötvös returned to Hungary in 1871 and soon became full professor at the University of Budapest. In 1876 he married, and eventually had two daughters who became delightful companions for him during his mountain climbs. Mountain climbing was his favorite hobby, and he became one of Europe's most famous climbers.

Eötvös conducted his famous work on determining surface tension at the University of Königsberg. These investigations were published in several papers between 1876 and 1886.

After 1886, Eötvös's research focused on the nature of gravitation. His paper published in 1890 reported on his work with torsion balances used to study the forces of attraction between masses. Torsion balances for measuring attraction between two masses existed before Eötvös's work—in fact his first devices were similar to those of scientists John Michell (1724–1793), Henry Cavendish (1731–1810), and Charles-Augustin de Coulomb (1736–1806)—but Eötvös refined his balance to gain added sensitivity.

The Eötvös balance became one of the best instruments for measuring gravitational fields at the surface of Earth and for predicting the existence of certain structures beneath the surface. In fact, although Eötvös focused on basic theory and research, his instruments later proved important for prospecting for oil and natural gas. Andrew L. Simon, author of *Made in Hungary*, notes that very soon after Eötvös's Heidelberg University studies under such professors as Gustav Kirchhoff (1824–1887), Robert Bunsen (1811–1899), and Hermann von Helmholtz (1821–1894),

> he became the greatest Hungarian scientist of theoretical and experimental physics.... The use of the Eötvös Torsion Balance has been a landmark in geological research and prospecting and was instrumental in the discovery of oil fields in Texas, Venezuela, the Zala oil fields in Hungary, and elsewhere.... Since Eötvös's work, his countrymen's studies in gravitation, geomagnetism and seismology have been notable.

Eötvös's measurements also showed that *gravitational mass* (the mass m in Newton's Law of Universal Gravitation $F = Gm_1m_2/r^2$) and *inertial mass* (the constant mass m responsible for inertia in Newton's Second Law, which we often write as $\mathbf{F} = m\mathbf{a}$) were the same—at least to an accuracy of about five parts in 10^9. In other words, Eötvös showed that the inertial mass (the measure of the resistance of an object to acceleration by an applied force) is the same as gravitational mass (the factor determining the weight of an object) to within a great deal of accuracy. This information later proved useful for Einstein when he formulated the General Theory of Relativity. Einstein cited Eötvös's work in Einstein's 1916 paper *The Foundation of the General Theory of Relativity*. The General Theory of Relativity suggests that gravitation is a consequence of curved space. One idea of General Relativity, called the Equivalence Principle, is that gravity pulling in one direction is equivalent to acceleration in the opposite direction. Many of the predictions of general relativity, such as the bending of starlight by gravity and the small shift in the orbit of the planet Mercury, have been confirmed by experiments.

Returning our attention to Eötvös's sensitive balance, I should note that this device was essentially the first instrument useful for gravitational gradiometry, that is, for the measurement of very local gravitational properties. For example, Eötvös's early measurements involved his mapping the second derivatives of the gravitational potential at different locations in his office and, shortly after, the entire building. Local masses in the rooms influence the values he obtained. The Eötvös balance could also be used to study the gravitational changes due to slow motions of massive bodies or fluids. According to Péter Király of the KFKI Research Institute for Particle and Nuclear Physics, "changes in the water level of the Danube could allegedly be detected from a cellar 100 meters away with a centimeter precision, but that measurement was not well documented."

Eötvös had many other related interests, including his studies involving: magnetic anomalies, the shape of Earth, paleomagnetic work on bricks that were thousands of years old, and variations in gravitational acceleration caused by the relative motion of an object with respect to Earth.

Mountaineering and stereoscopic photography were Eötvös's favorite hobbies. With his daughters, he made the first ascent of several peaks in the Dolomite Mountains of the Alps. When he was 68, shortly before his death, he climbed some of the highest peaks of the Tatra Mountains on the border of modern-day Poland and Slovakia.

In 1885, Eötvös and colleagues founded the Hungarian Society for Mathematics. From 1886 until his death, Eötvös taught at the University of Budapest, which in 1950 was renamed the Eötvös Loránd University in his honor. Biographer L. Marton writes of Eötvös's lasting effect on education in Hungary in the *Dictionary of Scientific Biography*:

> Eötvös's intensive research efforts did not prevent him from pursuing other interests. Shortly after his appointment as professor of physics, he became aware of the shortcomings of both high school and university instruction in Hungary, and from then on he devoted considerable effort to improving both.... [These efforts led to] a surprising increase in the number of outstanding Hungarian scientists during the twentieth century....

A lunar crater with a diameter of 99 kilometers was named after Eötvös and approved in 1970 by the International Astronomical Union General Assembly.

FURTHER READING

Király, P., "Eötvös and STEP," poster presented at the Satellite Test of the Equivalence Principle Symposium, Pisa, Italy, April 6–8, 1993. Published, without figures, in *Proceedings* (ESA WPP-115), R. Reinhard, editor (Noordwijk, The Netherlands: ESTEC, July 1996), pp. 399–406; see www.mek.iif.hu/porta/szint/tarsad/tudtan/eotvos/html/stepcikk.html or www.kfki.hu/eotvos/stepcikk.html.

Marton, L. "Roland Eötvös," in *Dictionary of Scientific Biography*, Charles Gillispie, editor-in-chief (New York: Charles Scribner's Sons, 1970).

Simon, Andrew L., *Made in Hungary: Hungarian Contributions to Universal Culture* (Safety Harbor, Fla.: Simon Publications, 1998).

INTERLUDE: CONVERSATION STARTERS

Anyone who cannot cope with mathematics is not fully human. At best he is a tolerable subhuman who has learned to wear shoes, bathe, and not make messes in the house.

—Robert A. Heinlein, *Time Enough for Love*

The secularization of the concept of nature's laws proceeds more slowly in England than on the continent of Europe. By the end of the eighteenth century, after the French Revolution, Laplace could boast that he had no need of the hypothesis of God's existence, and Kant had sought to ground the universality and necessity of Newton's laws not in God or nature, but in the constitution of human reason.... [Nonetheless] whether the laws of nature might be expressions of divine will was still much debated in the third quarter of the nineteenth century in Britain.... Not until Darwin's revolution had worked its way through British intellectual life did the laws of nature get effectively separated from God's will.

—Ronald N. Giere, *Science Without Laws*

No theory can be objective, actually coinciding with nature.... Each theory is only a mental picture of phenomena, related to them as sign is to designatum.... From this it follows that it cannot be our task to find an absolutely correct theory but rather a picture that is, as simple as possible and that represents phenomena as accurately as possible. One might even conceive of two quite different theories both equally simple and equally congruent with phenomena, which therefore in spite of their difference are equally correct.

—Ludwig Boltzmann, "On the Development of Methods of Theoretical Physics in Recent Times"

KOHLRAUSCH'S LAWS OF CONDUCTIVITY

Germany, 1874 and 1875. At low concentrations, the molar conductivities of strong electrolytes vary as the square root of electrolyte concentration. The conductivity of an electrolyte in solution depends upon the sum of the contributions from its individual ions.

CROSS REFERENCE: OSTWALD'S DILUTION LAW, ARRHENIUS'S LAW OF DISSOCIATION, AND OHM'S LAW OF ELECTRICITY.

> In 1874, the role of fluorides in preventing dental decay was discovered. British statesman Winston Churchill and Hungarian-born magician Harry Houdini were born. The Young Men's Hebrew Association in Manhattan was founded, and it still operates today as the "92nd Street Y." The first public zoo in the United States opened, in Philadelphia. German chemist Othmar Zeidler prepared DDT, but he did not discover its insecticidal properties. (It was not until 1939 when Swiss entomologist Paul Mueller discovered its use as an insecticide.)

Friedrich Kohlrausch was interested in understanding how electricity is conducted in solutions. This kind of study was difficult because direct current caused ions in the solution to collect near the electrodes and partially neutralize the electric potential. Additionally, products of decomposition tended to collect on the electrodes. In order to avoid this problem and achieve precise results, Kohlrausch decided to use alternating current instead of direct current. As a result, he found in the early 1870s that conductivity of solutions increased with increasing temperature. In 1876, he discovered that ions (charged particles) in dilute solutions did not interact much with one another, and that the water molecules were essentially the only particles that impede the flow of ions. He concluded that "in a dilute solution every electrochemical element has a perfectly definite resistance pertaining to it, independent of the compound from which it is electrolyzed." In a sense, he had found that some of the concepts relating to Ohm's Law of Electricity, related to resistors in circuits, could be used to understand electrical properties of solutions of electrolytes, which are substances that conduct electricity by facilitating the flow of ions in solutions.

As background to Kohlrausch's laws, note that the concept of electrical resistance may be used to study the motion of ions in solution. Some of the variables typically under study include the resistance R of a sample, which increases with its length l and decreases with its cross-sectional area A.

The conductivity κ is the inverse of the resistivity of the sample, or more precisely, $\kappa = 1/(RA)$. Λ_m is the molar conductivity, which is defined as $\Lambda_m = \kappa/c$, where c is the molar concentration of the electrolyte that is present in the solution.

Kohlrausch's findings can be stated as two laws: Kohlrausch's Square Root Law (1874) and Kohlrausch's Law on the Independence of Migrating Ions (1875).

KOHLRAUSCH'S SQUARE ROOT LAW (1874)

The conductivity of a solution depends on the number of ions in the solution. At low concentrations, the molar conductivities of strong electrolytes vary as the square root of concentration:

$$\boxed{\Lambda_m = \Lambda_m^\circ - Kc^{1/2}}$$

The constant Λ_m° is referred to as the *limiting molar conductivity*, that is, the molar conductivity at the limit of zero concentration. At this infinite dilution, the ions do not interact with one another. K is a constant and usually depends on the ratio of ions that compose an electrolyte.

An electrolyte is said to be strong if a high proportion of a solute dissociates to form free ions. For example, nitric acid (HNO_3) is a strong electrolyte and fully dissociates to ions according to $HNO_3 \rightarrow H^{+1} + NO_3^{-1}$.

KOHLRAUSCH'S LAW ON THE INDEPENDENCE OF MIGRATING IONS (1875)

Kohlrausch also found that the molar conductivity Λ_m of an electrolyte is the sum of the contributions from its individual ions, that is, from the cations or anions of the electrolyte. (A cation is an ion or group of ions having a positive charge. An anion is a negatively charged ion or group of ions.) The ions behave independently at dilute concentrations. If we denote the limiting molar conductivity of the cations as λ_+ and that of the anions as λ_-, then we have

$$\boxed{\Lambda_m^\circ = \nu_+\lambda_+ + \nu_-\lambda_-,}$$

where ν_+ and ν_- are the numbers of the cations and anions per formula unit of electrolyte. For example, $\nu_+ = 1$ and $\nu_- = 2$ for $BaCl_2$.

Kohlrausch's laws are perhaps the best known of the various solution conductivity laws that scientists have discovered. Several slightly less fundamental laws exist that concern the conductivity of solutions, for example, Ostwald's Dilution Law (1888):

$$\frac{1}{\Lambda_m} = \frac{1}{\Lambda_m^\circ} + \frac{\Lambda_m c}{K_a \left(\Lambda_m^\circ\right)^2},$$

where c is the solute concentration and K_a is the equilibrium constant for dissociation of the solute. The law was first suggested by the German chemist Wilhelm Ostwald (1853–1932). Ostwald's Dilution Law is a relationship that shows the concentration dependence of the molar conductivity, and the law is often used for weak electrolytes that are not fully ionized in solution—for example, weak acids and bases. The conductivity depends on the number of ions in solution and thus on the degree of ionization. We may use this equation for determining the limiting molar conductivity of the solution by plotting $1/\Lambda_m$ against c/Λ_m. The intercept at $c = 0$ will be $1/\Lambda_m^\circ$. An example of a weak electrolyte is acetic acid, which will not fully dissociate into its component ions when in an aqueous solution $HC_2H_3O_2 \leftrightarrow H^{+1} + C_2H_3O_2^{-1}$.

Another relationship, sometimes called Arrhenius's Law of Dissociation (developed around 1883–1887), states that the degree of electrolyte dissociation, a, to form ions when the electrolyte is dissolved in water is given by $a = \Lambda_m/\Lambda_m^\circ$, where Λ_m is the molar conductance at some concentration, and Λ_m° is the molar conductance at infinite dilution. This relationship holds only in situations where the ionic interaction effects are minimum. If one knows the degree of dissociation, one can easily calculate the dissociation constant. This law is named after Swedish physical chemist Svante Arrhenius (1859–1927).

Friedrich Wilhelm Kohlrausch (1840–1910), German physicist famous for his contributions to our understanding of the electrical conductivity of solutions.

CURIOSITY FILE: The Kohlrausch physicist-family triumvirate is composed of the physicist father Rudolph and two physicist sons Friedrich Wilhelm and Wilhelm Friedrich. Of the three, the son discussed in this entry is the most eminent, as judged from the amount of space allocated to each in the

Dictionary of Scientific Biography. • The Kohlrausch relaxation function in mathematics and physics is named in honor of Friedrich Kohlrausch's father, and it takes the form $F_s(t) = f \exp[-(t/\tau_r)^{\beta}]$, where β is a stretching parameter, $0 < \beta \leq 1$, and f is a scaling factor. A unique time τ_r characterizes the process. This formula, published by Kohlrausch in 1847, is often used today in the study of lubricants and the dynamics of viscous liquids and glasses. Friedrich Kohlrausch continued research in this area.

> *To some it may seem strange that the weaker [more dilute] the solution, the greater the flow of ions* (current), but this is what Kohlrausch's law states. His theory is correct only over a limited range of dilutions because if the dilution increases to the extent that no ions are available in the solution, no electric current will be conducted between the electrodes. In other words, absolutely pure water will not conduct electricity.
>
> —Robert E. Krebs, "Kohlrausch's Law," *Scientific Laws, Principles, and Theories*

Friedrich Wilhelm Kohlrausch was born in Rinteln, Germany. His father, Rudolph Kohlrausch (1809–1858), taught mathematics and physics and is famous for his work with Wilhelm Weber (1804–1891), the German physicist who, among other things, devised a logical system of units for electricity and, with Carl Friedrich Gauss (1777–1855), investigated terrestrial magnetism. In particular, Rudolph is remembered for his collaborations with Weber that demonstrated important relationships between electrostatic and electromagnetic units and the speed of light.

Friedrich, the son, received his doctoral degree at University of Göttingen in 1863, studying under his father's colleague Weber. Friedrich was appointed extraordinary professor at the University of Göttingen from 1866 to 1870. In 1866, Friedrich Wilhelm Kohlrausch collaborated with his brother, physicist Wilhelm Friedrich Kohlrausch, on the electrochemical properties of silver. (As you might guess, sometimes historians of science confuse the contributions of the three Kohlrausches!)

While at the University of Göttingen, Kohlrausch published his book *Leitfaden der praktischen Physik* (Guidelines to Practical Physics), which described a variety of experimental and measuring techniques and influenced German students for many years that followed.

In 1870, Kohlrausch became a professor at the Polytechnikum at Zurich. A year later, he moved to the Darmstadt University of Technology in Germany. At this time he showed, with his colleague Otto Grotrian, that the conductivity of solutions increased with increasing temperature. While at the Darmstadt University in 1874, he demonstrated that an electrolyte has a constant amount of electrical resistance. He calculated the transfer velocity of ions by studying the dependence of conductivity upon dilution.

Kohlrausch recognized that one convenient way to compare the conductance behavior of different electrolytes is to compare the limiting values of conductivity Λ_m as the concentration approaches zero. As discussed in Cooper H. Langford's *The Development of Chemical Principles*, for strong electrolytes, this limiting value at zero concentration, Λ_m°, is measured by performing a series of experiments and extrapolating experimental data in order to create a graph of Λ_m versus molar concentration. According to Langford, Kohlrausch was the first to make a systematic study of strong electrolytes and tabulate values for Λ_m°

Because Λ_m° is with respect to an infinite dilution solution, no interionic interactions would take place at such a dilution. Obviously, direct measurement of Λ_m° is impossible, because an infinitely dilute solution will have no conductivity contribution from the movement of ions. However, the simple extrapolation of experimental data allows scientists to make useful tables of values for Λ_m°.

In 1875, Kohlrausch worked at the University of Würzburg, and from 1875 to 1879, he examined numerous salt solutions and acids. This research led him to state his law of the independent migration of ions, in which each type of migrating ion has a specific resistance independent of what its original molecular combination may have been. The electrical resistance of a solution results only from the migrating ions of a given substances.

In 1888 Kohlrausch became director of the physical laboratory at Strasbourg University, and in 1895 he was president of the Physikalisch-Technische Reichsanstalt (Imperial Physical Technical Institute).

Kohlrausch can be considered the "measurement king," as a result of his improvements to a variety of measuring devices, which included the Kohlrausch bridge (for measuring conductivity), a tangent galvanometer (for determining the presence, direction, and strength of an electric current in a conductor), and a reflectivity meter. Kohlrausch was an important scientist in the history of electrochemistry for his two laws and because his work influenced so many other researchers. For example, Ostwald made use of Kohlrausch's methods and technologies in his own research, and Ostwald's Dilution Law is discussed above.

Ostwald himself was a fascinating person, independent of discussions of Kohlrausch. Ostwald—a viola player, landscape painter, and color theorist—is best known for his work on catalysis. During his life, he published 45 books, more than 500 articles, and thousands of reviews. In 1909, Ostwald won the Nobel Prize for Chemistry. He devoted a great amount of his time to promoting of world peace; however, his attempts to promote Ido, an improved version of the universal language Esperanto, did not achieve lasting results.

Friedrich Wilhelm Kohlrausch died in January of 1910 in Marburg, Germany. Iwan Rhys Morus eulogizes the life of Kohlrausch in *When Physics Became King*:

> Looking back over his career in 1900, Kohlrausch opined that "measuring nature is one of the characteristic activities of our age." A colleague... remarked that "no other physicist has surpassed Kohlrausch in the skill and care with which he used instruments and methods." He had [invented] dynamometers, galvanometers, magnetometers, and reflectometers.... Working with Weber, he had devoted almost forty years to working at determining the values of electrical and magnetic constants and units.... He [steered] the Reichsanstalt [scientific institution] towards helping ensure German domination of the expanding electrical industries.

FURTHER READING

Atkins, P. W. *Physical Chemistry*, 5th edition (New York: Freeman, 1996).

Dogra, S. K., *Physical Chemistry Through Problems* (New Dehli: New Age Publishers, 1984).

Drennan, Ollin, "Friedrich Kohlrausch," in *Dictionary of Scientific Biography*, Charles Gillispie, editor-in-chief (New York: Charles Scribner's Sons, 1970).

Hamann, Carl H., Andrew Hamnett, and Wolf Vielstich, *Electrochemistry* (Hoboken, N.J.: Wiley, 1988).

Katz, Eugenii, "Friedrich Wilhelm Georg Kohlrausch"; see chem.ch.huji.ac.il/~eugeniik/history/kohlrausch.htm.

Krebs, Robert, "Kohlrausch's Law," in *Scientific Laws, Principles, and Theories* (Westport, Connecticut: Greenwood Press, 2001).

Langford, Cooper H., *The Development of Chemical Principles* (New York: Dover, 1995).

Morus, Iwan Rhys, *When Physics Became King* (Chicago: University of Chicago Press, 2005).

Zeiss, Carl, "Kohlrausch's Laws," Carl Zeiss AG; see www.zeiss.de/C12567A100537AB9/Contents-Frame/9BBD454D0BCC325EC1256D0900331437.

INTERLUDE: CONVERSATION STARTERS

Mathematicians, astronomers, and physicists are often religious, even mystical; biologists much less often; economists and psychologists very seldom indeed. It is as their subject matter comes nearer to man himself that their antireligious bias hardens.

—C. S. Lewis, "Religion without Dogma?" *The Grand Miracle: And Other Selected Essays on Theology and Ethics from* God in the Dock

What is the status of claims that are typically cited as "laws of nature"—Newton's Laws of Motion, the Law of Universal Gravitation, Snell's Laws, Ohm's Law, the Second Law of Thermodynamics, the Law of Natural Selection? Close inspection, I think, reveals they are neither universal nor necessary—they are not even true.

—Ronald N. Giere, *Science Without Laws*, 1999

Mathematicians define things, otherwise they wouldn't have a clue what they are talking about. This is so because everything in mathematics was invented by people. Contrariwise, nothing of the substance of physics was invented by us: Nature is out yonder and there is not a shred of evidence that the Universe cares a fig about humans or other beings. The formulations of physics are uniquely ours, but those, of course, are mathematics. That is why, contrary to popular opinion, physicists never actually define anything physical.

—Vincent Icke, *The Force of Symmetry*

Science is not a system of certain, or well-established statements, nor is it a system which steadily advanced toward a state of finality.

—Karl Popper, *The Logic of Scientific Discovery*

CURIE'S MAGNETISM LAW AND THE CURIE-WEISS LAW

France, 1895, generalized in 1907. The magnetic susceptibility of paramagnetic materials is inversely proportional to the absolute temperature. A critical temperature (the Curie temperature) exists, above which the magnetic properties disappear.

> In 1895, the Armenians were massacred in Turkey. H. G. Wells published *The Time Machine.* Italian-Irish electrical engineer Guglielmo Marconi invented "radio telegraphy." German physicist Wilhelm Röntgen discovered X-rays. American businessman King Camp Gillette invented a safety razor with a disposable blade, along with the business model that eventually made him famous. Baseball superstar Babe Ruth and U.S. boxing champion Jack Dempsey were born. W. E. B. Du Bois became the first African American to receive a Ph.D. from Harvard University. Irish physicist George Fitzgerald suggested that distance contracts in the direction a body is traveling.

CURIE'S MAGNETISM LAW

French physical chemist Pierre Curie discovered the following law by fitting experimental results to a simple model. In particular, Curie's Law illuminates the relationship between the magnetization of certain kinds of materials and the applied magnetic field and temperature:

$$\boxed{M = C \cdot \frac{B_{\text{ext}}}{T}}$$

Here, M is the resulting magnetization, and B_{ext} is the magnetic flux density of the applied (external) field, measured in teslas. T is the absolute temperature measured in degrees kelvin, and C is the Curie point, a constant that depends on the material. According to Curie's Law, if one increases the magnetic field, one tends to increase the magnetization of a material in that field. As one increases the temperature while holding the magnetic field constant, the magnetization decreases.

Sometimes, Curie's Law is written as

$$\boxed{\chi = \frac{C}{T},}$$

where $\chi = M/B_{ext}$ is the magnetic susceptibility, that is, the degree of magnetization of a material in response to a magnetic field.

Curie's Law is applicable to paramagnetic materials, such as aluminum and copper, in which atomic magnetic dipoles have a tendency to align with an external magnetic field. These materials can become very weak magnets, and their attractive force can be measured by using sensitive instruments. In particular, when subject to a magnetic field, paramagnetic materials suddenly attract and repel like standard magnets. When there is no external magnetic field, the magnetic moments of particles in a paramagnetic material are randomly oriented, and the paramagnet no longer behaves as a magnet. (Generally speaking, "magnetic moment" refers to the magnetic strength and direction, or alignment of the magnetic poles, at a given time.) When placed in a magnetic field, the moments generally align parallel to the field, but this alignment may be counteracted by the tendency for the moments to be randomly oriented due to thermal motion.

Curie's Law holds only for a limited range of values of B_{ext}. In other words, Curie's Law works only for samples in which a relatively small fraction of the atoms are aligned with the magnetic field. When the aligned fraction becomes sufficiently large, Curie's Law no longer applies, and the magnetization M cannot increase indefinitely with increasing external magnetic field. Obviously, once the dipoles are nearly 100% aligned, further increases in the magnetization are impossible.

Paramagnetic behavior can also be observed in ferromagnetic materials, such as iron and nickel, that are above their Curie temperatures, T_c. The Curie temperature is a temperature above which the materials lose their ferromagnetic ability, that is, the ability to possess a net (*spontaneous*) magnetization even when no external magnetic field is nearby. For iron, the Curie temperature is 1,043°K (770°C). Ferromagnetism is responsible for most of the magnets you encountered at home, such as permanent magnets that may be sticking to your refrigerator door, or the horseshoe magnet you played with as a child.

As discussed above, paramagnetism is a "weaker form" of magnetism. Paramagnetic materials in magnetic fields act like magnets, but when the field is removed, thermal motion will quickly disrupt the magnetic alignment. For most atoms and ions, the magnetic effects of the electrons cancel so that a particle is not magnetic and does not exhibit paramagnetism. This is true for gases like neon and for the Cu^{+} ions (a copper atom from which one electron has been removed) that compose ordinary copper. For other atoms or ions, such as Mn^{2+}, the magnetic effects do not cancel, so such atoms have magnetic dipoles.

THE CURIE-WEISS LAW

The Curie-Weiss Law describes the magnetic susceptibility χ, defined above for Curie's Law, in the paramagnetic region above the Curie point:

$$\chi = \frac{C}{T - T_c}$$

Here, C is a material-specific Curie constant; T is absolute temperature measured in degrees kelvin, and T_c is the Curie temperature measured in degrees kelvin. At the temperature $T = T_c$, a spontaneous magnetization exists. Note that this law is an extension of Curie's Law and takes into account the Curie temperature, a specific temperature that changes depending on the nature of a substance and is approximately equal to 0 for paramagnets. Ferromagnetic substances have a large, positive value for T_c, indicative of their strong interactions. Substances called antiferromagnetics, such as chromium, have a negative value for T_c.

In other words, ferromagnetic materials show ferromagnetic behavior only when the temperature is less than T_c. When $T > T_c$ the ferromagnetic material has the standard paramagnetic behavior. The Curie-Weiss Law, which gives the magnetic susceptibility as a function of temperature, is only valid above the Curie temperature. Below T_c, the atomic magnetic moments tend to align in a common direction within the ferromagnetic material. The spontaneous magnetization below the Curie temperature arises from an internal magnetic field called the Weiss molecular field, in honor of French physicist Pierre Weiss (1865–1940). This field is proportional to the magnetization of small domains in the material.

A relative magnetic susceptibility of a material may be determined by placing a sample of the material inside a small coil and measuring the inductance of the coil with and without the sample. If the inductance is measured as a function of temperature from above to below the Curie temperature, the Curie-Weiss Law shown in the preceding formula can be used to determine the Curie temperature.

Let us review before proceeding to discussions of magnetic domains and the sources of magnetism. This Curie-Weiss Law comes from Weiss's theory, proposed for ferromagnetic materials, which incorporates the interactions between magnetic moments. T_c can be positive, negative, or zero. When $T_c = 0$, then the Curie-Weiss Law is the same as the Curie Law. When T_c is nonzero, then an interaction between neighboring magnetic moments exists, and the material is paramagnetic only above a certain transition temperature. If T_c is positive, then the material is ferromagnetic below the Curie temperature. The equation is valid only when the material is in a paramagnetic state.

To help us understand what these laws and terms really mean, it is useful to review the nature of magnetism more generally. Recall that electric currents generate magnetic fields. If a substance alters a magnetic field, we can visualize this alteration in terms of currents within the substance. Many atoms actually act like tiny bar magnets and can be imagined as containing tiny current loops. Some atoms—such as carbon, copper, and lead—because of the orientation of their electrons, have no permanent magnetic moments. Other atoms have permanent magnetic moments and may exhibit the property of paramagnetism. If we subject these atoms to a magnetic field, the field tends to align the magnetic moments with the field, and these atoms actually increase the magnetic field into which they are placed. As discussed, these paramagnetic effects are temperature dependent. Paramagnetic effects decrease with increasing temperature due to more energetic thermal motions that make alignment more difficult.

Iron, nickel, and cobalt react strongly in a magnetic field and are said to be ferromagnetic and have high values of χ. These atoms have large magnetic moments, and atoms in small regions of a substance, called domains, may align so that their magnetic moments are all in the same direction. For example, in an unmagnetized bar of iron, these domains exist, but the alignment direction of the magnetic moment in each domain may be random with respect to neighbor domains. Domains may be as large as a fraction of a millimeter in some materials. If an external magnetic field is applied to a piece of iron, the domains aligned with the field direction grow at the expense of unaligned domains. This causes an augmentation of the applied field. Domains actually grow by the motion of domain boundaries. At very high applied field, almost all the atoms may be aligned with the field. However, if the temperature is raised sufficiently high (beyond the Curie temperature), the atoms will break loose from each other's influence, and the alignment is lost. The Curie temperatures for several ferromagnetic materials are listed in table 10. As mentioned, above the Curie temperature, these materials behave like other paramagnetic materials.

All elements can be classified in terms of their magnetic behavior and fall into one of several categories according to their bulk magnetic susceptibility (see table 11). Most elements are either diamagnetic or paramagnetic at room temperature. Ferrimagnetic properties are observed in compounds. Only the elements iron, cobalt, and nickel are ferromagnetic

TABLE 10 Curie Temperatures for Several Ferromagnetic Materials

Material	T_c (°K)
Iron	1,043
Nickel	633
Iron and nickel alloy (50% each)	803
Gadolinium	293
Gadolinium chloride ($GdCl_3$)	2.2

From D. H. Martin's *Magnetism in Solids.*

at and above room temperature. At temperatures above T_c, the susceptibility varies according to the Curie-Weiss Law.

Pierre Curie (1859–1906) and Pierre Weiss (1865–1940), French physicists responsible for the Curie-Weiss law showing the dependence of magnetic susceptibility on temperature.

CURIOSITY FILE: Pierre Curie considered himself to have a feeble mind and never went to elementary school. He later shared the Nobel Prize with his wife, Marie. Marie subsequently received a second Nobel Prize, making her the first person to win or share two Nobel Prizes. • The Curie's elder daughter and her husband also received the Nobel Prize. • The 2006 Nobel Prize awarded to Roger Kornberg for his work on the transcription of genetic information made him and his father the sixth father–son pair to win Nobel Prizes since the prizes were first awarded in 1901 and the eighth set of parent and child laureates.

> Every [scientific] discovery, however small, is a permanent gain.
>
> —Pierre Curie's 1984 letter to Marie, urging her to join him in "our scientific dream"

> We must eat, drink, sleep, be idle, love, touch the sweetest things of life and yet not succumb to them. It is necessary that, in doing all this, the higher thoughts to

TABLE II Materials and Their Magnetic Behaviors

Type	Susceptibility χ	Magnetic Behavior	
		Magnetic Moments of Atoms[a]	Magnetic Field Effect
Diamagnetism	Small, negative (e.g., $\chi = -2.7 \times 10^{-6}$ for gold)	Atoms have no magnetic moment	As the external field increases, the material's magnetization decreases
Paramagnetism	Small, positive (e.g., $\chi = 21.0 \times 10^{-6}$ for platinum)	Atoms have randomly oriented magnetic moments	As the external field increases, the material's magnetization increases
Ferromagnetism	Large, positive; a function of applied field; microstructure dependent (e.g., $\chi \sim 100{,}000$ for iron)	Atoms have aligned magnetic moments, facing one direction	As the external field increases, the material's magnetization increases but rapidly reaches a plateau value for magnetization
Antiferromagnetism	Small, positive (e.g., $\chi = 3.6 \times 10^{-6}$ for chromium)	Atoms have aligned magnetic moments	As the external field increases, the material's magnetization increases
Ferrimagnetism	Large, positive; a function of applied field; microstructure dependent (e.g., $\chi \sim 3$ for barium ferrite)	Atoms have aligned magnetic moments	As the external field increases, the material's magnetization increases but rapidly reaches a plateau value for magnetization

[a] Characteristics in this column refer to the material in the absence of a magnetic field.

Adapted from the University of Birmingham's Applied Alloy Chemistry Group, "Classification of Magnetic Materials."

> which one is dedicated remain dominant and continue their unmoved course in our poor heads. It is necessary to make a dream of life, and to make of a dream a reality.
>
> —Pierre Curie, quoted in Marie Curie's *Pierre Curie*

> Radium could become very dangerous in criminal hands, and here the question can be raised whether mankind benefits from knowing the secrets of Nature....
>
> —Pierre Curie's Nobel Prize speech, "Radioactive Substances, Especially Radium"

The Curie Law and more general Curie-Weiss Law reflect the combined efforts of Pierre Curie and Pierre Weiss. As discussed above, the transition from ferromagnetic to paramagnetic properties at the Curie point is related to a change in the relationship of the magnetic susceptibility to the temperature. In 1895, Curie asserted that above the Curie point, the susceptibility varies inversely as the absolute temperature. In 1907, Pierre Weiss showed this was not generally true, and so he modified Curie's Law to account for the susceptibility of paramagnetic substances above the Curie point—so that the susceptibility varies inversely as the *excess* of the temperature about that point. The Curie-Weiss Law does not operate at or below the Curie point.

Pierre Curie is famous for his research in the fields of radioactivity, crystallography, magnetism, and piezoelectricity, which is the ability of certain crystals to generate a voltage when the crystal is mechanically stressed. He became so devoted to science and scientific research that he essentially spent most of his life in a laboratory. The line between his private and scientific life became totally blurred.

Curie was born in Paris, where his father was a physician. His father provided Curie's early education at home before Curie entered the Faculty of Sciences at the Sorbonne. Of Curie's younger years, his wife, Marie, would later write in her 1923 biography *Pierre Curie*:

> Pierre passed his childhood entirely within the family circle; he never went to the elementary school....His earliest instruction was given him first by his mother and was then continued by his father and his elder brother....Pierre's intellectual capacities were not those which would permit the rapid assimilation of a prescribed course of studies. His dreamer's spirit would not submit itself to the ordering of the intellectual effort imposed by the school.
>
> He himself believed that he had this slow mind and often said so....It seems to me, rather, that...it was necessary for him to

concentrate his thought with great intensity upon a certain definite object, in order to obtain a precise result, and that it was impossible for him to interrupt or to modify the course of his reflections to suit exterior circumstances.

As a young teenager, Curie had a love of mathematics, especially spatial geometry, which would later be of value to him in his work on crystallography. In 1880, Curie and his brother Jacques demonstrated that electricity was produced when certain crystals were compressed—a phenomenon now called "piezoelectricity." Their demonstrations involved such crystals as tourmaline, quartz, and topaz. In 1881, the brothers demonstrated the reverse effect, namely, that electric fields could cause some crystals to deform. Although this deformation is small, it was later found to have practical applications in the production and detection of sound and the focusing of optical assemblies. Piezoelectric applications have been used in the design of phonograph cartridges, microphones, and ultrasonic submarine detectors. Today, piezoelectric applications are encountered commonly in electric cigarette lighters, which use a piezoelectric crystal to produce a voltage in order to ignite the gas from the lighter. The U.S. military is exploring the possible use of piezoelectric materials in soldier's boots for generating power in the battlefield.

In 1884, Curie published a memoir on subjects that included crystal symmetry. Another article on symmetry and its repetitions appeared in 1885. While working on his exhaustive studies of the groups of symmetry that might exist in nature, Curie sought wider application of his ideas and wrote, as quoted in Marie Curie's *Pierre Curie*, "I think it is necessary to introduce into physics the ideas of symmetry familiar to crystallographers."

In 1895, Curie obtained his Doctor of Science degree and was appointed professor of physics. His doctoral thesis was on various forms of magnetism. His early work with his brother Jacques focused on crystallography and piezoelectric effects. Later, he turned his attention to magnetism, and he built many sophisticated apparatuses, which made use of balances, electrometers, and piezoelectric crystals.

Curie designed a sensitive torsion balance for measuring various magnetic effects, and his equipment proved useful to other researchers. Curie's interest in magnetism led him to his discovery of Curie's Law, discussed above. The law shows how a paramagnetic material is sensitive to temperature. The material constant in Curie's Law is known as the Curie constant. He also discovered the transition temperature (or Curie point), above which ferromagnetic materials lose their ferromagnetic behavior.

His doctoral thesis work on the magnetic properties of materials at different temperatures actually started around 1891. His precise words relating to his objectives are found in his wife's biography:

> From the point of view of their magnetic properties, bodies may be divided into two groups: *diamagnetic* bodies, bodies only feebly magnetic, and *paramagnetic* bodies. At first sight the two groups seem entirely separate. The principal aim of this research has been to discover if there exist transitions between these two states of matter, and if it is possible to make a given body pass progressively through them. To determine this I have examined the properties of a great number of bodies at temperatures differing as much as possible, in magnetic fields of varying intensities.
>
> My experiments failed to prove any relation between the properties of *diamagnetic* and those of paramagnetic bodies. And the results support the theories which attribute magnetism and diamagnetism to causes of a different nature. On the contrary, the properties of *ferro-magnetic* bodies and of bodies *feebly magnetic* are intimately united.

Curie's experimental work presented many challenges, because it necessitated the measuring of very minute forces within a container subjected to very high temperatures.

In 1894, Curie met his future wife when she was living in Paris and studying at the Sorbonne. His affection is evident in his love letter to Marie when she had gone back to her beloved Poland to visit her father:

> We have promised each other (is it not true?) to have, the one for the other, at least a great affection. Provided that you do not change your mind! For there are no promises which hold; these are things that do not admit of compulsion.
>
> It would, nevertheless, be a beautiful thing in which I hardly dare believe, to pass through life together hypnotized in our dreams: your dream for your country; our dream for humanity; our dream for science. Of all these dreams, I believe the last, alone, is legitimate.

Marie later noted that Curie was so dedicated to scientific study that he did not marry until he was 36 because he did not believe that a marriage could meet his strict requirements. Pierre and Marie married in 1895. Pete Moore writes of their relationship in *E=mc²: The Great Ideas That Shaped Our World*:

> Shortly after graduating, Marie married renowned physicist Pierre Curie. It was a relationship that was initially based on mutual interest in science, but which prospered. Curie wrote once: "I have the

best husband one could dream of. I could never have imagined finding one like him. . . . ”

In 1896, French physicist Antoine Henri Becquerel (1852–1908) discovered that a uranium compound, when placed upon a photographic plate covered with black paper, develops the plate in a manner similar to light. The Curies grew intrigued, and they collaboratively studied radioactive substances, making use of poor laboratory equipment while at the same time attempting to manage a heavy teaching load in order to earn a living. In fact, they conducted most of their pioneering research in a room that had previously been used as a small storeroom and machine shop. Later, they worked in a wooden shed with an unfinished floor and a leaky glass roof. The room had no hoods to carry away the poisonous gases produced during their experiments.

In 1898, the Curies announced the discovery of radium and polonium and later conducted landmark studies into the properties of radioactivity and radioactive byproducts. (Polonium was named after Poland, Marie's birthplace.) Pierre Curie and one of his students observed heat production from radium particles, which, in effect, was the first discovery of nuclear energy. He and his colleagues were the first to report on the decay of radioactive materials that was accompanied by skin burns. By using magnetic fields, he discovered that some radioactive particles had no charge while others were either positively or negatively charged.

During 1899 and 1900, the Curies published a memoir on the discovery of the radioactivity produced by radium—along with accounts on its luminous, electrical, and chemical properties. The Curies, along with Becquerel, were awarded the Nobel Prize for Physics in 1903 “in recognition of the extraordinary services they have rendered by their joint researches on the radiation phenomena. . . . ” The Curies were also awarded the Davy Medal of the Royal Society of London in 1903.

Pierre Curie's work was published in many journals such as the *Comptes Rendus de l'Académie des Sciences*, the *Journal de Physique*, and the *Annales de Physique et Chimie*. In 1905, Curie was elected to the Academy of Sciences.

In 1906, Pierre Curie was killed in a Paris street, his head crushed under a carriage wheel during a rainstorm. Had he not died in this manner, he possibly would have succumbed to the radiation poisoning that later killed his beloved Marie, who died from leukemia caused by her exposure. In fact, while working with radium, Curie had voluntarily exposed his arm to radium for several hours to understand its effects. This exposure had created a lesion resembling a burn that required several months to heal.

Barbara Goldsmith, author of *Obsessive Genius: The Inner World of Marie Curie*, writes of the carelessness with which radioactivity was treated, as judged by the standards of today. Here, she describes physicist Ernest Rutherford's 1903 visit to the Curies:

> After the last toast, the group strolled out into the garden. In the dark of the night, Pierre reached in his vest pocket and drew forth a glass tube of radium bromide. Its magnificent luminosity gleamed as he held it up, illuminating an expression of rapture on Marie's face. Rutherford observed that it also illuminated the cracked flesh and burned skin of Pierre's irrevocably destroyed fingers.

After Pierre Curie's death, the Faculty of Sciences of Paris asked Marie to take her husband's place in order to assure the continuance of his work. The French Society of Physics posthumously honored Pierre Curie in 1908 by issuing a complete publication of his works, a single volume of about 600 pages.

Marie received the Nobel Prize in Chemistry in 1911 "in recognition of her services to the advancement of chemistry by the discovery of the elements radium and polonium, by the isolation of radium and the study of the nature and compounds of this remarkable element." She was the first person to win or share two Nobel Prizes.

The Curies' elder daughter, Irène (1897–1956), married French physicist Jean Frédéric Joliot (1900–1958), and the husband and wife team received the Nobel Prize for Chemistry in 1935. The younger daughter, Eve, wrote *Madame Curie*, a famous biography of her mother, published in 1938. As mentioned above, Marie Curie had earlier written a short biography of her husband, Pierre, in which she described the struggling scientist who had earlier been denied financial and other support for his work.

As a result of working with the uranium-rich ore called pitchblende, Marie Curie's lab assistant, Blanche Wittman, suffered from radiation poisoning. Both of her legs and one arm had to be amputated. She used a wagon to get around and lived in Marie's Paris apartment, where she died in 1913. Per Olov Enquist's *The Book About Blanche and Marie*, a "reality-based" novelization of their relationship, was published in 2006. Luan Gaines, reviewer for *Curled Up with a Good Book*, eloquently describes the plight of Wittman:

> Think of Blanche and Marie, heads bent over their shining experiments, the poisonous element that spells the ruin of Blanche's extraordinary beauty and deforms Curie's right hand, the two basking in an island of friendship and mutual admiration while hovering over luminescent death. Picture Blanche near the end of her days, ensconced in her wooden box, reduced to a torso with only a right

arm and hand to pen her thoughts, dissecting the nature of love and woman, radiation the fusion of all yet the instrument of her death.

In the mid-1990s, Marie Curie's publications, diaries, and workbooks were sent to the Bibliothèque Nationale, a national library in Paris. Because the papers were still highly radioactive, before they could be studied they had to be decontaminated through a two-year process.

Pierre Weiss was born in Mulhouse, France, and is famous for his theories of ferromagnetism that accounted for various observations such as the sudden disappearance of ferromagnetism above the Curie temperature. Weiss's father owned a clothing shop in France. In 1887, Weiss graduated, first in his class, from the Polytechnikum in Zurich with a degree in engineering. In 1902, he became a professor at the Polytechnikum, where he directed the physics laboratory. As discussed above, in 1907 Weiss modified Curie's Law so that it accounted for the susceptibility of paramagnetic substances above the Curie point. Also in 1907, Pierre Weiss postulated the first modern theory of permanent magnets, which he suggested contained tiny individual magnetic domains. In 1918, Weiss discovered "magnetocaloric effects" and showed how thermodynamics can be used to calculate the temperature variation of a magnetic substance placed in a varying field.

We know little of Weiss's personal life except for his 1898 marriage to Jane Rances, whose mother was of English origin. Weiss was tall and distinguished looking with his large moustache. The Weisses had one daughter, Nicole, who in 1938 married Henri Cartan (born 1904), a famous French mathematician who made fundamental advances in the theory of analytic functions. After Weiss's wife died in 1919, he waited three years and then married Marthe Klein, an X-ray technician instructor and a university-level certified physics teacher. Marthe is perhaps best remembered for a letter she wrote in 1919, which shows the difficulty that women had in rising through the scientific establishment:

> *On m'a donné un service où j'ai à enseigner de l'histoire naturelle à des gamines de 10 à 14 ans! C'est pour prouver que ce que j'ai fait ici ne devait pas servir à mon enseignement. J'ai même à faire des cours d'économie domestique*!
>
> [They have given me a job in which I am to teach Natural History to girls between 10 and 14 years old. It's to prove that what I have done in here mustn't be used for my teaching activity. I even have to take some classes in Home Economics!]

Weiss was influential in his ideas about the small magnetic domains that exist within magnets. According to Etienne Du Tremolet de Lacheisserie and colleagues in *Magnetism: Fundamentals*:

> Pierre Weiss assumed that at the macroscopic scale (typically for sizes larger than one micrometer), a ferromagnetic material is spontaneously divided into domains. Each domain has spontaneous magnetization, but from domain to domain the resulting magnetization does not have the same direction. Thus, at the macroscopic level, there is no resulting moment.... [Today] magnetic domains are named *Weiss domains*. They are separated by walls, the *Bloch walls*, which consist of a certain number of atomic planes in which the orientation of magnetic moments more or less progressively passes from that of one domain to that of one another.

I conclude this entry by returning to the death of Pierre Curie. In his biography, Marie Curie published extracts from published appreciations of her husband that she received after his accident. Perhaps the most moving are the words of French mathematician Henri Poincaré (1854–1912):

> On the night preceding his death ... I sat next to him and he talked with me of his plans and his ideas. I admired the fecundity and the depth of his thought, the new aspect which physical phenomena took on when looked at through that original and lucid mind. I felt that I better understood the grandeur of human intelligence—and the following day, in an instant, all was annihilated. A stupid accident brutally reminded us how little place thought holds in the face of the thousand blind forces that hurl themselves across the world without knowing whither they go, crushing all in their passage....
>
> In foreign countries the most illustrious scientists joined in trying to show the esteem in which they held our compatriot, while in our own land there was no Frenchman, however ignorant, who did not feel more or less vaguely what a force his nation and humanity had lost.... True physicists, like Curie, neither look within themselves, nor on the surface of things, but they know how to look through things.... He never separated the worship of this ideal from what he rendered to science, and he gave us a shining example of the high conception of duty that may spring from a simple and pure love of truth. It matters little in what God he believed; it is not the God, but faith, that performs miracles.

A lunar crater with a diameter of 151 kilometers was named after Pierre Curie and approved in 1970 by the International Astronomical Union General Assembly.

FURTHER READING

Applied Alloy Chemistry Group, "Classification of Magnetic Materials," University of Birmingham; see www.aacg.bham.ac.uk/magnetic_materials/type.htm#Diamagnetism#Diamagnetism.

Curie, Marie, *Pierre Curie*, Charlotte and Vernon Kellogg, translators (New York: Macmillan, 1923); see etext.lib.virginia.edu/toc/modeng/public/CurPier.html.

De Lacheisserie, Etienne Du Tremolet, Damien Gignoux, and Michel Schlenker, *Magnetism: Fundamentals* (New York: Springer, 2004).

Enquist, Per Olov, *The Book about Blanche and Marie* (New York: Overlook Press, 2006).

Gaines, Luan, "Book Review of *The Book about Blanche and Marie*," *Curled Up with a Good Book* (e-zine); see www.curledup.com/blanchem.htm.

Goldsmith, Barbara, *Obsessive Genius: The Inner World of Marie Curie* (New York: W. W. Norton, 2004).

Martin, D. H., *Magnetism in Solids* (Cambridge, Mass.: MIT Press, 1967).

Moore, Pete, *$E = mc^2$: The Great Ideas That Shaped Our World* (New York: Sterling Publishing, 2005).

Nobel Foundation, "Pierre Curie Biography," in *Nobel Lectures, Physics 1901–1921* (Amsterdam: Elsevier Publishing Company, 1967); see nobelprize.org/physics/laureates/1903/pierre-curie-bio.html.

Perrin, Francis, "Pierre Weiss," in *Dictionary of Scientific Biography*, Charles Gillispie, editor-in-chief (New York: Charles Scribner's Sons, 1970).

Prepost, R., "Ferromagnetism—the Curie Temperature of Gadolinium," University of Wisconsin; see www.hep.wisc.edu/~prepost/407/curie/curie.pdf.

Wyart, Jean, "Pierre Curie," in *Dictionary of Scientific Biography*, Charles Gillispie, editor-in-chief (New York: Charles Scribner's Sons, 1970).

INTERLUDE: CONVERSATION STARTERS

> If a lunatic scribbles a jumble of mathematical symbols, it does not follow that the writing means anything merely because to the inexpert eye it is indistinguishable from higher mathematics.
>
> —Eric Temple Bell, quoted in J. R. Newman's *The World of Mathematics*

> Laws of nature are those fundamental laws of physics which hold everywhere within the universe. . . . There are

only a few of them, [e.g.] $F(x,t) = m(x) \cdot d^2 s(x,t)/dt^2$. Another kind of law of nature are *special force laws*, e.g. the classical laws for gravitational force or electrical force.... Laws of nature are strictly true—but at the cost of not per se being applicable to real systems, because they do not specify which forces are active. *System laws*, in contrast ... refer to particular systems of a certain kind in a certain time interval Δt. They contain or rely on a *specification* of all the forces.... Examples of system laws in classical physics are Kepler's laws of elliptic planetary orbits ... and the classical wave equations....

—Gerhard Schurz, "Normic Laws, Non-monotonic Reasoning, and the Unity of Science"

There are whole walls in libraries covered by countless shelves which are bending under the books on electronics and quantum electrodynamics. But in none of those books will you find a proper definition of an electron, for the very good reason that we haven't the foggiest idea what an electron "is."

—Vincent Icke, *The Force of Symmetry*

Not only does God play dice with the universe but, if he did not, the complex universe we see around us would not exist at all. We owe everything to randomness.... Quantum events happen randomly, for no reason at all.... The diversity we see around us is created by quantum theory exploring all possible evolutions forward in time...."

—Marcus Chown, "It's All Down to a Roll of the Dice," *New Scientist* (excerpts of interview with Stephen Hsu and Nick Evans)

1900 AND BEYOND

It seems we all face a fundamental paradox in that it's impossible to think about the universe except in terms of its relation to humans. You can't make sense of language, or even scientific laws or mathematics, without the concept of an observer, and yet at the same time we know perfectly well that humans are a very late addition: the universe was here long before us and will be here long after us.

—Michael Frayn, quoted in "All the World's a Stage," *New Scientist*, September 23, 2006

The universe might actually be able to fine-tune itself. If you assume the laws of physics do not reside outside the physical universe, but rather are part of it, they can only be as precise as can be calculated from the total information content of the universe. The universe's information content is limited by its size, so just after the big bang, while the universe was still infinitesimally small, there may have been wiggle room, or imprecision, in the laws of nature.

—Patrick Barry, "What's Done Is Done," *New Scientist,* September 30, 2006

Nature, and Nature's laws lay hid in night: God said: "*Let Newton be!*" and all was light.

—Alexander Pope, "Epitaph Intended for Sir Isaac Newton"

It did not last: the Devil howling "*Ho! Let Einstein be!*" restored the status quo.

—John Collings Squire (1884–1958), British journalist, "In Continuation of Pope on Newton"

PLANCK'S LAW OF RADIATION

Germany, 1900. The amount of energy at a particular wavelength radiated by a blackbody depends on the temperature of the body and the wavelength. Planck's formulation is notable because it incorporates the earliest known practical application of quantum theory.

CROSS REFERENCE: ALBERT EINSTEIN, RUDOLF CLAUSIUS, KIRCHHOFF'S LAW OF THERMAL RADIATION, WIEN'S DISPLACEMENT LAW, THE STEFAN-BOLTZMANN RADIATION LAW, THE RAYLEIGH-JEANS LAW, AND WIEN'S RADIATION LAW.

> In 1900, Sigmund Freud published *The Interpretation of Dreams*. Hawaii officially became a U.S. territory and was granted self-governance. American inventor Cornelius J. Brosnan filed for a U.S. patent for a paperclip. He called his invention the "Konaclip."

Quantum theory, which suggests that matter and energy have the properties of both particles and waves, had its origin in pioneering research concerning hot objects that emit radiation. For example, imagine the coil on an electric heater that glows brown and then red as it gets hotter. More particularly, Planck's Law of Radiation quantifies the amount of energy emitted (at a particular wavelength) by a particular kind of hot glowing object called a blackbody. A blackbody, discussed in "Kirchhoff's Electrical Circuit and Thermal Radiation Laws" (see part III), is an object that emits and absorbs the maximum possible amount of radiation at any given wavelength and at any given temperature. I describe blackbodies further at the end of this section.

Thermal radiation is radiant energy emitted by an object as a result of the temperature of the object. The spectrum of the thermal radiation from a hot body is continuous over a range of wavelengths. Usually, the amount of radiation at any given frequency is different than at other frequencies. Many of the objects that we encounter in our daily lives emit a large portion of their radiation spectrum in the infrared, or far-infrared, portion of the spectrum, and this radiation is not visible to our eyes. However, as the temperature of a body increases, the dominant portion of its spectrum shifts so that we can see a glow from the object.

Even though such thermal radiation spans a range of wavelengths, scientists are often interested in the radiation emitted per unit wavelength, called monochromatic radiation, and the amount of such radiation depends on the temperature. For example, the Sun has an effective surface temperature of about 5,800°K and emits most of its energy below a wavelength of 3 micrometers. Earth, which is much cooler, emits most of its energy at longer wavelengths. In particular, Earth has a surface temperature of about 290°K, and it emits virtually all of its radiation at wavelengths greater than 3 micrometers.

We can use the term $E_{b\lambda}d\lambda$ to denote the radiant energy emission per unit time and per unit area from a blackbody at wavelength λ in the wavelength range $d\lambda$. $E_{b\lambda}$ is known as the monochromatic blackbody emissive power. Planck's Law, discovered in 1900, shows how this emissive power is distributed among different wavelengths for a perfect radiator at temperature T. Thus, an ideal radiator, or blackbody, emits radiation according to Planck's Law:

$$\boxed{E_{b\lambda} = \frac{C_1}{\lambda^5(e^{C_2/\lambda T} - 1)},}$$

where $E_{b\lambda}$ is the monochromatic emissive power of a blackbody at absolute temperature T (units of W/m^3), λ is the wavelength (m), T is the absolute temperature of the body (°K), C_1 is the first radiation constant (3.7415×10^{-16} $W{\cdot}m^2$), and C_2 is the second radiation constant 1.4388×10^{-2} (m·K, where the K indicates degrees kelvin). (Another way of writing this law, involving Planck's constant and Boltzmann's constant, is described shortly.)

It is instructive for students to draw plots of this function for different values of wavelength. At temperatures below about 5,800°K, the emission of radiation occurs to a great extent between wavelengths of 0.2 and 50 micrometers. The wavelength for which the emissive power is maximum, $E_{b\lambda}(\lambda_{max}, T)$, decreases with increasing temperatures. In general, only a small amount of the total emitted radiation is visible to the eye. At about 800°K, an object glows with a dull red color.

WIEN'S DISPLACEMENT LAW

One of the earliest laws concerning blackbodies is Wien's Displacement Law, which was discovered in 1893 by German physicist Wilhelm Wien (1864–1928). Wien's Displacement Law can actually be derived from Planck's Law by setting the derivative of the formula to 0 in order to reveal the relationship between the wavelength λ_{max} at which the emissive power is maximum and the absolute temperature:

$$\boxed{\lambda_{max} T = 2.898 \times 10^{-3}\ \mathrm{m \cdot K}}$$

In other words, as the temperature of an object increases, the wavelength of emitted radiation decreases. The product of the absolute temperature and the wavelength corresponding to the peak of the emissivity power curve is a constant. Wien was awarded a Nobel Prize for work relating to his law.

If we were to use a value of 5,780°K for the Sun's surface temperature, we would calculate a peak emission at 500 nm, according to Wien's Displacement Law. Electromagnetic radiation of 500 nm is roughly in the middle of the visual spectrum. (Yellow light has a wavelength of about 570 nm. Sodium lamps, often used in parking lots, emit a yellow light of wavelength 589 nm.)

This formula makes clear the origins of a star's color because the color relates to different surface temperatures as specified by Wien's relationship. Red stars are cool, with a surface temperature of around 3,000°K, yellow ones are closer to 6,000°K, and white stars are about 10,000°K. Blue stars are extremely hot at about 20,000°K. Wien's Displacement Law also means that we can simply look at a star and estimate its temperature.

STEFAN-BOLTZMANN RADIATION LAW

German physicist Gustav Kirchhoff (1824–1887) had realized that the emission from a blackbody (a concept he created and a word he coined) is a function of the wavelength and the temperature, but he could not precisely articulate the function. Kirchhoff had shown that the heat radiation that occurs from a blackbody at a particular temperature was independent of the nature of the material, and that it depended only on temperature and wavelength. In fact, if one considers the entire wavelength spectrum, the total emission $E_b(T)$ of radiation per unit surface area of a blackbody per unit time is given by the Stefan-Boltzmann Radiation Law:

$$\boxed{E_b(T) = \sigma T^4}$$

This expression gives the total radiated power per unit area of the blackbody, summed over all wavelengths. Joseph Stefan (1835–1893)—the Slovene-Austrian physicist, mathematician, and poet—empirically found his law in 1879 through experiments, and Ludwig Boltzmann (1844–1906) derived the law in 1884 from theoretical principles involving thermodynamics. Here, T is the absolute temperature of the area in degrees kelvin, and σ is the Stefan-Boltzmann constant [5.67×10^{-8} W/($\mathrm{m}^2 \cdot \mathrm{K}^4$)]. The constants C_1 and C_2 in Planck's Law are related to the Stefan-Boltzmann constant σ as follows:

$$\sigma = \left(\frac{\pi}{C_2}\right)^4 \frac{C_1}{15} = 5.670 \times 10^{-8}\ \mathrm{W/(m^2 \cdot K^4)}$$

Interestingly, Stefan was able to determine an approximate temperature of the Sun's surface using his law, because he had a reasonable estimate of the Sun's energy flux. The value he obtained was 5,700°K for the temperature of the Sun—close to the currently accepted value of 5,780°K.

Note that the Stefan-Boltzmann constant is *universal* in the sense that it is independent of the material or temperature of a body. Because actual hot objects radiate less efficiently than do perfect blackbodies, scientists sometimes add another quantity ε to the Stefan-Boltzmann Law. ε is the *emissivity* and has a value between 0 and 1, where 1 corresponds to a perfect blackbody and 0 to a perfect reflector.

The precise value usually depends on the material, the surface, and the temperature. Thus, the Stefan-Boltzmann Law may be recast as follows for real objects:

$$E_b(T) = \varepsilon\sigma T^4$$

Note that although ε is called the emissivity, scientists also use it to describe the absorption of radiation by a material.

Using the Stefan-Boltzmann Radiation Law, we can predict the total energy a blackbody emits, which is proportional to the fourth power of the temperature of the object. A star that is the same size and four times as hot as our Sun radiates 4^4 or 256 times more energy than the Sun. A spherical blackbody (e.g., a star) will produce a luminosity that depends on the surface area of the star times the fourth power of its temperature.

HOW TO MAKE A BLACKBODY

In the laboratory, a blackbody can be approximated by a large hollow, rigid object such as a sphere. Imagine that we use a nail to poke a hole in the sphere. Any radiation entering the hole bounces around against the inner sphere walls. With each bounce, a part of the radiation is absorbed by the sphere. Some of the radiation is reflected, and in the next bounce, additional radiation is absorbed. By the time the radiation has bounced around a number of times and exits through the same nail hole, its intensity is so weak that it is negligible. This means that the nail hole may be considered a blackbody because nearly all radiation that enters the hole is absorbed by the hollow sphere. (For a creative analogy, imagine a fly that enters the hole, bangs against the inner walls of the sphere, and finally becomes so tired that it is too weak to escape.) The radiation emitted by the inner surface of the sphere depends on the temperature of the inner surface. The amount of radiation produced in the spherical cavity whose walls are at temperature T is equal to the emissive body of a blackbody at temperature T. With idealized blackbodies, the radiation-emitting properties of the body are independent of the material of the body and vary in a simple way with temperature.

If we construct cavity radiators (blackbodies) using different metals, heat each cavity to the same temperature (e.g., 2,000°K), and observe the light emitted in a dark room, we would find that the radiance of the hole in the cavity is the same for different blackbodies, even though the radiances of the outer surfaces may be different. As suggested above, the radiancy of hole varies according to Stefan-Boltzmann Law $E_b(T) = \sigma T^4$. However, the radiancy of the outer surfaces does vary with both material and temperature.

RAYLEIGH-JEANS LAW

Several related blackbody laws have preceded Planck's radiation law. In 1900, British physicist Lord Rayleigh (1842–1919) proposed a law based on a classical physics analysis of standing waves within a three-dimensional cavity, such as the spherical cavity just discussed. The Rayleigh-Jeans Law gives the energy density of a blackbody radiation as a function of wavelength λ and temperature:

$$f(\lambda, T) = 8\pi k \frac{T}{\lambda^4}$$

Here, T is the temperature in degrees kelvin, and $k = 1.3806505 \times 10^{-23}$ joules/kelvin is Boltzmann's constant. Notice that Rayleigh derived a fourth-power dependence on wavelength, and a more complete derivation was presented by Rayleigh and British physicist Sir James Hopwood Jeans (1877–1946) in 1905. Their law works well and agrees with experiments performed at long wavelengths, but strongly disagrees with experimental results at short wavelengths. In fact, notice that as λ approaches zero, the value of the function becomes infinite! This result was known as the "ultraviolet catastrophe" because the deviation from experiment occurs in the ultraviolet range and shorter wavelengths.

ADDITIONAL RENDITIONS OF THE LAWS

In 1900, Planck announced that he had made a modification to the classical calculation so that the function $f(\lambda, T)$ agreed with experimental data at all wavelengths. Planck's more accurate radiation formula with its more complicated dependence on wavelength has been presented above. Here is another way of writing the previous Planck's equation, which is frequently found in textbooks:

$$\boxed{f(\lambda, T) = \frac{8\pi hc\lambda^{-5}}{e^{hc/\lambda kT} - 1}}$$

Planck's equation approaches zero as λ approaches zero. The constant h is Planck's constant and equals $6.6260693 \times 10^{-34}$ J·s, k is Boltzmann's constant, and c is the speed of light. Similarly, we may also write the Stefan-Boltzmann constant mentioned above in terms of other constants of nature:

$$\sigma = \frac{2\pi^5 k^4}{15c^2h^3} = 5.670 \times 10^{-8}\mathrm{J/(s \cdot m^2 \cdot K^4)}$$

Planck had initially found the formula for his law empirically as he fit curves to the data. In some sense, his initial work was inspired guesswork, and there was no substantial theory behind it. However, he soon found that he could derive this equation by modifying the calculations of the energy per wave in the cavities that Rayleigh contemplated. In particular, he modeled the cavity walls as a collection of little electromagnetic oscillators. In order to arrive at his formula, Planck assumed that the energy of oscillators is discrete and could assume only certain values of nhf. Here, n is an integer now called a quantum number, f is the frequency of the oscillator, and h is Planck's constant. These oscillators both emit energy into the cavity and absorb energy from it via discrete jumps or packages called quanta. [Later research has revealed that the correct formula for the oscillator involves $(n + 1/2)hf$, but this adjustment makes no change to Planck's conclusions.]

At the time, Planck's model of a discrete system was unconventional, and Planck did not fully appreciate it for years, writing in *Scientific Autobiography and Other Papers*, "My futile attempts to fit the elementary quantum of action [i.e., the quantity nh] somehow into the classical theory continued for a number of years, and they cost me a great deal of effort." Planck originally believed that the quantization of tiny oscillators in the cavity walls (now known to be atoms) was simply a mathematical convenience that yielded accurate predictions of the blackbody spectrum. Other physicists also considered such quantization too radical and a mere mathematical artifice. Although Planck could not articulate the reason for the quantization, it fit the data so precisely that he came to realize that it must be valid. Note that the extremely small value of Planck's constant h causes the difference between two adjacent values of nhf to be tiny.

Helge Kragh writes in "Max Planck: The Reluctant Revolutionary":

> If a revolution occurred in physics in December 1900, nobody seemed to notice it. Planck was no exception, and the importance ascribed to his work is largely a historical reconstruction. Whereas Planck's radiation law was quickly accepted, what we today consider its conceptual novelty—its basis in energy quantization—was scarcely noticed. Very few physicists expressed any interest in the justification of Planck's formula, and during the first few years of the 20th century no one considered his results to conflict with the foundations of classical physics.

It was not until Einstein showed, in 1905, that Planck's quantum could be applied to different phenomena that physicists realized that the traditional continuous wave theory of light gave an inadequate picture of reality. In particular, Einstein proposed a model whereby light was emitted and absorbed in packets or photons; thus, it was Einstein who was one of the first to recognize the essence and implications of quantum theory.

Planck's quantum (discrete oscillator energy) approach to theoretically derive his radiation law finally led to his 1918 Nobel Prize. In fact, Planck's Law actually let him calculate values for both h and k from experimental data. Today, we set the birth date of quantum physics to December 14, 1900—the day Planck described his theory to the Berlin Physical Society. His model was just the tip of the iceberg, a hint that a large portion of physics was still unknown. In many ways, his findings were the birth of modern physics, which is eloquently described in Henry Hooper and Peter Gwynne's *Physics and the Physical Perspective*:

> Starting in the 1890s, a series of astonishing experimental discoveries and theoretical concepts started to open up the inside of the atom to the figurative gaze of wide-eyed physicists.... The key advance of what the late cosmologist George Gamow described as "thirty years that shook physics" was the development of what we term the quantum theory of radiation—the idea that electromagnetic energy, light, and other forms of radiation come in the form of discrete packets....

Before closing this section, I should note that another attempt prior to Planck's Law was Wien's Radiation Law (also known as Wien's Distribution Law), formulated by Wien in 1896:

$$E_{b\lambda} = \frac{C_1}{\lambda^5 (e^{C_2/\lambda T})}$$

However, Wien's Radiation Law was based solely on his attempts to fit the data and was not exact. It also failed experimentally at very long wavelengths. However, the law was the stimulus for Planck, who discovered his Law of Radiation in 1900 in an attempt to improve upon and derive Wien's Radiation Law.

The essential difference between some of the key radiation laws can be summarized as follows:

- Planck's law is accurate at all wavelengths.
- Wien's Law is a good approximation at short wavelengths.
- The Rayleigh-Jeans Law is a good approximation at large wavelengths; however, at short wavelengths the radiation value becomes huge, and we have the ultraviolet catastrophe.

Max Planck (1858–1947), German physicist famous for his foundational work in quantum theory and for his blackbody radiation law.

CURIOSITY FILE: Planck was the first famous physicist to enthusiastically support Einstein's 1905 Special Theory of Relativity. • Planck so detested Hitler's policies with respect to race and religion that Planck went directly to Hitler to voice his concerns. Hitler simply responded with a tirade against the Jews. Planck's son was executed by the Nazis in 1945. • Planck and Einstein often played chamber music together, Planck at the piano and Einstein playing violin.

> Science cannot solve the ultimate mystery of nature. And that is because, in the last analysis, we ourselves are part of nature and therefore part of the mystery that we are trying to solve. We have no right to assume that any physical laws exist, or if they have existed up until now, that they will continue to exist in a similar manner in the future.
>
> —Max Planck, "The Mystery of Our Being"

In the year 1900, Max Planck solved a problem which had been vexing physicists for years, and in so doing opened a Pandora's Box filled with surprises for mankind.

—Frederick Bueche, *Introduction to Physics for Scientists and Engineers*

As a man who has devoted his whole life to the most clear headed science, to the study of matter, I can tell you as a result of my research about atoms this much: There is no matter as such. All matter originates and exists only by virtue of a force which brings the particle of an atom to vibration and holds this most minute solar system of the atom together. We must assume behind this force the existence of a conscious and intelligent mind. This mind is the matrix of all matter.

—Max Planck, acceptance speech for 1918 Nobel Prize for Physics

We can consider the whole universe to be like the interior of a huge oven; the oven's temperature is the average temperature of the universe. [Using Planck's law of radiation for blackbodies] and knowing the wavelength distribution of the electromagnetic radiation striking us from outer space, we can determine the [average] temperature of ... the universe ... is close to 3 K.

—Frederick Bueche, *Introduction to Physics for Scientists and Engineers*

Reality is ... just a very specific, narrow slice of that vast range of what our thoughts try to encompass.

—Max Planck, 1923 lecture on the law of causality and free will

Planck lived almost ninety years. He witnessed the two world wars, two Reichs, and the Weimar Republic. He saw the great German scientific establishment, which he had helped build, destroyed by Nazi anti-Semitic racial polices and other insanities. He deplored everything the Nazis did, but chose to remain in Germany....

—William H. Cropper, *Great Physicists*

Max Planck was born in Kiel, a city in northern Germany. His father was professor of constitutional law at the University of Kiel. Planck was his family's sixth child. As a youngster, he enjoyed climbing mountains with his family. He did well at school and excelled in music, but he did not display any particular genius in mathematics and science at an early age.

As a teenager, Planck was sufficiently talented in piano playing that he considered a musical career. He had perfect pitch and especially enjoyed playing the works of Franz Schubert and Johannes Brahms.

Planck entered the University of Munich in 1874. Here, his interest in physics steadily grew, despite the fact that one professor told him that the field of physics was well understood, with little more to discover, and to expect no significant future developments in the field! Planck explained his nascent fascination with physics in his *Scientific Autobiography and Other Papers*, published in 1949: "The outside world is something independent from man, something absolute, and the quest for the laws which apply to this absolute appeared to me as the most sublime scientific pursuit in life." He also noted that "the laws of human reasoning coincide with the laws governing the sequences of impressions we receive from the world about us; that, therefore, pure reasoning can enable man to gain an insight into the mechanisms of the [world]."

In 1877, Planck continued his education at the University of Berlin, where he said that his famous teacher Gustav Kirchhoff was rather boring as a lecturer. While in Berlin, he studied and was intrigued by the research papers of Rudolf Clausius (1822–1888), on thermodynamics. In 1879, Planck received his doctorate for his own work on the Second Law of Thermodynamics. His thesis was titled "On the Second Law of Mechanical Theory of Heat." In 1885, he was appointed extraordinary professor of theoretical physics in Kiel, and in 1887 he married Marie Merck.

In 1888, he was appointed as an extraordinary professor of theoretical physics at the University of Berlin. He never gave up his love of music and held concerts in his home, using his customized harmonium, an organlike keyboard instrument that employs metal reeds that vibrate as a result of air forced from a bellows.

In 1897, Planck published his *Vorlesungen über Thermodynamik*, a popular collection of fundamental papers on thermodynamics. This classic work of modern physics was to become a standard reference in the field of thermodynamics for nearly fifty years, and it included his 1879 thesis and all his important papers until 1896.

He also maintained a continuing interest in philosophy and religion and suggested that science should have as its goal "the establishment of a single grand connection among all the forces of nature." In an 1899 paper published in *Sitzungsberichte der Preußischen Akademie der Wissenschaften*, he mentioned his quest for "natural" scientific units that are independent of particular objects so that the units would "retain their meaning for all times and for all cultures, including extraterrestrial and nonhuman ones."

Today, physicists sometimes make use of "Planck units" to honor Planck's desire for finding natural units to describe all of nature. Planck units are units of measurement for length, mass, time, electric charge, and absolute temperature, expressed in such a way that five fundamental physical constants become 1 when expressed in these units. The five fundamental constants are the speed of light in a vacuum c; the gravitational constant G; the reduced Planck's constant $\hbar = h/(2\pi)$, where h is Planck's constant; the Coulomb force constant $1/(4\pi\varepsilon_0)$, where ε_0 is the permittivity in vacuum; and the Boltzmann constant k. Some popular authors refer to Planck units as God's units, because they eliminate the more common but arbitrary human conventions with respect to units. At the time that Planck envisioned his natural units, he had not yet discovered his blackbody radiation law in which h made its first appearance.

John D. Barrow and Frank Tipler list various Planck units in *The Anthropic Cosmological Principle* and note that the Planck length and time are both extremely small, many orders of magnitude smaller even than nuclear sizes and times. On the other hand, the Planck mass is roughly equal to the mass of a small grain of sand:

$$l_{\mathrm{P}} = \left(\frac{G\hbar}{c^3}\right)^{1/2} \approx 10^{-33}\mathrm{cm}$$

$$t_{\mathrm{P}} = \left(\frac{G\hbar}{c^5}\right)^{1/2} \approx 5 \times 10^{-44}\mathrm{s}$$

$$m_{\mathrm{P}} = \left(\frac{c\hbar}{G}\right)^{1/2} \approx 10^{-5}\mathrm{gm}$$

Henrik Smith in *Quantum Mechanics* describes how quickly Planck arrived at his blackbody radiation formula while in Berlin:

> One Sunday afternoon on October 7, 1900, the Planck family had guests for tea. After hearing about these new [radiation] results [from a guest], Planck attempted later in the day to construct an empirical formula, which would combine the regions of long and short wavelengths corresponding respectively to Rayleigh's and Wien's expressions. The result, Planck's radiation law, was made public only twelve days later.... It contained a new fundamental constant, which today we call the Planck constant. Two months later, Planck published a derivation of the radiation law [which] contained assumptions that were quite foreign to the physics of the nineteenths century.

Planck's theoretical explanation of the observed spectra made use of a theory that involved quanta, or packets, of energy, and he presented his concepts at a meeting of the *Physikalische Gesellschaft* in Berlin. As mentioned above, this radiation law and explanation eventually led Planck to receive the Nobel Prize for Physics in 1918. In his 1920 Nobel Lecture, he said,

> Either the quantum of action was a fictional quantity, then the whole deduction of the radiation law was essentially an illusion representing only an empty play on formulas of no significance, or the derivation of the radiation law was based on a sound physical conception. In this case the quantum of action must play a fundamental role in physics, and here was something completely new, never heard of before, which seemed to require us to basically revise all our physical thinking, built as this was, from the time of the establishment of the infinitesimal calculus by Leibniz and Newton, on accepting the continuity of all causative connections. Experiment decided it was the second alternative.

According to Hooper and Gwynne:

> In the first several months after his announcement of the quantum inspiration, Planck plowed a lonely furrow. Scientists always tend to be somewhat suspicious of new ideas, particularly those without any obvious fundamental foundation, and many decided to wait for the other shoe to drop, in the form of a successful application of Planck's theory to some other branch of science. Their wait was short one.

In 1905, just a few years after Planck announced his Law of Radiation, Albert Einstein used the idea of radiation quantization to explain the results of another experiment that had confounded physicists for years. The experiment involved the photoelectric effect, for which Einstein provided evidence suggesting that light may be thought of as discrete particles. Later in life, Planck remarked in *Scientific Autobiography and Other Papers*:

> An important scientific innovation rarely makes its way by gradually winning over and converting its opponents: it rarely happens that Saul becomes Paul. What does happen is that its opponents gradually die out and that the growing generation is familiarized with the idea from the beginning.

Planck's wife Marie died in 1909, leaving him with four children. This was the first of many tragedies for Planck. For example, one of his children, Karl, died in 1916 in World War I. Both daughters died in childbirth (1917 and 1919).

In 1911, he married the niece of his first wife. Planck was Secretary of the Mathematics and Natural Science Section of the Prussian Academy of Sciences from 1912 until 1943. In 1943, the city of Frankfurt wanted to honor Planck with the Goethe Award; however, Joseph Goebbels forbade the award, "because Planck has been repeatedly speaking for the Jew Albert Einstein."

Alan A. Grometstein writes in *The Roots of Things: Topics in Quantum Mechanics*:

> As the Nazi Party rose to power, Planck chose to remain in Germany rather than escape to freedom as other scientists had.... He despised Hitler, viewing the Third Reich as a cancer upon the nation to which he continued to give allegiance, but believed that his presence would act as a moderating and civilizing influence. Perhaps it did; he is known to have aided and sheltered vulnerable colleagues.

During World War II, in 1944, Planck's home in a suburb of Berlin was demolished by fire after an air raid. Many of his most important scientific records were destroyed. In 1945, the Gestapo tortured his son Erwin to death, because he was suspected of being involved in the plot to assassinate Hitler. Most likely, Erwin simply knew someone involved in a plot to assassinate Hitler. When Planck was 85, he fled and lived in a farmhouse, and when the farm had to be evacuated, Planck, age 87, camped with his wife in a forest.

Through his very long life, Planck received numerous awards in addition to his Nobel Prize, including election as a Fellow of the Royal Society in 1926, the Royal Society Copley Medal awarded in 1929, and election Fellow of the Royal Society of Edinburgh in 1937. In the same year, in his lecture "Religion and Naturwissenschaft," Planck said that he believed that God is omnipresent and that "the holiness of the unintelligible Godhead is conveyed by the holiness of symbols."

In his later years, Planck, like Einstein, was not convinced that ultimate reality had a quantum nature at its very core. He believed the world was not "really" the indeterminate, statistical entity that other scientists such as German physicists Niels Bohr (1885–1962) and Werner Heisenberg (1901–1976) believed it to be. Planck thought that reality had an independent existence, and he never accepted that the observer and observed were so strongly coupled, as suggested by physicists such as Bohr.

During the war, Americans had taken Planck to Göttingen, when they learned that his refuge near the Elbe River had been destroyed and that he was caught between the Allied armies and the retreating German forces. Planck died in Göttingen at the age of 89.

Ian Duck and E. C. G. Sudarshan write of Planck's discoveries in *100 Years of Planck's Quantum*:

> When in the year 5000 people look back three thousand years to our era, we all hope that they will find some epochal events as myth-making for them as the Trojan War for us. Many events of such lasting significance are to be found among the achievements of our twentieth century scientific revolution in physics. The very first of these revolutionary events . . . is Planck's invention of the energy quantum in 1900. . . . Planck opened the door to an utterly new, totally unanticipated, wonderfully strange and mysterious but absolutely necessary ultimate reality of the world. . . . We cannot know where it will lead and we cannot believe it will end.

A lunar crater with a diameter of 314 kilometers was named after Planck and approved in 1970 by the International Astronomical Union General Assembly.

In 2003, researchers at Sandia National Laboratories created a material that exceeded the predicted output of Planck's blackbody radiation law for a special material. In particular, they used filaments created from tungsten lattices that, when heated, emitted much more energy in certain regions of the near-infrared spectrum than solid tungsten filaments did. The researchers hope that this lattice material may someday provide a superior energy source. The radiation of the lattice can pump energy into wavelengths used by photovoltaic cells that customarily convert light into electricity.

The latticelike materials, also known as photonic crystals, do not actually break Planck's Law, which gives the emission for a solid blackbody and not a photonic lattice. The researchers note,

> A photonic lattice apparently subjects energies passing through its links and cavities to more complex photon-tungsten interactions than Planck dreamt of when he derived his system that successfully predicted the output energies of simple heated solids. And a lattice's output is larger than a solid's only in the frequency bands the lattice's inner dimensions permit energy to emerge in.

I discussed Planck's interest in fundamental physical constants. Although these fundamental constants, such as the speed of light in a vacuum, are usually assumed to be unvarying, twenty-first century scientists have occasionally claimed to detect subtle changes in their values. For example, in 2006, a team of physicists and astronomers in the Netherlands,

Russia, and France suggested that the constant μ that represents the ratio of the mass of the proton to that of an electron may have decreased slightly since the birth of the universe. Not without controversy, their evidence involves comparison of light absorption patterns of distant hydrogen molecules in outer space to molecules in the laboratory. Because looking far into space is equivalent to looking back in time, scientists may be able to use such comparisons to search for changes in the constants of nature. Further evidence would be required for such findings to become generally accepted by the scientific community.

FURTHER READING

Barrow, John D., and Frank Tipler, *The Anthropic Cosmological Principle* (New York: Oxford University Press, 1988).

Duck, Ian, and E. C. G. Sudarshan, *100 Years of Planck's Quantum* (Singapore: World Scientific, 2000).

Grometstein, Alan A., *The Roots of Things: Topics in Quantum Mechanics* (New York: Springer, 1999).

Hooper, Henry, and Peter Gwynne, *Physics and the Physical Perspective* (New York: Harper and Row, 1977).

Kragh, Helge, "Max Planck: The Reluctant Revolutionary," *PhysicsWeb*, December 2000; see www.physicsweb.org/articles/world/13/12/8/1.

Kreith, Frank, and Mark S. Bohn, *Principles of Heat Transfer*, 6th edition (Pacific Grove, Calif.: Brooks/Cole, 2001).

Planck, Max, *Scientific Autobiography and Other Papers* (New York: Philosophical Library, 1949).

Singer, Neal, "Energy Emissions Far Greater Than Predicted by Planck's Law: Revolutionary Tungsten Photonic Crystal Could Provide More Power for Electrical Devices," DOE/Sandia National Laboratories, July 8, 2003; see www.newmaterials.com/news/2245.asp.

Smith, Henrik, *Quantum Mechanics* (Singapore: World Scientific, 1991).

INTERLUDE: CONVERSATION STARTERS

> God exists since mathematics is consistent, and the devil exists since we cannot prove the consistency.
>
> —Morris Kline, *Mathematical Thought from Ancient to Modern Times*

> It doesn't make much difference whether this determinism is due to an omnipotent God or to the laws of science. Indeed, one could always say that laws of science are the expression of the will of God.
>
> —Stephen Hawking, *Black Holes and Baby Universes*

The laws of Nature give a fundamental role to certain entities. We are not really sure what they are, but at the present level of understanding they seem to be the elementary quantum fields. They are highly simple because they are governed by symmetries. These are not objects with which we are familiar. In fact, our ordinary notions of space and time, causation, composition, substance and so on really lose their meaning on that scale. But it is just at that scale, at the level of the quantum fields, that we are beginning to find a certain satisfying simplicity.

—Steven Weinberg, "Is Science Simple?" in *The Nature of the Physical Universe*

Physics is a mixture of discovery and invention. For instance, Newton's laws of motion and of gravitation are mathematical models of the actual universe and hence a discovery of sorts, but the equations were Newton's invention. The fact that Newton's laws hold only approximately suggests that they have not "always existed," but are instead an imperfect description of the universe. The universe is as it is; human theories of it are always inventions and are discoveries to the extent that they reflect reality. Further, I suspect we will only ever have models of the universe; we will never understand it from the inside out.

—Nick Hobson, 2006, personal communication with the author

BRAGG'S LAW OF CRYSTAL DIFFRACTION

England, 1913. The angles at which radiation produces the most intense reflections from crystals depends on the spacing of atomic planes in the crystals and the wavelength of the incident radiation.

In 1913, Mahatma Gandhi, leader of the Indian Passive Resistance Movement, was arrested. French author Albert Camus was born, as was Richard Nixon, thirty-seventh President of the United States. The Woolworth Building and Grand Central Terminal opened in New York. Niels Bohr formulated his theory of atomic structure. Zippers became popular on clothing. Henry Ford pioneered assembly-line methods for building cars in a factory. U.S. company R.J. Reynolds Tobacco introduced Camel cigarettes.

Discovered by the English physicists Sir W. H. Bragg and his son Sir W. L. Bragg in 1913, Bragg's Law explains the results of experiments involving the diffraction of electromagnetic waves from crystal surfaces. Bragg's Law provides a powerful tool for studying crystal structure. For example, when X-rays are aimed at a crystal surface, they interact with atoms in the crystal, causing the atoms to reradiate waves that may interfere with one another. The interference is constructive (reinforcing) for integer values of n according to Bragg's Law:

$$\boxed{n\lambda = 2d\sin(\theta),}$$

where λ is the wavelength of the incident electromagnetic waves (e.g., X-rays), d is the spacing between the planes in the atomic lattice of the crystal, and θ is the angle between the incident ray and the scattering planes. For example, X-rays travel down through crystal layers, reflect, and travel back over the same distance before leaving the surface. The distance traveled depends on the separation of the layers and the angle at which the X-ray entered the material. For maximum intensity of reflected waves, the waves must stay in phase to produce the constructive interferences. Two waves stay in phase, after both are reflected, when n is a whole number. For example, when $n = 1$, we have a "first-order" reflection; for $n = 2$, we have a "second-order" reflection. If only two rows were involved in the diffraction, then as the value of θ changes, the transition from constructive to destructive interference is gradual. However, if interference from many rows occurs, then the constructive interference peaks become sharp, with mostly destructive interference occurring between the peaks.

Bragg's Law can be used for measuring wavelengths and for calculating the spacing between atomic planes of crystals. The observations of X-ray wave interference in crystals, commonly known as X-ray diffraction, provided direct evidence for the periodic atomic structure of crystals that was postulated for several centuries.

William Henry Bragg (1862–1942) and his son William Lawrence Bragg (1890–1971), British physicists famous for using X-rays to determine the structure of minerals.

CURIOSITY FILE: The younger Bragg discovered a new cuttlefish, which was named *Sepia braggi* in his honor. He was also the youngest person ever to receive the Nobel Prize in Physics.

> Physicists use the wave theory on Mondays, Wednesdays, and Fridays, and the particle theory on Tuesdays, Thursdays, and Saturdays.
>
> —William Henry Bragg, "Electrons and Ether Waves"

> Sadly, the joint work and its recognition created lasting problems between father and son; both rather reserved men, they had difficulty in discussing the matter, and the younger inwardly resented the way in which outsiders tended to give too much credit to his father.
>
> —Anthony C. T. North, "*Light Is a Messenger*—Book Review"

> Sir Lawrence Bragg, who died on July 1, 1971, aged 81, had the unique distinction of having himself created the science to which he devoted his life's work, and lived long enough to experience its revolutionary impact, first on inorganic chemistry and mineralogy, then on metallurgy, and finally on organic chemistry and biochemistry.
>
> —Max Perutz, "A Hundred Years and More of Cambridge Physics: Sir Lawrence Bragg"

> Everything that has already happened is particles, everything in the future is waves. The advancing sieve of time coagulates waves into particles at the moment "now."
>
> —William Lawrence Bragg, quoted in Ronald Clark's *Einstein: The Life and Times*

Father and son Braggs were awarded the Nobel Prize in Physics in 1915 for their pioneering studies in determining crystal structures for substances including table salt (NaCl), zinc sulfide (ZnS), and diamond. William Henry Bragg, the father, was born in Westward, Cumberland, a county in northern England. He came from a family that included farmers and merchant seamen.

The elder Bragg did well in school, and in 1875 he attended King William's College, where he developed a near-phobia for the Bible due to horrifying stories of eternal punishment in the afterlife that were described to him along with other scary Bible tales. Later, Bragg wrote in his private notes, and in spoke in his 1941 lecture at Cambridge titled "Science and Faith," "For many years, the Bible was a repelling book, which I shrank from reading. . . . I am sure that I am not the only one to whom . . . the literal interpretation of Biblical texts cause years of acute misery and fear." He never lost his faith entirely, but refused to take the idea of Hell literally. Later, he softened his position and came to believe, as expressed in *World of Sound*, "From religion comes a man's purpose; from science his power to achieve it."

While at Cambridge, he focused his energy on mathematics to the near exclusion of other subjects. For three years, he read mathematics throughout the morning, from about five to seven in the afternoon, and around an hour every evening. As a result, he did exceptionally well on a mathematics exam, and he wrote, as quoted in Paul Forman's "William Lawrence Bragg," "I never expected anything so high . . . I was fairly lifted in a new world. I had new confidence: I was extraordinarily happy." In 1855, he became a professor of mathematics and physics at the University of Adelaide in Australia.

For many of his early academic years, Bragg published very little, preferring to focus on classroom teaching and encouraging young minds. He was an excellent golfer, and his oldest son William Lawrence caddied for his father.

Forman comments about Bragg's "late start" in research science in the *Dictionary of Scientific Biography*:

> This is not the sort of life that brings election to the Royal Society of London (1907), the Bakerian lectureship (1915), the Nobel Prize in physics (1915), the Rumford Medal of the Royal Society (1935–1940), and membership in numerous foreign academies. . . . The new life began, at age forty-one, in 1903–1904.

I would be interested in hearing from readers who know of other famous scientists who seemed to flower so late in their lives. Note that E. M. C.

Andrade and K. Lonsdale make a similar observation in Bragg's obituary, published in 1943:

> Bragg had an astonishing career. Up to the age of forty he never showed any desire to carry out original experiment. He then straightaway embarks upon a perfectly precise and important piece of work and within a few years his name is known wherever physics is seriously studied.... He then himself conclusively demonstrates, by the work with which his name will always be associated, the wave nature of X-rays. He starts life as an extremely shy and retiring youth, never, apparently, quite at home in Cambridge, and in his old age becomes a national figure, at ease in all surroundings, whose personal appeal is known all over England.

In the spring of 1904, Bragg began experiments on the absorption of alpha particles (positively charged helium nuclei, He^{2+}). For the next two years, Bragg continued his investigations, publishing a paper every few months. By 1912, many scientists believed that X-rays were very short electromagnetic waves, with wavelengths of the order of a fraction of an angstrom, that is, about one-thousandth of the wavelength of visible light. However, the world awaited experimental verification of this hypothesis.

In 1912, William Lawrence, Bragg's oldest child, finally showed how recent X-ray observations by German physicist Max von Laue (1879–1960) and others might be best understood as reflections of electromagnetic radiation from planes in a crystal, and he derived Bragg's Law, $n\lambda = 2d \sin(\theta)$, as described above. William Lawrence presented his derivation of Bragg's Law at a meeting of the Cambridge Philosophical Society on November 11, 1912, and his paper on the subject, titled "The Diffraction of Short Electromagnetic Waves by a Crystal," was published in 1913 in the *Proceedings of the Cambridge Philosophical Society*. Of this formula and its implications, Austrian-British molecular biologist Max Perutz (1914–2002) wrote in his essay "Sir Lawrence Bragg":

> Why did this twenty-two-year-old student succeed in correctly interpreting the diffraction pattern predicted and discovered by an accomplished theoretician [Max von Laue] eleven years his senior? Bragg himself modestly attributes it to a "concatenation of fortunate circumstances" but his paper soon convinces you that its success owed more to Bragg's astute powers of penetrating through the apparent complexities of physical mechanisms to their underlying simplicity.

Von Laue did suggest in 1912 that a thin crystalline plate should act as a three-dimensional diffraction grating so that X-rays passing through a crystal might produce a diffraction pattern similar to that obtained when visible light is passed through an optical grating; however, Bragg was the first to determine the law or formula to describe the precise effect. For his own work in the field, von Laue received the Nobel Prize in 1914. Not only did von Laue's work provide striking evidence that X-rays were indeed electromagnetic waves with definite wavelengths, but he also provided support that the geometrical properties of crystals reflect an arrangement of atoms in a three-dimensional lattice.

William Henry followed his son's work with excitement, and used an inverted formula so that given a known wavelength, he could determine d, the distance between atomic planes. The X-ray spectrometer that he built helped to determine the atomic positions for a variety of crystals. The Braggs understood that the X-ray reflection from crystals was not simply a surface phenomenon and that only a small portion of the energy of an incident wave front would be reflected from the first layer of atoms. They reasoned that additional reflections must also come from deeper layers or atomic planes. Whether these additional reflected waves interfere with each other and other reflected waves in a constructive or destructive manner depends on the distance between atomic planes and the wavelength of the waves.

Here are William Lawrence's own words in his December 1912 paper in *Nature*:

> The spots in Laue's crystallographs can be shown to be due to partial reflection of the incident beam in sites of parallel planes in the crystal on which the atoms centers may be arranged, the simplest of which are the actual cleavage planes of the crystal. This is merely another way of looking at diffraction.... Only a few minutes' exposure to a small X-ray bulb sufficed to show the effect....

During and after World War I, the elder Bragg worked on antisubmarine devices and the detection and measurement of sound for locating submarines. He also established a school of crystallographic research at University College in London. Bragg enjoyed conveying the wonders of science to a broader public, and his "Christmas Lectures" for children were very popular.

William Lawrence Bragg, William Henry's son, was born in Adelaide, South Australia, and like his father, he particularly excelled in mathematics at school. William Lawrence was a quiet child, preferring solitary hobbies such as shell collecting. As mentioned above later in life he even discovered a new cuttlefish (a squidlike creature), which was named *Sepia braggi* in his honor.

At age 5, William Lawrence fell from his tricycle and broke his arm. His father had read about Roentgen's recent experiments with X-rays and used them to examine his son's arm. This is the first recorded medical use of X-rays in Australia.

In 1909, the younger Bragg went to England to enter Trinity College, Cambridge, where he received a large scholarship in mathematics, despite taking the exam while in bed with pneumonia. During the summer vacation of 1912, his father began to discuss the nature of X-ray diffraction with him, and after several experiments, William Lawrence published the Bragg equation, which describes the angular behavior of X-rays striking a crystal. As discussed above, this equation is basic to X-ray diffraction, which is used to understand crystal structures. His father had designed the X-ray spectrometer to make exact measurements of X-ray wavelengths.

Father and son spent vacations together in order to determine the atomic arrangement of numerous minerals such as diamond. Their collaboration led some to believe that William Henry had actually initiated the research that led to the Bragg equation, which bothered William Lawrence. However, his delight was evident when he became the youngest person ever to receive the Nobel Prize in Physics at the age of 25. Many accounts suggest that the mutual scientific interest William Lawrence and his father shared was enjoyable for both of them. Over the next two years they collaborated in their X-ray crystallography research and coauthored the book *X Rays and Crystal Structure*, which was published in 1915.

In 1921, William Lawrence Bragg married the daughter of a doctor, and they eventually had four children. During World War I, Bragg helped the British army develop methods to determine the distance of enemy artillery from the sounds of their firing. From 1937 to 1938, Bragg was director of the British National Physical Laboratory.

After World War I, the close collaboration between William Lawrence and his father ended. They agreed to focus on different areas of X-ray crystallography. William Lawrence was to focus on inorganic compounds, metals, and silicates; William Henry, on organic compounds. After his father's death, William Lawrence became fascinated by the challenge of using X-ray crystallography to determine the structures of biological macromolecules, and he made contributions to the study of the phase relationships between the X-ray reflections from hemoglobin crystals. Later, his group at the Royal Institution of Great Britain, London, played key roles in determining the structures of myoglobin and hemoglobin, the first two proteins whose structures were determined by X-ray crystallography.

William Lawrence played an important role in the 1953 discovery of the structure of DNA, due to the support he provided to scientists Francis Crick (1916–2004) and James D. Watson (born 1928) who worked under his guidance at the Cavendish Laboratory.

William Lawrence felt that science students everywhere should be provided with sufficient time to have a general education and not be so narrow in their studies that they could not appreciate a fuller life and outlook. Thus, in 1954, when he became director of the Royal Institution, following in his father's footsteps, he suggested that the institution give back to the community and hold a series of lectures for schoolchildren. The idea was a huge success, and thousand of children attended these lectures over the years.

William Lawrence loved gardening throughout his life. When he had moved to London, he missed having a garden to such an extent that he worked as a part-time gardener. His employer did not recognize the fame of his gardener until a guest at the home expressed shock at seeing him laboring in the garden!

Perutz wrote of William Lawrence personality:

> So often men of genius were hellish to live with, but Bragg's creativity was sustained by a happy home life; typically one would find him tending his garden, with Lady Bragg, children and grandchildren somewhere in the background, and before getting down to business he would proudly demonstrate his latest roses. To the present young generation Bragg was an avuncular figure who showed them that science can be fun.... Bragg's superb powers of combining simplicity with rigor, his enthusiasm, liveliness and charm of manner, and his beautiful demonstrations all conspired to make him one of the best lecturers on science that ever lived.

In an obituary for the elder Bragg, Andrade and Lonsdale comment on the work of both Braggs and its effect on science:

> The work of Bragg and his son Lawrence in the two years 1913, 1914 founded a new branch of science of the greatest importance and significance, the analysis of crystal structure by means of X-rays. If the fundamental discovery of the wave aspect of X-rays, as evidenced by their diffraction in crystals, was due to Laue and his collaborators, it is equally true that the use of X-rays as an instrument for the systematic revelation of the way in which crystals are built was entirely due to the Braggs.

FURTHER READING

Andrade, E. N. C., and K. Lonsdale, "William Henry Bragg, 1862–1942," *Obituary Notices of Fellows of the Royal Society*, 4(12): 276–300, November 1943; Reprinted in P. P. Ewald, *Fifty Years of X-Ray Diffraction* (New York: Springer, 1962); see www.iucr.org/iucr-top/publ/50YearsOfXrayDiffraction/.

Bragg, William Henry, "X-rays and Crystals," *Nature*, 90: 219 and 360–361, 1912.

Bragg, William Lawrence, "The Specular Reflection of X-rays," *Nature*, 90: 410, 1912.

Clark, Ronald, *Einstein: The Life and Times* (New York: HarperCollins, 1984).

Bragg, William Henry, *The World of Sound, Six Lectures Delivered before a Juvenile Auditory at the Royal Institution, Christmas, 1919* (London: Bell, 1921).

Forman, Paul, "William Lawrence Bragg," in *Dictionary of Scientific Biography*, Charles Gillispie, editor-in-chief (New York: Charles Scribner's Sons, 1970).

Hunter, G. K., *Light Is a Messenger—the Life and Science of William Lawrence Bragg* (New York: Oxford University Press, 2004).

North, Anthony C. T., "Book Review: *Light Is a Messenger*," *Acta Crystallographica Section A: Foundations of Crystallography*, 61: 262–264, March 2005; see journals.iucr.org/a/issues/2005/02/00/pf0015/pf0015bdy.html.

Perutz, M. F., "Sir Lawrence Bragg," in *A Hundred Years and More of Cambridge Physics* (Cambridge, U.K.: Cambridge University Physics Society, 1974; reprinted 1995); see www.phy.cam.ac.uk/cavendish/history/years/bragg.

"Sir William Henry Bragg"; see www.nobel-winners.com/Physics/william_henry_bragg.html.

INTERLUDE: CONVERSATION STARTERS

> The mathematical life of a mathematician is short. Work rarely improves after the age of twenty-five or thirty. If little has been accomplished by then, little will ever be accomplished.
>
> —Alfred Adler, "Mathematics and Creativity," *New Yorker* magazine, 1972

> Without regularities embodied in the laws of physics, we would be unable to make sense of physical events; without regularities in the laws of nature, we would be unable to discover the laws themselves.
>
> —Gerd Baumann, *Symmetry Analysis of Differential Equations with Mathematica*, 2000

> Could it be that there are no laws of Nature at all? Perhaps all the order we see is a manifestation of that peculiar type of total lawlessness and independence that leads to predictability?
>
> —John D. Barrow, *The Universe That Discovered Itself*

Laws of Nature are generalizations of observations. They are discovered, not invented. The Laws of Nature do not "run the universe" or tell it how to behave—they describe its observed behavior. Laws of Nature do not explain, they describe. Theories explain—for example, the law of gravity is not a theory of gravity. Theories are "invented," not discovered. The fact that the current formulation of a given law may be only a locally applicable approximation of a more general (and as yet undiscovered) formulation doesn't mean that the current formulation is "wrong." 3.14 is not an incorrect description of pi; it is merely a less precise description than, for example, 3.14159.

—Bill Gavin, 2006, personal communication with the author

HEISENBERG'S UNCERTAINTY PRINCIPLE

 *Germany, 1927.* The position and the velocity of an object cannot both be known with high precision, at the same time. Specifically, the more precise the measurement of position, the more imprecise the measurement of momentum, and vice versa.

In 1927, American inventor Philo Farnsworth transmitted the first experimental electronic television pictures. The Academy of Motion Picture Arts and Sciences was founded. Saudi Arabia became independent of the United Kingdom. An experiment confirmed French physicist Louis de Broglie's hypothesis that subatomic particles behave like waves.

The Heisenberg Uncertainty Principle states that it is impossible to precisely measure the values of certain pairs of physical quantities for a single particle. The most common expression of this principle depicts the relations between the position x and the momentum p of a particle in space:

$$\Delta x \Delta p \geq \frac{\hbar}{2}$$

Here, Δx corresponds to the uncertainty of the position measurement; Δp corresponds to the uncertainty of the momentum measurement, and $\hbar$ is the reduced Planck's constant, $h/2\pi$. Notice that as Δx becomes smaller (i.e., the more precisely we know the position of the particle), the larger the uncertainty becomes for the momentum, Δp. (Recall that the momentum of a particle is its velocity times its mass.)

Until this law was discovered, most scientists believed that the precision of any measurement was limited only by the accuracy of the instruments being used. Werner Karl Heisenberg showed that even if we could construct an infinitely precise measuring instrument, we still could not accurately determine both the position and momentum of a particle. Because the formula indicates that the product of the position and momentum uncertainties is equal to or greater than about 10^{-35} J·s, the uncertainty principle becomes significant only at the small size scales of atoms and subatomic particles.

Some writers have erroneously suggested that the uncertainty principle concerns itself with the degree to which the measurement of the position of a particle may disturb the momentum of a particle. However, this is not a correct interpretation of the principle. Note also that we could measure a particle's position x to a high precision, but as a consequence, we could know little about the momentum.

The uncertainty principle also applies to measurements of energy E and time t:

$$\boxed{\Delta E \Delta t \geq \frac{\hbar}{2}}$$

In this formulation, ΔE is the uncertainty of our knowledge of the energy of a particle, and Δt is the time interval during which the particle had the energy; alternatively, we can think of Δt as the uncertainty in the time interval during which the measurement is made.

Although Heisenberg developed the relationship that involves uncertainties in position and momentum in 1927, the energy–time uncertainty relationship was not proven until 1945 by Russian scientists Leonid Mandelshtam (1879–1944) and Igor Tamm (1895–1971).

For those scientists who accept the Copenhagen interpretation of quantum mechanics, the Heisenberg uncertainty principle means that the physical universe literally does not exist in a deterministic form but is rather a collection of probabilities. Similarly, the path of an elementary particle such as a photon cannot be predicted, even in theory, by an infinitely precise measurement.

Werner Karl Heisenberg (1901–1976), German physicist famous for his formulations of quantum mechanics and his uncertainty principle.

CURIOSITY FILE: On the TV show *Star Trek*, the "Heisenberg compensator" was a device used in the transporter system to compensate for the Heisenberg Uncertainty Principle, which allegedly would have otherwise rendered transporters theoretically impossible. • In computer science, a Heisenbug is a software error that disappears or alters its behavior when one attempts to probe or isolate it further. • Consider the most common Heisenberg joke on the Web: A policeman pulls Heisenberg to the side of the road for speeding. "Do you know how fast you were going?" the policeman asks. "No," replies Heisenberg, "but I know exactly where I am!"

> Only those quantities that can be measured have any real meaning in physics. If we could focus a "super" microscope on an electron in an atom and see it moving around in an orbit, we would declare that such orbits have meaning. However, we shall show that it is fundamentally impossible to make such an observation—even with the most ideal instrument that could conceivably be

constructed. Therefore, we declare that such orbits have no physical meaning.

—David Halliday and Robert Resnick, *Physics*

All of my meager efforts go toward killing off and suitably replacing the concept of the orbital path which one cannot observe.

—Werner Heisenberg, 1925 letter to Wolfgang Pauli

I learned optimism from Sommerfeld, mathematics at Göttingen, and physics from Bohr.

—Werner Heisenberg, quoted in Laurie M. Brown, Abraham Pais, and A. B. Pippard, *Twentieth Century Physics*

Even the Uncertainty Principle isn't "merely" philosophy: it predicts real properties of electrons. Electrons jump at random from one energy state to another state which they could never reach except that their energy is momentarily uncertain. This "tunneling" makes possible the nuclear reactions that power the sun and many other processes. Physicists have put some of these processes to practical use in microelectronics.

—David Cassidy, "Werner Heisenberg and the Uncertainty Principle"

The more I think about the physical portion of Schrödinger's theory, the more repulsive I find it.... What Schrödinger writes about the visualizability of his theory "is probably not quite right"; in other words it's crap.

—Werner Heisenberg, 1926 letter to Wolfgang Pauli

One moonlit night we walked all over Hainberg Mountain [near Göttingen], and [Heisenberg] was completely enthralled by the visions he had, trying to explain his newest discovery to me. He talked about the miracle of symmetry as the original archetype of creation, about harmony, about the beauty of simplicity, and its inner truth. It was a high point of our lives.

—Elisabeth Heisenberg, *Inner Exile*

Werner Heisenberg was born in Würzburg, Germany. His father was an Evangelical Lutheran who taught classical languages and ruled his family in a stiff, domineering manner. However, in private, Heisenberg's mother and father admitted that they were not religious, did not believe in the supernatural, and only followed Christian ethics.

As a boy in school, Heisenberg excelled in subjects such as mathematics, physics, and religion, and his overall scholastic record was generally excellent. One of his school teachers remarked, "He is more developed toward the side of rationality than of fantasy and imagination." He also enjoyed chess, read many mathematical books, and tried to prove Fermat's Last Theorem after studying an advanced book on number theory. He taught himself calculus when his parents asked him to tutor a college student for her final exams. His father, after observing young Heisenberg's passion for mathematics, became worried that Heisenberg was neglecting his Latin studies, so the father brought home old math papers and books written in Latin.

In 1920, together with his fellow student Wolfgang Pauli (1900–1958), Heisenberg began to study theoretical physics under Arnold Sommerfeld (1868–1951) at the University of Munich. Of his studies, Heisenberg wrote in *Physics and Beyond: Encounters and Conversations*:

> My first two years at Munich University were spent in two quite different worlds: among my friends of the youth movement and in the abstract realm of theoretical physics. Both worlds were so filled with intense activity that I was often in a state of great agitation, the more so as I found it rather difficult to shuttle between the two.

He finished his doctoral dissertation in 1923 in just three years. His thesis was essentially a 59-page calculation titled "On the Stability and Turbulence of Fluid Flow." The work focused on the challenges of understanding the nature of the precise transition of a smoothly flowing fluid (laminar flow) to a turbulent flow. The problem was so difficult mathematically that Sommerfeld had written, "I would not have proposed a topic of this difficulty as a dissertation to any of my other pupils."

Heisenberg's passion for theory over experiment nearly ended his scientific career in 1923 when he sought approval of his doctoral thesis from four key professors. Pete Moore writes in *$E = mc^2$: The Great Ideas That Shaped Our World*:

> Sommerfeld asked questions about theoretical mathematics, and these were answered with ease.... Wilhelm Wein was more concerned about practical physics and [checked] that Heisenberg understood the details behind the experimental work. Apparently, he didn't. The result was a raging argument between Sommerfeld and Wein, the one wanting to pass him with flying colors, the other

wanting to fail him. In the end, a compromise was reached, and Heisenberg was given a mediocre pass.

Heisenberg followed Pauli to the University of Göttingen, where he studied under German physicist Max Born (1882–1970). In 1924, he went to the Institute for Theoretical Physics in Copenhagen to study under Danish physicist Niels Bohr (1885–1962). While in Copenhagen, Heisenberg met Albert Einstein for the first time.

In 1925, Heisenberg had one of his greatest breakthrough ideas with respect to quantum mechanics and atomic theory. David Cassidy of Hofstra University, along with the American Institute of Physics, describe Heisenberg's findings at a popular Heisenberg tribute Web site:

> Since the electron orbits in atoms could not be observed, Heisenberg tried to develop a quantum mechanics without them. He relied instead on what can be observed, namely the light emitted and absorbed by the atoms. By July 1925 Heisenberg had an answer.... Heisenberg handed a paper on the derivation to his mentor, Max Born, before leaving on a month-long... trip. After puzzling over the derivation, Born finally recognized that the unfamiliar mathematics was related to the mathematics of arrays of numbers known as "matrices." Born sent Heisenberg's paper off for publication. It was the breakthrough to quantum mechanics.

In 1925, Heisenberg had invented a way to express quantum mechanics in terms of matrices after considering sets of quantized probability amplitudes. These amplitudes formed a noncommutative algebra, for example, an algebra in which $A \times B$ does not equal $B \times A$. In 1926, Heisenberg developed the concept of "matrix mechanics" further in a paper coauthored with Born and German physicist Pascual Jordan (1902–1980). The approach was able to account for many of the properties of atomic events. In matrix mechanics, individual terms of a matrix correspond to probabilities of occurrences of states and to transitions among states. Heisenberg used the new matrix mechanics to interpret the spectrum of the helium atom. Thomas Powers, in *Heisenberg's War*, writes of Heisenberg's new formulations:

> The solutions that [Heisenberg's matrix mechanics] provided came only with agony and labor and it demanded difficult concessions—for example, giving up the idea of "orbits" within the atom. This aroused the wasp in Pauli: The moon, like an electron, occupied a stationary state, and yet it moved in an orbit. If nature made a place for orbits among the spheres, why did Heisenberg ban them from

> the atom and insist only on "observables"? "Physics is decidedly confused at the moment," Pauli remarked in 1925. "In any event, it is much too difficult for me, and I wish I... had never heard of it."

In 1926, Heisenberg was appointed Lecturer in Theoretical Physics in Copenhagen, where he continued to work with Bohr. In 1927, Heisenberg was appointed to a chair at the University of Leipzig. In 1932, he was awarded the Nobel Prize in Physics for his matrix description of quantum mechanics. Professor Henning Pleijel, Chairman of the Nobel Committee for Physics of the Royal Swedish Academy of Sciences, said that Heisenberg had viewed the problem in quantum mechanics

> from the very beginning, from so broad an angle that it took care of systems of electrons, atoms, and molecules.... Heisenberg's quantum mechanics has been applied by himself and others to the study of the properties of the spectra of atoms and molecules, and has yielded results which agree with experimental research.

Heisenberg is today perhaps best known for the Uncertainty Principle, discovered in 1927, which states it is not possible, even in theory, to determine the precise position and momentum of a particle. (Most physicists interpret this fact to mean that a particle does not simultaneously have a precise position and momentum.) These errors are negligible in our daily lives but become important when studying atomic and subatomic phenomena.

As discussed above, the uncertainty formulas involve the uncertainties in the measurements of position and momentum. This "uncertainty" is sometimes called the "imprecision" of the measurement in other areas of science and may be thought of as applying to a range of the results of repeated measurements. For example, suppose you measure the length of a pea with a ruler. You measure the pea to be 8 millimeters in diameter. But because your ruler has a limited precision, another measurement of the pea might yield 8.5 millimeters or 7.5 millimeters. In fact, if you measure the pea a hundred times, you will get a bell curve of measurements centered on an average value, such as 8 millimeters. The spread of the bell curve indicates the uncertainty of the measurement. Note that some physics text books list the uncertainty principle as $\Delta x \Delta p \geq \hbar$ instead of $\Delta x \Delta p \geq \hbar/2$, simply because their definitions are slightly different with respect to what the Δ refers. For example, in older text books Δ may have indicated the full width within which resides 50% of the probability.

According to some interpretations of the Uncertainty Principle, every explanation of reality has a meaning only in terms of the experiments that can be used to measure an aspect of reality under discussion. If this is true, as Heisenberg seemed to believe, things that cannot be measured have no meaning in physics. Heisenberg also believed that we cannot predict the future based on the past because the future of a single particle cannot be known, even if we know all the forces acting on the particle. We cannot know the precise position and momentum of a particle at a given instant. The best we can do is make statistical predictions.

Later in life, Heisenberg remarked that the time period roughly between 1927 and 1932 was the "golden age of atomic physics... in which the great obstacles that had occupied all our efforts in the preceding years had been cleared... and fresh fruits seemed ready for the picking" (quoted in Laurie M. Brown, Abraham Pais, and A. B. Pippard, *Twentieth Century Physics*). In 1928, Heisenberg published *The Physical Principles of Quantum Theory*, a standard introduction to the underlying mathematical formalism of quantum mechanics.

In 1935, Heisenberg was a logical choice to replace his former mentor Sommerfeld at the University of Munich. Alas, the Nazis required that "German physics" must replace "Jewish physics," which included quantum theory and relativity. As a result, Heisenberg's appointment to Munich was blocked even though he was not Jewish. One Nazi functionary is quoted as saying, "The concentration camp is obviously the most suitable place for Herr Heisenberg!" A 1937 SS newspaper said that "Heisenberg is only one example of many others.... They are all representatives of Judaism in German spiritual life who must all be eliminated just as the Jews themselves." After a frightening investigation, the SS finally cleared Heisenberg of any accusations.

Heisenberg was a superb pianist, and it was through his passion for music that he met his other life's passion, Elisabeth Schumacher. He had met Elisabeth at a concert in which he was performing. She was 22, and he was 35. They married three months later in 1937 and eventually had seven children.

During World War II, Heisenberg led the unsuccessful German nuclear weapons program. Today, historians of science still debate as to whether the program failed because of lack of resources, lack of the right scientists on his team, Heisenberg's lack of experimental skills, or his lack of a desire to give such a powerful weapon to the Nazis. Whatever the case, it is true that Hitler's military eventually decided to concentrate on rockets and jet aircraft and give less financial support to the nuclear weapons effort.

Heisenberg does appear to have been committed to at least some of Germany's extreme nationalism, as indicated in this portion of his 1943 letter to Dutch scientist Hendrik Casimir, published in Dan Kurzman's *Blood and Water*:

> History legitimizes Germany to rule Europe and later the world. Only a nation that rules ruthlessly can maintain itself. Democracy cannot develop sufficient energy to rule Europe. There are, therefore, only two possibilities: Germany and Russia, and perhaps a Europe under German leadership is the lesser evil.

Dutch-American physicist Samuel Goudsmit (1902–1978) expressed his reservations about Heisenberg in his August 27, 1948, letter to Michael Perrin, an American official who was involved with the detention of German scientists who had worked on the Nazi atom bomb project:

> Heisenberg doesn't seem to be willing even now to condemn the Nazis openly. Instead, he tries to impress upon the world how excellent the quality of German scientific work was, even under the Nazis, and how, after all, their intentions were only peaceful. The only mildly anti-Nazi article I have seen by Heisenberg is a speech to the students at Göttingen, in which he points out that science has nothing to do with race or religion. I think his speech would have been much stronger if he had given examples of the destructive influence of the Nazi doctrine.

At the end of World War II, the Allies captured Heisenberg and detained him and several other German scientists for several months at an English country manor, where their private conversations were secretly recorded.

In 1946, Heisenberg was appointed director of the Max Planck Institute for Physics and Astrophysics at Göttingen. He continued to hold this post when the institute moved to Munich, until he resigned in 1970. Heisenberg died of cancer in Munich in 1976.

Two of his famous books that dealt with the philosophy of physics were *Physics and Philosophy* (1962) and *Physics and Beyond* (1971). David Cassidy writes of Heisenberg's impact in "Werner Heisenberg: An Overview of His Life and Work":

> As a physicist, Heisenberg ranks with Niels Bohr, Paul Dirac, and Richard Feynman in his contributions and impact upon contemporary physics. He was a key player in the development of quantum mechanics. . . . He went on to formulate a quantum theory of ferromagnetism, the neutron-proton model of the nucleus, the S-matrix theory in particle scatting. . . . During his lifetime, Heisenberg produced nearly 600 original research papers, philosophical essays, and explanations for general audiences.

FURTHER READING

Brown, Laurie M., Abraham Pais, and A. B. Pippard, *Twentieth Century Physics* (Boca Raton, Fla.: CRC Press, 1995).

Cassidy, David C., *Uncertainty: The Life and Science of Werner Heisenberg* (New York: W. H. Freeman & Company, 1991).

Cassidy, David, "Werner Heisenberg: An Overview of His Life and Work," in *100 Years, Werner Heisenberg: Works and Impact*, Dietrich Papenfuß, Dieter Lüst, and Wolfgang P. Schleich, editors (New York: Wiley, 2002).

Cassidy, David C., and the Center for History of Physics of the American Institute of Physics, "Werner Heisenberg and the Uncertainty Principle"; see www.aip.org/history/heisenberg/.

Heisenberg, Werner, *Physics and Beyond: Encounters and Conversations* (London: G. Allen & Unwin, 1971).

Kurzman, Dan, *Blood and Water: Sabotaging Hitler's Bomb* (New York: Henry Holt, 1997), 35.

Moore, Pete, *$E = mc^2$: The Great Ideas That Shaped Our World* (New York: Sterling Publishing, 2005).

Powers, Thomas, *Heisenberg's War: The Secret History of the German Bomb* (New York: Da Capo Press, 1993).

Rose, Paul, *Heisenberg and the Nazi Atomic Bomb Project, 1939–1945: A Study in German Culture* (Berkeley, Calif.: University of California Press, 1998).

INTERLUDE: CONVERSATION STARTERS

> Philosophers and great religious thinkers of the last century saw evidence of God in the symmetries and harmonies around them—in the beautiful equations of classical physics that describe such phenomena as electricity and magnetism. I don't see the simple patterns underlying nature's complexity as evidence of God. I believe *that* is God. To behold [mathematical curves], spinning to their own music, is a wondrous, spiritual event.
>
> —Paul Rapp, "Get Smart: Controlling Chaos," *Omni*

> If our abstractions of nature are mathematical, in what sense can we be said to understand the universe? For example, in what sense does Newton's Law explain *why* things move?
>
> —Lawrence M. Krauss, *Fear of Physics*

> Imagine that the world is something like a great chess game being played by the gods, and we are observers of the game.... If we watch long enough, we may eventually catch on to a few of the rules.... However, we might not be able to understand why a particular move is made in the game, merely because it is too complicated and our

minds are limited. . . . We must limit ourselves to the more basic question of the rules of the game. If we know the rules, we consider that we "understand" the world.

—Richard Feynman, *Feynman Lectures on Physics*

Every scientific theory has its domain of applicability, every theory has realms where their approximations work, and realms where their approximations break down. We don't use Newtonian gravity to build buildings on the Earth (unless the building is *very* tall), we use Galileo's model of gravity. We don't use Einstein's theory of gravity for navigating the space shuttle when Newton's theory works to the level of precision needed for the task. The relevant question is "Could we have learned the greater understanding revealed by Einstein without the two centuries of observations, analysis, and *experience* developed under Newton's ideas?" I think the answer is probably "no."

—William T. Bridgman, "The Cosmos in Your Pocket: How Cosmological Science Became Earth Technology. I."

HUBBLE'S LAW OF COSMIC EXPANSION

United States, 1929. The greater the distance a galaxy is from an observer on Earth, the faster it recedes. The distances between galaxies, or galactic clusters, are continuously increasing and, therefore, the universe is expanding.

CROSS REFERENCE: THE DOPPLER EFFECT, THE GENERAL THEORY OF RELATIVITY, AND BUYS-BALLOT'S WIND AND PRESSURE LAW.

> In 1929, Ernest Hemingway published *A Farewell to Arms*. The U.S. Stock Exchange collapsed. Popular songs included "Singing in the Rain" and "Tiptoe Through the Tulips." The comic strip character Popeye made his debut. Bandleader Guy Lombardo played "Auld Lang Syne" for the first time. Yasser Arafat, Palestinian leader and recipient of the Nobel Peace Prize, was born.

In 1929, American astronomer Edwin Hubble discovered a linear relationship between recessional velocity and distance of galaxies:

$$\boxed{v = H \cdot D}$$

Here, v is the recessional velocity (e.g., the movement of a galaxy away from an observer on Earth), H is Hubble's constant, and D is the distance of a galaxy from an observer or, more precisely, the proper distance that the light had traveled from the galaxy in the rest frame of the observer, that is, the reference frame in which the observer is at rest. Today, we now know that Hubble's constant can be used to characterize a time scale for the evolution of the cosmos and probably changes through time. Thus, H is sometimes written as H_0 in formulations of the law, which signifies that the value is not static.

The Hubble constant is thought to be the same throughout the universe for a given time. Today, astronomers have determined that H is approximately 71 (km/s) per megaparsec. A megaparsec is about 3 million light years—and a light year is the distance that light travels in a vacuum in one year, approximately 9.46 trillion (9.46×10^{12}) kilometers or 5.88 trillion (5.88×10^{12}) miles. To get a feeling for the magnitude of a megaparsec, which is often abbreviated Mpc, consider that the diameter of our Milky Way galaxy is about 0.02 megaparsecs and the distance from our galaxy to the Andromeda galaxy is 0.77 megaparsecs.

For many galaxies, the velocity v can be estimated from the redshift of a galaxy, which is an observed increase in the wavelength of electromagnetic radiation received by a detector on Earth compared to that emitted by the source. Such redshifts occur because galaxies are moving away from our own galaxy at high speeds due to the expansion of space itself. The change in the wavelength of light that results from the relative motion of the light source and the receiver is an example of the Doppler effect (see "Buys-Ballot's Wind and Pressure Law" in part III for more information on this effect). Other methods also exist for determining the velocity of faraway galaxies.

Systems that exert their own nearby gravitational influences, such as stars within a single galaxy, are not subject to Hubble's Law. This is why the stars within our Milky Way are not expanding away from one another in the same manner in which intergalactic distances are generally increasing. This recession sometimes applies only to clusters of galaxies rather than to galaxies themselves, because those galaxies that exist within a group are bound together by gravity. Our local cluster of galaxies, for example, is not expanding, but our cluster is moving farther away from other clusters.

Although an observer on Earth finds that all distant galactic clusters are flying away from Earth, our location in space is not special. Observers in another galaxy would also see the galactic clusters flying away from their position, because all of space is expanding.

As far back as 1917, before scientists really knew much about galaxies, researchers noticed that most galaxies had their spectral lines Doppler-shifted to the red, which, as discussed above, suggested that they were moving away from us at high speeds. The more distant the galaxy, the faster it moved. This is one of the main lines of evidence for the Big Bang that created the universe and the subsequent expansion of space. However, it is important to remember that the galaxies are not like the flying debris from a bomb that has just exploded. Space itself is expanding. The distances between galaxies are increasing in the same way that black dots painted on the surface of a balloon move away from one another when the balloon is inflated. It doesn't matter on which dot you reside in order to observe this expansion. Looking out from any dot, the other dots appear to be receding.

The Hubble constant helps scientists gain insight into the universe because it establishes a time scale for the universe. In particular, it provides a rough measure for the time since the Big Bang. The age of expansion for the universe is given by $1/H$. For example, if we set $H = 71$ km/s/Mpc, the expansion age of the universe is about 13.7 billion years. However, this assumes that the universe has expanded without any acceleration or deceleration. If the universe is accelerating, as suggested by many scientists today, then the universe could be older than that determined by the simple estimate provided by $1/H$.

In 1929, Hubble had found his relationship of the distance of a galaxy from Earth and the redshift of a galaxy by plotting a straight line, using data associated with 46 galaxies. The line yielded a Hubble constant of 500 km/s/Mpc. The value we use today is much lower because various errors were associated with Hubble's distance measurements. When Hubble presented his plots of recessional velocity versus distance for his 46 galaxies, although there was a clear linear relationship, the data points were somewhat scattered partly because all galaxies have some additional residual motion in addition to the motion resulting from the expansion of the universe. In 1958, American astronomer Allan Sandage (born 1926) published a value of $H_0 = 75$ km/s/Mpc.

In the 1990s, the Hubble Space Telescope allowed researchers to perform careful studies of Cepheid variables—stars whose distinctive pulsations give information as to the intrinsic brightness and distance of the stars. By observing these kinds of stars in 31 galaxies, scientists computed H_0 to a precision of about 10%. Armed with this value for H_0 and with measurements of the cosmic microwave background, the age of the universe was estimated to be 13.7 billion years. Astronomers have also used the Hubble Space Telescope, along with other observations, to determine that a mysterious dark energy seems to compose about three-quarters of the total energy density of the universe. About five billion years ago, this dark energy started to cause the rate of expansion of the universe to accelerate.

Edwin Hubble (1889–1953), U.S. astronomer famous for his theories on the expansion of the universe.

CURIOSITY FILE: Astronomer Milton Humason (1891–1972) played a major role in assisting Edwin Hubble when formulating Hubble's Law by performing various spectroscopic studies. Humason started out as a janitor at the Mount Wilson Observatory in 1917! • When Hubble died, his wife did not have a funeral for him. She never revealed what was done with his body. • While in England, Hubble ran track and played on one of the first baseball teams in the British Isles.

> Arguably the most important cosmological discovery ever made is that our Universe is expanding. Its stands, along with the Copernican Principle—that there is no preferred place in the Universe, and Olbers' paradox—that the sky is dark at night, as one of the cornerstones of modern cosmology. It forced cosmologists to dynamic models of the Universe, and also implies the existence of a timescale or age for the Universe. It was made

possible ... primarily by Edwin Hubble's estimates of distances to nearby galaxies.

—John P. Huchra, "The Hubble Constant"

Hubble's astronomical triumphs earned him worldwide scientific honors and made him the toast of Hollywood during the 1930s and 1940s—the confidant of Aldous Huxley and a friend to Charlie Chaplin, Helen Hayes and William Randolph Hearst.

—Michael D. Lemonick, "The Time 100: Edwin Hubble"

Equipped with his five senses, man explores the universe around him and calls the adventure Science.

—Edwin Powell Hubble, *The Nature of Science*, 1954

As he rose through the social and economic strata, Hubble reinvented himself.... There was a discontinuity between one Hubble, with an ordinary Midwestern background, and the other, a wealthy Anglophile who mingled with the Hollywood greats. Along the way, Hubble partly disowned his family members, not allowing any of them to meet his wife or her family.

—William H. Cropper, *Great Physicists*

Hubble was born in Marshfield, Missouri, and attended high school in Chicago. His father, a religious man, was involved in the insurance business. According to legends, at Hubble's high school graduation ceremony, the school principal approached him and said, "Edwin Hubble, I have watched you for four years and I have never seen you study for ten minutes." The principal paused, and then said, "Here is a scholarship to the University of Chicago."

At age 16, Hubble started the University of Chicago, where he received a B.S. in mathematics and astronomy—and was also a heavyweight boxer. A sports promoter desperately wanted to train Hubble to fight the current heavyweight champion of the world, but Hubble decided that a more cerebral career behooved him, and he continued his studies at Queen's College, Oxford, in 1910, as a Rhodes Scholar. Hubble's father, who was strongly against the drinking of alcohol, made Hubble vow that he would not touch alcohol while in England.

He received a B.A. from Oxford in jurisprudence in 1912. In 1913, Hubble is alleged to have opened a law office in Kentucky, but conflicting stories exist with respect to his life. Bill Bryson in *A Short History of Nearly Everything* writes:

> For all his gifts, Hubble was an inveterate liar.... Though he later *claimed* to have passed most of the second decade of the century practicing law in Kentucky, in fact he worked as a high school teacher and basketball coach in New Albany, Indiana, before belatedly attaining his doctorate.

Rocky Kolb in *Blind Watchers of the Sky* writes of this matter in more diplomatic fashion:

> After three years at Oxford, Hubble returned to Louisville, Kentucky, where his family had moved while he was in England. But rather than practice law... like most Rhodes scholars trained in jurisprudence, he decided to do something noble with his life, and he became a high school teacher.

Having boxed both in the United States and while at Oxford, it was time to try something new again. In 1914, he went to the Yerkes Observatory of the University of Chicago, where he was awarded a Ph.D. in astronomy in 1917.

When the United States entered World War I, Hubble enlisted and eventually rose to the rank of major. He returned to the United States in 1919, when he began work at the Mount Wilson Observatory and studied nebulae, interstellar dust, and gas that are often visible as luminous patches. At Mount Wilson, he discovered that some nebulae were outside our Milky Way galaxy and were actually separate galaxies in their own right. Michael D. Lemonick writes in *Time*:

> [In the 1920s], most of Hubble's colleagues believed the Milky Way galaxy, a swirling collection of stars a few hundred thousand light-years across, made up the entire cosmos. But peering deep into space from the chilly summit of Mount Wilson, in Southern California, Hubble realized that the Milky Way is just one of millions of galaxies that dot an incomparably larger setting.

In 1925, Hubble classified nebulae and galaxies according to their shapes and brightness patterns. For example, he determined that most galaxies had a rotational symmetry and a central core. Spiral galaxies had the appearance of swirling around a dense core of stars.

During his galactic observations, he determined that these galaxies are receding from ours, and the more distant galaxies were receding from us faster than the nearby galaxies. Up to this point in the history of science, the universe was often considered a rather static object. Hubble established that the universe was expanding and that the ratio of the galactic speed to their separation distances is a constant. Although his calculation of the Hubble constant was incorrect and suggested an age for the universe that was far too young, subsequent astronomers obtained more accurate data and validated Hubble's theory.

According to John Huchra in "The Hubble Constant":

> Hubble is generally given credit for the discovery of the expansion, even though papers by Georges Lemaitre and H. P. Robertson using Hubble's data on the velocity-distance relation preceded his 1929 landmark, because it was his systematic program of measuring galaxy distances and his 1924 discovery of Cepheid variable stars in M31 [which could be used to compute distances] and his actual plot of the relation that finally convinced the community at large.

After Einstein learned of Hubble's discovery, Einstein gave up his work on the cosmological constant, which he had hypothesized in order to allow for a static solution to his equations. He later said that his work on the cosmological constant was one of his "greatest blunders." Einstein's General Theory of Relativity had initially suggested the universe was either expanding or contracting, but because astronomers had told Einstein that the universe was static, he had added a fudge factor to his equations—a cosmological constant that acted like an antigravity force to prevent the universe from contracting. Although Hubble was never confident that the model of the expanding universe was actually correct, his discovery eventually resulted in George Gamow (1904–1968) and Fred Hoyle's (1915–2001) Big Bang theory.

In 1942, Hubble attempted to enlist in the army, but the U.S. War Department appointed him to chief of ballistics and director of the Supersonic Wind Tunnel Laboratory in Maryland.

Near the end of his life, Hubble suffered from heart problems, and he died suddenly in 1953 from a cerebral thrombosis (blood clot in a blood vessel supplying the brain). His wife never revealed what was done with his body, and I believe that to this day, the location of his remains is still a mystery.

A lunar crater with a diameter of 80 kilometers was named after Hubble and approved in 1964 by the International Astronomical Union General Assembly. Sandage, writing on the centennial of the birth of Edwin Hubble, notes that Hubble's name has been attached to numerous astronomical theories, constants, and devices:

> There is Hubble's zone of avoidance, the Hubble galaxy type, the Hubble sequence, the Hubble luminosity law for reflection nebulae, the Hubble luminosity profile for E galaxies, the Hubble constant,

> the Hubble time, the Hubble diagram, the Hubble redshift-distance relation, the Hubble radius for the Universe, and now the Hubble Space Telescope. It seems appropriate in this centennial year to celebrate the memory of a scientist whom some have called the greatest astronomer (in changing paradigms) since the times of Galileo, Kepler and Newton.

Alexander S. Sharov and Igor D. Novikov in *Edwin Hubble* speak in glowing terms of Hubble's achievements that also stimulated other scientists to build upon his important observations:

> Hubble opened the world of galaxies for science when he proved that the nebulae outside the Milky Way are gigantic stellar systems [similar to] the galaxy which includes our Sun and its planets. However, the most important discovery was that of the red-shift in spectra of galaxies... the Universe was smaller in the past.... The explosive origin of the Universe determined its subsequent evolution [that] gave rise to the human race.... This is why astronomers rank Edwin Hubble with Copernicus and Galileo Galilei.

John Gribbin, author of *The Birth of Time*, emphasizes that Hubble was much more of an observer than a theoretician:

> Edwin Hubble never really subscribed to any theory about the Universe at all, in spite of the association made today between his name and the theory of the Big Bang. Hubble was an observer, and he reported the observations he made almost entirely without any trappings of theoretical interpretation, leaving that for others to do.... Hubble always exaggerated his own social status and achievements outside astronomy.

Bryson agrees with Gribbin's assessment:

> Hubble was a much better observer than a thinker and didn't immediately appreciate the full implications of what he had found.... At all events, Hubble failed to make theoretical hay when the chance was there. Instead, it was left to a Belgian priest-scholar (with a Ph.D. from MIT) named Georges Lemaître to [formulate] his own "fireworks theory," which suggested that the universe began as a geometrical point... which burst into glory and had been moving apart ever since.

Before leaving the entry on Hubble, I should point out that our concept and view of the universe has changed dramatically over the last four centuries. Back in the time of Galileo Galilei (1564–1642), many learned people still believed that Earth was the center of the universe. As this geocentric theory faded away like a dying flower, new theories evolved that suggested the Sun was the center of the universe, and when this theory died, some believed that the Milky Way galaxy was the center of everything. Today, we realize that our modern telescopes have only begun to reveal the immense numbers and variety of stars, and we know that the Sun is just an ordinary star in our galaxy, which contains roughly 200 billion stars. As I discuss in *The Stars of Heaven*, the galactocentric view of the universe, which placed our Milky Way at the center of the universe, turned out to be just another centrism that finally died in the 1920s, when Hubble proved that the Milky Way was not the only galaxy in the universe.

In our observable universe, we know there is roughly one galaxy for every star in the Milky Way. This fact certainly would have disturbed some of the scientists in the last four centuries and destroyed some of their notions about the heavens. Is it possible that the next four centuries will bring equally radical changes in our cosmological theories? Each of our centrisms died hard, and each demise was aided by new tools, new images, and new maps. One possibility is that our universe is not the only universe. I believe that new knowledge gleaned in the twenty-first century about other universes will destroy the centrism that our universe is all that there is.

FURTHER READING

Bryson, Bill, *A Short History of Nearly Everything* (New York: Random House, 2003).

Christianson, Gale E., *Edwin Hubble: Mariner of the Nebulae* (Chicago: University of Chicago Press, 1996).

Gribbin, John R., *The Birth of Time: How Astronomers Measured the Age of the Universe* (New Haven, Conn.: Yale University Press, 2001).

Hall, Stephen, *Mapping the Millennium* (New York: Random House, 1992), p. 21.

Huchra, John P., "The Hubble Constant," Harvard-Smithsonian Center for Astrophysics; see cfa-www.harvard.edu/~huchra/hubble/.

Kolb, Rocky, *Blind Watchers of the Sky* (New York: Basic Books, 2006).

Lemonick, Michael D., "The Time 100: Edwin Hubble," *Time*, March 29, 1999; see www.time.com/time/time100/scientist/profile/hubble.html.

Livio, Mario, "Hubble's Top 10," *Scientific American*, 295(1): 43–38, July 2006.

Pickover, Clifford, *The Stars of Heaven* (New York: Oxford University Press, 2001).

Sandage, Allan, "Edwin Hubble 1889–1953," *Journal of the Royal Astronomical Society of Canada*, 83(6), December 6, 1989; see antwrp.gsfc.nasa.gov/diamond_jubilee/1996/sandage_hubble.html.

Sharov, Alexander S., and Igor D. Novikov. *Edwin Hubble, the Discoverer of the Big Bang Universe* (New York: Cambridge University Press, 2005).

Whitrow, G. J., "Edwin Hubble," in *Dictionary of Scientific Biography*, Charles Gillispie, editor-in-chief (New York: Charles Scribner's Sons, 1970).

INTERLUDE: CONVERSATION STARTERS

The miracle of appropriateness of the language of mathematics for the formulation of the laws of physics is a wonderful gift which we neither understand nor deserve. We should be grateful for it, and hope that it will remain valid for future research, and that it will extend, for better or for worse, to our pleasure even though perhaps also to our bafflement, to wide branches of learning.

—Eugene Wigner, "The Unreasonable Effectiveness of Mathematics"

The tooth fairy is real, the laws of physics are real, the rules of baseball are real, and the rocks in the field are real. But they are real in different ways. What I mean when I say the laws of physics are real is that they are real in pretty much the same sense ... as the rocks in the field, and not in the same sense ... as the rules of baseball. We did not create the laws of physics or the rocks in the field.... I am making an implicit assumption ... that our statements about the laws of physics are in a one-to-one correspondence with aspects of objective reality.... If we ever discover intelligent creatures on some distant planet and translate their scientific works, we will find that we and they have discovered the same laws.

—Steven Weinberg, "Sokal's Hoax," *The New York Review of Books*, August 8, 1996

We see reality according to our thought. Therefore thought is constantly participating both in giving shape and form and figuration to ourselves, and to the whole of reality. Now, thought doesn't know this. Thought is thinking that it isn't doing anything. I think this is really where the difficulty is. We have got to see that thought is part of this reality and that we are not merely thinking about it, but we are thinking it. Do you see the difference?

—David Bohm, *On Creativity*

Astronomy provides a laboratory for extreme physics, a window into environments at extremes of distance, temperature and density that often can't be reproduced in Earth laboratories, or at least not right away. A surprising amount of the science we understand today started out solutions to problems in astronomy.... When it comes to discoveries in fundamental science, few of the discoverers have any inkling of the eventual consequences of their discoveries.

—William T. Bridgman, "The Cosmos in Your Pocket: How Cosmological Science Became Earth Technology. I."

THE GREAT CONTENDERS

At once philosophy and genial fantasy, practical physics and terrifying weapon, $E = mc^2$ has become metonymic of technical knowledge writ large. Our ambitions for science, our dreams of understanding and our nightmares of destruction find themselves packed into a few scribbles of the pen.

—Peter Galison, "The Sextant Equation" in Graham Farmelo's *It Must Be Beautiful*

God [could] vary the laws of Nature, and make worlds of several sorts in several parts of the universe.

—Isaac Newton, "Questions" from *Opticks*

This section of the book contains a large panoply of eponymous scientific laws. Some of these laws are slightly more obscure than the laws in the main section of the book, and I also include a few favorite biological laws in this section. Some of the "great contenders," although quite fundamental, do not have simple formulas for their expression or are not referred to as "laws" in many literature references.

As I researched the following laws, I often discovered dozens of interesting books that explained facets of these laws, and I often indicate these book resources within each short entry. Sometimes, I give definitions of the laws from several different authors' perspectives. In the interest of brevity, the "Great Contenders" section is much less formal and less detailed than the main entries of this book and thus may serve simply as a launch pad for

TABLE 12 Time Distribution of Great-Contender Laws

Time Period	Number
1600–1700	2 (4.25%)
1700–1800	2 (4.25%)
1800–1900	28 (59.5%)
1900–2000	15 (32%)

further study. Perhaps a future book will permit me to give the following contenders a fuller treatment.

As with the main entries, most of the following laws fall in the period between 1800 and 1900, and most are associated with scientists from Western Europe. Table 12 indicates the historical distribution of the laws in this section. Table 13 catalogues these laws by country of birth, or primary country affiliation, for the various lawgivers. Sometimes the determination of country affiliation is a judgment call when a lawgiver is born in one country but works in another. Thus, table 13 should be used only as a rough indicator. (If a lawgiver is associated with more than one law, he is counted only once.)

1600–1700

Mersenne's Law of Vibration: Physics, 1626

Marin Mersenne (1588–1648), French mathematician.

For a small-amplitude vibration, the fundamental frequency of vibration of a uniform string is proportional to the square root of the tension of the string, the reciprocal of the square root of mass per unit length of the string, and the reciprocal of the length of the string. Using these relationships, a piano maker can achieve a range of frequencies. Of course, the piano maker cannot rely solely on using strings of different length to change frequencies, because the longest piano string would have to be more than 150 times the length of the shortest to achieve the desired sound! Thus, piano strings also have different weights and tensions as required by Mersenne's Law. (See *Science and Music* by James Jeans.)

TABLE 13 Country Distribution of Great-Contender Laws

Country	Number
France	15
Britain	12
Germany	10
United States	7
Netherlands	3
Italy	2
Austria	2
Russia	2
Switzerland	1
India	1
Ireland	1
Denmark	1

Mersenne was of many famous mathematicians who were deeply religious. He was a French theologian, philosopher, number theorist, priest, and monk. He argued that God's majesty would not be diminished had He created just one world instead of many because the one world would be infinite in every part. Mersenne first publications were theological and argued against atheism and skepticism.

Torricelli's Law of Efflux: Physics, 1643

Evangelista Torricelli (1608–1647), Italian mathematician and physicist who was also the inventor of the barometer.

Torricelli's Law states that the speed of efflux of liquid from a hole is the same speed that a body would achieve in falling freely from the free surface of the liquid to the hole. If a hole is at a distance h below the surface of the liquid, the square of the speed of efflux v from the hole is $v^2 = 2gh$, where g is the gravitational acceleration constant. The law assumes that viscous effects can be ignored. The quantity of the efflux, or discharge, is given by $q = av$, where a is the cross-sectional area of the hole.

Note that $q = av$ is sometimes called Leonardo da Vinci's Law. In particular, around 1500, da Vinci noticed that water in a river moved faster where the river narrowed. He also found that the area of a cross section of a river multiplied by the velocity of the water flowing through that section is constant at any point in the river, thus conserving the mass of the fluid. This is sometimes known as the Law of Continuity, expressed as $av =$ constant. In particular, da Vinci wrote in 1502, "A river in each part of its length in an equal time gives passage to an equal quantity of water, whatever the width, the depth, the slope, the roughness, and the tortuosity (sinuosity)" (quoted in Robert Philip Benedict's *Fundamentals of Temperature, Pressure, and Flow Measurements*). Although he did not give the actual formula $q = av$, many have ascribed this equation to da Vinci. This continuity equation was actually formulated in 1628 by Italian monk and hydraulics expert Benedetto Castelli (1578–1643). In practice, the continuity equation allows researchers to follow changes in velocity and cross-sectional area at different locations along a river. Many sources suggest that this law was first discovered by Hero of Alexandria (c.10–70), centuries before da Vinci and Castelli. Hero was a Greek engineer and geometer in Alexandria in Hellenistic Egypt.

Various authors have considered the effect of the shape of the hole on the velocity of flow in Torricelli's Law. Holes that are square or in which the fluid approaches the hole with a high velocity cannot be treated by Torricelli's Law unless corrections are made for these special conditions (see *A History and Philosophy of Fluid Mechanics* by G. A. Tokaty).

Aside from his work in physics, Torricelli is also famous for the discovery of a mathematical object whose surface area is infinite, but whose volume is finite. This object, also called Torricelli's trumpet, is a horn-like object created by revolving $f(x) = 1/x$ for $x \in$ of $[1, \infty)$ about the x-axis. John dePillis in *777 Mathematical Conversation Starters* explains that, mathematically speaking, pouring red paint into Torricelli's trumpet could fill the trumpet funnel. In filling the trumpet, you could paint the entire inside, which is an infinite surface—even though you have a finite number of paint molecules. This seeming paradox can be partly resolved by remembering that Torricelli's trumpet is actually a mathematical construct, and our finite number of paint molecules that "fills" the horn is an approximation to the actual finite volume of the horn.

Torricelli was astounded by this object that seemed to be an infinitely long solid with an infinite-area surface and a finite volume. He thought it was a paradox and unfortunately did not have the tools of calculus to fully appreciate and understand the object. Today, Torricelli is remembered for the telescopic astronomy he did with Galileo.

Readers who become fascinated by Torricelli's trumpet may enjoy considering the following problem: For what values of a does $f(x) = 1/x^a$ produce a horn with finite volume and infinite area?

1700–1800

Maupertuis's Rule of Least Action: Physics, 1746

Pierre-Louis Moreau de Maupertuis (1698–1759), French mathematician.

Maupertuis said, "Nature is thrifty in all its actions," and this rule is sometimes called the Principle of Least Action. For example, Maupertuis deduced, from this principle, the laws of reflection and refraction of light "with the important principle by which nature, when realizing its actions, always goes along the simplest path." In 1746, he published his universal law of motion and equilibrium: "When a change occurs in nature, the quantity of action necessary for this change is least possible. The quantity of action is the product of the masses of the bodies by their speeds and by the distance over which they move." (See *Encyclopaedia of Mathematics*, edited by Michiel Hazewinkel.)

Maupertuis believed his principle applied to the universe when he wrote (as quoted in Morris Kline's *Mathematics and the Physical World*):

> The laws of movement and of rest deduced from this principle being precisely the same as those observed in nature, we can admire the application of it to all phenomena. The movement of animals, the vegetative growth of plants... are only its consequences; and the spectacle of the universe becomes so much the grander, so much more beautiful, the worthier of its Author, when one knows that a small number of laws, most wisely established, suffice for all movements.

Richter's Law of Chemical Reactions: Chemistry, 1791

Jeremias Richter (1762–1807), German chemist.

An equivalent weight of an acid will exactly neutralize an equivalent weight of a base. More generally, Richter discovered that the ratio by weight of the compounds consumed in a chemical reaction was always the same. For example, 615 parts by weight of magnesia (MgO) always neutralize 1,000 parts by weight of sulfuric acid.

Richter published two books that laid the foundations of stoichiometry (the quantitative relationship between reactants and products in a chemical reaction). He was an engineer in the department of mines in Silesia who later was appointed chemist to the porcelain works in Berlin. In his *Anfangsgründe der Stöchiometrie*, he set forth his belief that all of chemistry might be reduced to a mathematical system. (See *The Development of Modern Chemistry* by Aaron J. Ihde.)

1800–1900

Malus's Law of Polarization: Optics, 1809

Etienne Louis Malus (1775–1812), French physicist.

The intensity of transmitted light I produced when a polarizer is placed in front of an incident beam of plane-polarized light of intensity I_0 is given by $I = I_0\cos^2\theta$. The experiment is usually conducted using two polarizers. Unpolarized light is passed through a first polarizer. A second polarizer, called the analyzer, is inserted into the light path. Malus's Law describes the intensity of light transmitted through the analyzer when the analyzer is placed at an angle θ with respect to the first polarizer.

Today, we can only guess at the reasoning that Malus used to formulate his law. According to Jed Buchwald's *The Rise of the Wave Theory of Light: Optical Theory and Experiment in the Early Nineteenth Century*, Malus did not discover his law through calculations based on particles and forces. He did not seem to discover it simply by correlating observations at different refractive indices, nor did he have a way to measure intensity. Malus died too soon to present the full details of his theory.

The Bell-Magendie Law of Nerve Function: Neurophysiology, 1811

Sir Charles Bell (1774–1842), Scottish anatomist, and ***François Magendie (1783–1842),*** French physiologist.

Ventral roots of the spinal nerves (toward the belly of a creature) serve a motor function, and the dorsal roots (toward the back) serve a sensory function. In order to formulate his law, Bell performed experiments on live animals in which he cut certain nerves and noted their reactions.

Irwin A. Brody and Robert H. Wilkins, in *Neurological Classics*, state the law as follows: Sensory nerves enter the spinal cord by the posterior (dorsal) roots, and motor nerves leave it by the anterior (ventral) roots.

Edward S. Reed, in *From Soul to Mind: The Emergence of Psychology, from Erasmus Darwin to William James*, suggests that this law quickly led to the idea that the brain received and interpreted the spinal input signals and ordered whatever output signals were needed to return down the spine. Hence, the mind could still be located in the brain.

The experiments that led to the Bell-Magendie Law also led to the establishment in 1824 of the first Society for Prevention of Cruelty to Animals in England.

von Humboldt's Law of Tree Lines: Biology, 1817

Alexander von Humboldt (1769–1859), German naturalist and explorer.

The upper tree limit or "tree line" occurs at increasingly lower elevations as observations are made farther from the Equator. (The tree line, also called the "timberline," is the upper "edge" of the topography at which trees are capable of growing, due mostly to restrictions imposed by cold temperatures and wind.) The tree line finally reaches sea level above the Arctic Circle.

Between 1799 and 1804, von Humboldt explored South and Central America, and was the first naturalist to describe his observations in these regions from a scientific point of view. Over 21 years, he wrote much of his description in a huge set of volumes. As a child, he collected and labeled everything he could find, with an emphasis on shells, plants, and insects.

The Fresnel-Arago Laws of Optics: Optics, 1819

Augustin Fresnel (1788–1827) and Dominique Arago (1786–1853), French physicists.

This set of laws enumerates the conditions under which two rays of polarized light produce interference fringes. For example, two beams of light that are plane polarized in perpendicular planes do not produce interference fringes. Two beams polarized in the same plane interfere in the same manner as ordinary light.

Mitscherlich's Law of Isomorphism: Chemistry, 1821

Eilhard Mitscherlich (1794–1863), German chemist.

Substances that crystallize in isomorphous (identical) forms have similar chemical compositions. The law can be used to suggest the likely formulas of chemical compounds. For example, chromium oxide is isomorphous with Fe_2O_3, which suggests correctly that the formula for chromium oxide is Cr_2O_3.

Mitscherlich was a Renaissance man and had devoted himself as a young man to the study of ancient texts and languages, and he became an expert on the Persian language.

Hamilton's Principle of Dynamical Systems: Physics, 1835

Sir William Rowan Hamilton (1805–1865), Irish mathematician.

Jennifer Bothamley, in *Dictionary of Theories*, states this law as follows: The evolution of a dynamical system from time t_1 to a time t_2 is such that the action $S(t_1, t_2)$ is a minimum with respect to arbitrary small changes in the trajectory. Dare A. Wells in *Schaum's Outline of Lagrangian Dynamics* suggests that Hamilton's principle played an important role in the development of quantum mechanics.

Hamilton was a child prodigy. At the age of 7 he spoke Hebrew, and by 13 he had mastered many classical and modern European languages such as Farsi, Arabic, Hindustani, Sanskrit, and Malay. His oldest son, William Edwin Hamilton, noted that his father (as quoted in Peter Guthrie Tait's *Scientific Papers*) used to carry on long trains of algebraic and arithmetical calculations in his mind, during which he was unconscious of the earthly necessity of eating; we used to bring in a "snack" and leave it in his study, but a brief nod of recognition of the intrusion of the chop or cutlet was often the only result, and his thoughts went on soaring upwards.

Babinet's Principle of Diffraction: Physics, 1838

Jacques Babinet (1794–1872), French physicist.

According to Jennifer Bothamley's *Dictionary of Theories*, this principle involves the effects of complementary diffracting screens on electromagnetic radiation. Consider screens S_1 and S_2 that are complementary because S_2 is the screen obtained by making the opaque parts of screen S_1 transparent and the transparent parts opaque. Complementary diffraction screens produce identical intensity distributions.

Let me reiterate the meaning of this principle by presenting the perspective of several authors. According to Samuel Silver's *Microwave Antenna Theory and Design*, the optical Babinet's Principle states that the sum of the two complementary fields at any point is equal to the initial wave amplitude at the point in the absence of any screen.

Dipak K. Basu, in *Dictionary of Pure and Applied Physics*, writes:

> Suppose that we have a plane opaque screen in the x, y-plane; A_m is the area covered by the screen and A_0 is the aperture area. The complementary screen is defined to be that covering the area A_0 and having aperture area A_m. In both cases, let there be an initial field u_i arising from sources in the negative z-region of space, and let u_1 and u_2 be the diffraction field produced in the positive z-region by the respective screens. The optical Babinet's principle states that the sum of the two complementary fields at any point is equal to the initial wave amplitude at the point in the absence of any screen.

Babinet's Principle in the context of acoustic or electromagnetic waves asserts that the problem of diffraction by a plane screen S containing apertures L is related to a complementary diffraction problem in which L is a screen with apertures S (see Christopher M. Linton and Phillip McIver's *Handbook of Mathematical Techniques for Wave/Structure Interactions*).

As a final perspective on the law, consider two complementary screens, A and B. Screen A may just have a hole and screen B a stop the same size as the hole. If added together, no light may pass. Babinet's Principle tells us that complementary screens generate similar diffraction patterns. (See Karl Dieter Möller's *Optics: Learning by Computing, with Model Examples Using MathCad, MATLAB, Mathematica, and Maple.*)

Hess's Law of Constant Heat Summation: Chemistry, 1840

Chemist Germain Henri Hess (1802–1850), Swiss-born Russian.

Hess's Law is used to predict the change in heat content, or enthalpy (ΔH), in chemical reactions. The overall enthalpy change is the same whether it takes place in one or several steps. The enthalpy change of the overall chemical reaction is simply the sum of the enthalpy changes of each reaction step.

Randall K. Noon, in *Engineering Analysis of Fires and Explosions*, writes that the consequences of Hess's Law are important because they imply that heat-of-formation equations and heat-of-combustion equations can be manipulated together algebraically in order to accurately calculate the heats of formation or combustion of substance whose values have not been previously determined experimentally.

J. A. McLean and G. Tobin, in *Animal and Human Calorimetry*, note that in the field of calorimetry, Hess's Law states that the heat released by a chain of reactions is independent of the chemical pathways, and dependent only on the end products. In effect, this law and the law of conservation of energy together ensure that the heat evolved in the complex biochemical pathways that describe food digestion is exactly the same as the heat produced during measurements when the same food is converted into the same end products by simple combustion in a lab experiment.

Bergmann's Rule of Species Size: Biology, 1847

Christian Bergmann (1814–1865), German biologist.

Within a given species or genus, the average size of an animal tends to be smaller when the animal lives in warmer regions on Earth, and larger when in colder regions. Larger bodies have a smaller surface-to-volume ratio than do smaller bodies, which is useful for keeping an animal warm. The rule is especially applicable to bird species, but many mammals also follow this rule.

Jim Zumbo, in *Elk Hunting*, notes that elk that live in more northerly latitudes should be bigger in body size because of Bergmann's Rule.

Zumbo says that the larger elk size allows less heat loss per square inch of body surface.

Carrol L. Henderson, in *Field Guide to the Wildlife of Costa Rica*, notes that raccoons, cougars, and the white-tailed deer demonstrate Bergman's Rule in Costa Rica. Deer near the Equator have a smaller body than do deer of the same species in colder northern temperate climates.

The Gladstone-Dale Law of Refraction: Physics/Chemistry, 1858

John Hall Gladstone (1827–1902) and T. P. Dale, English chemists.

For a given temperature, the refractive index n of a gas is related to its density ρ by $(n - 1)/\rho = k$, where k is a constant.

Gladstone was very religious but believed that Christianity did not conflict with science. Referring to the Bible, he wrote in his "Points of Supposed Collision Between the Scriptures and Natural Science":

> The store houses of natural science have often been ransacked for weapons against the old book; the defenders of the faith have sometimes shrieked with alarm, and the assailants have sung their paean in anticipation of victory. Earthworks which form no part of the original fortress have been easily carried, but the citadel itself has remained unshaken and the very vigor of these repeated attacks has proved how impregnable are its valuable walls.

Kopp's Law of Heat Capacity: Chemistry, 1864

Hermann Franz Moritz Kopp (1817–1892), German chemist.

The molar heat capacity of a solid compound is approximately equal to the sum of the atomic heat capacities of its constituents. Related laws dealing with heat capacities include the Neumann-Kopp Law (1831) and Neumann's Law (1831), named after German chemist Franz Neumann (1798–1895).

Some have called Kopp the first great historian of chemistry. He published a comprehensive *History of Chemistry*, in four volumes, to which three supplements were added. He also wrote *The Development of Chemistry in Recent Times* and published a two-volume work titled *Alchemy in Ancient and Modern Times*.

Matthiessen's Rule of Electrical Resistivity: Physics, 1864

Augustus Matthiessen (1831–1870), English physicist.

The electrical resistivity of a metal that contains atomic impurities is almost always greater than that of the pure metal. According to Gerald D. Mahan's *Many-Particle Physics*, in metals, electrical resistivity arises from both electron scattering by impurities and electron scattering by phonons, the latter effect becoming large at high temperatures. Matthiessen's Rule says that these two contributions to the resistivity are additive; however, several deviations from the rule are known.

Maxwell's Law of Gas Viscosity: Physics, 1866

James Clerk Maxwell (1831–1879), Scottish physicist.

The coefficient of viscosity of a gas is independent of its density and of its pressure. The law assumes that the experiment is performed at constant temperature, and it is most accurate at pressures that are not extremely low or high.

According to Florian Cajori's *A History of Mathematics*:

> Maxwell predicted that so long as Boyle's Law is true, the coefficient of viscosity and the coefficient of thermal conductivity remain independent of the pressure. His deducing that the coefficient of viscosity should be proportional to the square root of the absolute temperature . . . induced him to alter the very foundation of his kinetic theory of gases by assuming between the molecules a repelling force varying inversely as the fifth power of their distances.

The Berthelot-Thomsen Principle of Chemical Reactions: Chemistry, 1867

Marcellin Pierre Eugene Berthelot (1827–1907), French physical chemist, and ***Hans Peter Jurgen Julius Thomsen (1826–1909),*** Danish chemist.

Consider all possible chemical reactions that may take place for a given set of reactants. The reaction that leads to the greatest release of energy will occur. Carl W. Hall's *Laws and Models* describes the law as follows:

Of the various possible low-temperature nonendothermic reactions that can proceed without the aid of external energy, the process takes place accomplished by the greatest evolution of heat.

Hans-Dieter Jakubtke and Hans Jeschkeit suggest in *Concise Encyclopedia of Chemistry* that the affinity of a chemical reaction (i.e., the tendency of a reaction to occur) is proportional to the heat released by the reaction.

Mendeleyev's Periodic Law of Elements: Chemistry, 1869

Dimitri Mendeleyev (1834–1907), Russian chemist.

The chemical and physical properties of the elements are periodic functions of the atomic weights. Mendeleyev was so confident that his periodic law was correct that he left empty cells in his periodic table of the elements for undiscovered elements that he thought should exist in these table locations. Five years later, gallium was discovered, and its atomic weight and chemical properties matched Mendeleyev's predictions almost exactly. Scientists were not sure as to the precise reasons for the period pattern in properties of the elements until fifty years after Mendeleyev's discovery, when quantum mechanics could be used to explain why elements form groups with similar properties. (See Paul Fleisher's *Matter and Energy: Principles of Matter and Thermodynamics.*)

The Lorentz-Lorenz Law of Refractive Indices: Physics, 1870

Hendrik Lorentz (1853–1928), Dutch physicist, and ***Ludwig Lorenz (1853–1928),*** Danish physicist.

At a constant temperature, the refractive index n of all states of a dielectric is related to the density of the medium ρ as $(n^2 - 1)/(n^2 + 2) = k \times \rho$, where k is a constant. (A dielectric material is a substance, such as glass, that is a poor conductor of electricity but an efficient supporter of electrostatic fields.) This formula was published by Ludwig Lorenz in 1869 and by Hendrick Lorentz (who discovered it independently) in 1870.

Lorentz won the Nobel Prize for Physics in 1902 for his theory of electromagnetic radiation. He also suggested that moving bodies that approach the velocity of light contract in the direction of motion, which helped give rise to Albert Einstein's Special Theory of Relativity.

Coppet's Law of Freezing Point Lowering: Chemistry, 1871

Louis Cas de Coppet (1841–1911), French physicist.

The degree of lowering of the freezing point of a solution is proportional to the amount of solute dissolved in the solution. The English chemists Charles Blagden (1748–1820) and Richard Watson (1737–1816) discovered similar laws. In 1788, Blagden, the secretary of British scientist Henry Cavendish (1731–1810), had shown that the freezing-point depression was proportional to the amount of substance (solute) in a given volume of solvent, and François Marie Rauolt (1830–1901) was made aware of Blagden's Law from an investigation made by Coppet in 1871 in which "atomic depressions" were calculated. This law helps to explain the use of salt to melt ice on roads (see William H. Brock's *The Chemical Tree: A History of Chemistry*).

Boltzmann's Distribution Law: Physics, 1871

Ludwig Boltzmann (1844–1906), Austrian physicist.

This law describes the statistical distribution of velocities and energies of gas molecules at thermal equilibrium. (See Jennifer Bothamley's *Dictionary of Theories.*) The law states that the natural logarithm of the ratio of the number of particles in two different energy states is proportional to the negative of their energy separation. Scottish physicist James Clerk Maxwell first suggested this distribution in 1859, using probabilistic arguments. Boltzmann generalized Maxwell's findings in 1871. This distribution law is sometimes referred to as the Maxwell-Boltzmann Distribution Law.

Dipak K. Basu's *Dictionary of Material Science and High Energy Physics* describes the Boltzmann Distribution Law as a law of statistical mechanics that states that the probability of finding a system at temperature T with an energy E is proportional to $e^{-E/KT}$, where K is Boltzmann's constant.

According to Stephen G. Brush's *Kinetic Theory of Gases: An Anthology of Classic Papers with Historical Commentary*, Boltzmann complained that Maxwell's derivation of the velocity distribution law was difficult to understand because of its brevity. Thus, Boltzmann devoted the first part of a 44-page memoir to explaining with concrete examples the steps that Maxwell had skipped.

Abney's Law of Luminosity: Physics, 1877

Sir William de Wiveleslie Abney (1844–1920), English chemist.

Let me express this law from several authors' perspectives:

- The luminous power of a source is the sum of the powers of the components of any spectral decomposition of the light (Richard A. Matzner's *Dictionary of Geophysics, Astrophysics, and Astronomy*).
- The luminances of differently colored lights add linearly. The luminosity of a compound stimulus is equal to the sums of the component luminances (Ian P. Howard and Aan P. Howard's *Binocular Vision and Stereopsis*).
- Light arriving at a surface is the sum of the light arriving from all sources to which the surface is exposed (Marc Schiler's *Simplified Design of Building Lighting*).

Allen's Rule of Body Form: Biology, 1877

Joel Asaph Allen (1838–1921), American zoologist.

According to Robert B. Eckhardt's *Human Paleobiology*, this law expresses a correlation of body form with temperature, stating that in warm-blooded species, the relative sizes of anatomical parts projecting from the body (e.g., limbs, tails, and ears) decrease with declining annual temperatures. This tendency correlates with the need to conserve heat in cold locations and radiate heat in warm locations.

Roger Lewin's *Human Evolution: An Illustrated Introduction* expresses the law this way: Populations of a geographically widespread species living in warm regions will have longer extremities (arms and legs) than those living in cold regions. The rule is reflected by the observation that people from tropical areas tend to have longer, thinner limbs, which maximize heat loss, while people from higher latitudes tend to have shorter limbs.

According to Paul B. Weisz's *The Contemporary Scene: Readings on Human Nature, Race, Behavior, Society, and Environment*, Allen's Rule explains the lean body build of desert folk.

Nernst's Law of Electrode Potentials: Chemistry, 1880

Walther Nernst (1864–1941), German chemist.

According to Jennifer Bothamley's *Dictionary of Theories*, this law specifies the concentration dependence of a reversible equilibrium potential of a working electrode. In particular, the law states that the zero-current electrode potential E of a reversible electrode immersed in a solution of an ion of valence z is given by $E = E^0 + [RT/zF]\cdot\ln a$, where E^0 is the standard electrode potential, R is the gas constant, T is the temperature, F is the Faraday constant, and a is the activity of the ion.

Andrew W. Batchelor, Loh Nee Lam, and Margam Chandrasekaran, in *Materials: Degradation and Its Control by Surface Engineering*, succinctly write that the Nernst Law describes the effect of electrolyte concentrations and temperature on the electrochemical potential.

For his work in thermochemistry, Nernst received the Nobel Prize in Chemistry in 1920. In addition to being a theoretician, Nernst was an inventor. His improved electric light, the Nernst lamp, made use of a ceramic body. His electrical piano replaced the sounding board with radio amplifiers. Nernst's two sons were both killed in the First World War.

Raoult's Law of Vapor Pressures: Chemistry, 1882

François Raoult (1830–1901), French chemist.

The vapor pressure of each component in an ideal solution depends on the vapor pressure of the individual components and the mole fraction of each component in the solution. In particular, $P_t = P_1 X_1 + P_2 X_2 + \ldots$, where P_t is the total vapor pressure of the solution; P_i are the vapor pressure of the pure components, and X_i are the mole fractions for each component.

According to *Safety and Health in Confined Spaces* by Neil McManus, liquids occurring in mixtures may be completely miscible, partly miscible, or completely immiscible in each other. In an ideal solution containing completely miscible liquids, interactions between molecules of solute and solvent are the same as solute–solute and solvent–solvent interactions. Raoult's Law provides a basis for predicting vapor pressures of components of ideal solutions in equilibrium states.

Raoult's Law is the basis of the techniques of distillation, used to separate substances with different boiling points or volatilities (see *SAT Subject Tests: Chemistry 2005–2006* by Kaplan, Inc.).

van't Hoff's Law of Osmotic Pressure: Chemistry, 1885

Jacobus van't Hoff (1852–1911), Dutch chemist.

The osmotic pressure of a solution (explained below) depends on the concentration of osmotically active particles according to the expression $\pi V = nRT$ or $\pi = C \times RT$, where π is the osmotic pressure, V is the volume of solution, n is the number of moles of solute in solution, C is the concentration of solute, R is the gas constant, and T is the absolute temperature. When the solute concentration increases, the osmotic pressure also increases. In other words, the osmotic pressure of a solution is directly proportional to the concentration of solute in dilute solution. The higher the osmotic pressure of a solution, the greater the water flow into the solution from the surroundings.

Osmotic pressure can be understood by considering two solutions that are separated by a semipermeable membrane. If the two solutions have different effective osmotic pressures (e.g., because a solute concentration is higher in one than the other), water will flow from the solution with lower osmotic pressure into the solution with the higher osmotic pressure.

Due to osmotic pressure, when a cell is placed in pure water, water may rush into the cell and cause it to burst. π should not be thought of as the pressure of the solute but rather the pressure that must be applied to the solution to keep solvent from flowing in through a semipermeable membrane. (See *BRS Physiology* by Linda S. Costanzo.)

The Ramsay-Young Law of Vapor Pressures: Chemistry, 1885

Sir William Ramsay (1852–1916), Scottish chemist, and ***Sidney Young (1857–1937),*** Ramsay's assistant at Bristol University in England.

According to a relationship proposed by Ramsay and Young, the ratio of the absolute boiling points of two substances A and B at various vapor pressures varies linearly with temperatures. F. C. Kracek expresses this as follows in his 1930 *Journal of Physical Chemistry* article "Vapor Pressures of Solutions and the Ramsay-Young Rule": $(T_A/T_B)_p = (T_A/T_B)_{p0} + c(T - T_0)_B$, where p corresponds to the vapor pressures.

Jennifer Bothamley, in *Dictionary of Theories*, states the law more succinctly: If two chemically similar compounds have the same vapor pressure at different absolute temperatures, the ratio of those temperatures is independent of this vapor pressure.

Ramsay is famous for his discovery of the noble gases, and he named the gas "argon" and discovered neon, krypton, and xenon. He received the Nobel Prize in Chemistry in 1904.

The Cailletet-Mathias Law of Density: Chemistry, 1886

Louis Cailletet (1832–1933), French physicist, and ***Emile Mathias (1861–1942),*** French chemist.

The mean density of a liquid and its saturated vapor is a linear function of temperature. Liquid metals are known to deviate from this law at certain temperatures. Carl W. Hall, in *Laws and Models: Science, Engineering, and Technology*, states the law in the following manner. A linear function exists between the arithmetical average of the densities of a pure unassociated liquid and its saturated vapor and the temperature of the liquid: $(d_l + d_v)/2 = A + Bt$. Here d_l and d_v are the densities of liquid and vapor, respectively. A and B are constants that depend on the liquid under study, and the value of B is negative. t is the temperature in °C. In 1900, the law was modified by British chemist Sidney Young (1857–1937). The law is also referred to as the Cailletet-Mathias Rectilinear Diameter Law.

Dollo's Law of Evolution: Biology, 1890

Louis Dollo (1857–1931), French-born Belgian paleontologist.

Evolution is not reversible. In other words, structures and functions that are lost through time can never be reacquired exactly in the same way. A tenet of evolutionary theory, Dollo's Law states that organisms do not re-evolve along lost pathways but can find alternate routes through a chain of random mutational events. For example, whales will never again walk on land with re-evolved pelvic appendages that derive from the current remnant structures that correspond in us to legs. (See *The Long and the Short of It: More Essays on the Fiction of Gene Wolfe* by Robert Borski.)

According to *The Blind Watchmaker: Why the Evidence of Evolution Reveals a Universe Without Design* by Richard Dawkins, Dollo's Law is just a statement about the statistical improbability of following exactly the same evolutionary trajectory twice (or, indeed, any *particular* trajectory), in either direction. A single mutational step can easily be reversed, but the chances of reversal of large numbers of mutational steps are vanishingly small.

According to *Laws and Models: Science, Engineering, and Technology* by Carl W. Hall, evolution is reversible in that features gained can be lost, but irreversible in that features lost cannot be regained. Several researchers have hypothesized that some evolutionary changes may result from genes being switched off. If these silent genes were switched back on, lost traits might reappear. Possible evolutionary throwbacks include the occasional appearance of extra nipples or breasts in humans, as well as the appearance of web fingers and toes. (See Michael Le Page, "The Ancestor Within All Creatures," *New Scientist*, January 13, 2007.

Sutherland's Law of Gas Viscosity: Chemistry/Physics, 1893

William Sutherland (1859–1911), Scottish theoretical physicist.

This law specifies a relationship between the viscosity (η_T) of a gas at temperature T and the viscosity (η_0) at a reference temperature T_0:

$$\frac{\eta_T}{\eta_0} = \frac{T_0 + S}{T + S}\left(\frac{T}{T_0}\right)^{3/2},$$

where S is the Sutherland constant. The temperature is in degrees Kelvin.

Sutherland was born in Glasgow, Scotland. When he was a young child, his family immigrated to Australia. Another example of simultaneous occurrences in science, Sutherland also derived a relation linking the diffusion coefficient to the viscosity of a solvent and the diameter of the diffusing molecule. Soon afterward, in 1905, Einstein published the same equation in his paper on Brownian motion, having arrived at it by exactly the same line of reasoning. Sutherland reported his relation in 1904 at a conference in New Zealand and published it 1905. Today, what might have been called the "Sutherland-Einstein Diffusion Equation" is usually referred to simply as the "Einstein Diffusion Equation," and Sutherland's work is not widely recognized.

Lorentz Force Law: Physics, 1895

Hendrik Lorentz (1853–1928), Dutch physicist.

A particle of charge q and velocity $\mathbf{v}$ residing in a magnetic field of induction $\mathbf{B}$ and an electric field of strength E experiences a force $\mathbf{F}$ given by

$\mathbf{F} = q(E + \mathbf{v} \times \mathbf{B})$. Here, the $\times$ symbol is the cross product. In other words, when both electric and magnetic fields are present, the total electromagnetic force on the charge is **F**. This law has been verified experimentally numerous times, and it is found to be true even for particles moving at speeds close to the speed of light. Thus, electric charge is relativistically invariant, which means that it does not change with velocity. This equation, along with Maxwell's Equations, unifies electrodynamics. (See *Electrodynamics of Solids and Microwave Superconductivity* by Shu-Ang Zhou.)

1900 AND BEYOND

Grüneisen's Law of Thermal Expansion: Physics, 1908

Eduard A. Grüneisen (1877–1949), German physicist.

Jennifer Bothamley in *Dictionary of Theories* expresses the law as follows: The ratio of the expansivity of a metal to its specific heat capacity at constant pressure is a constant over a wide range of temperatures.

According to *Polymer Properties at Room and Cryogenic Temperatures* by Gunther Hartwig, Grüneisen's Law states that the coefficient of thermal expansion α is proportional to the specific heat c. The Grüneisen relation is given by $\alpha = (1/3)(\rho/K)\gamma c$, where ρ is the density, K the bulk modulus, and γ is the Grüneisen parameter.

Sabine's Law of Acoustics: Physics, 1910

Wallace Sabine (1868–1919), British scientist.

The reverberation time of a hall is proportional to the volume of the hall divided by the total absorption of the hall, where reverberation time is usually defined as the time required for the sound intensity to fall to 10^{-6} of its initial value. Another definition of reverberation time is the number of seconds required for the intensity of the sound to decay by 60 decibels.

The Sabine experiments started in 1885, when Harvard University discovered that its new Fogg Art Museum had severe acoustical problems. The head of the university asked Sabine, a young physics professor, to "do something" about the difficulties. Sabine began to experiment with various local lecture rooms and used seat cushions as his portable means for experimenting with sound absorption. He used organ pipes to produce sounds, a stopwatch to time the reverberation, and his own keen hearing to make the measurements before he finally arrived at Sabine's Law. (See *Sound System Engineering* by Carolyn Davis and Don Davis.)

Child's Law of Diode Current: Physics, 1911

Clement Dexter Child (1868–1933), American physicist.

The space charge-limited current (SCLC) in a diode varies as the three-halves power of the anode voltage and inversely as the square of the distance separating the cathode and the anode. (Readers interested in learning more about such currents may consult *Dielectrics in Electric Fields* by Gorur G. Raju.)

The Geiger-Nuttall Rule of Particle Energy: Physics, 1911

Hans Wilhelm Geiger (1882–1945), German physicist, and ***John Nuttall (1890–1958),*** British physicist.

The energy of alpha particles emitted from different nuclides in a given radioactive series depends on the half-life of the nuclide according to $\ln(\lambda + c) = \ln R$, where λ is the decay constant for a nuclide, c is a constant for a particular radioactive series, and R is the range of the alpha particles in a medium. (The term "nuclide" usually refers to an atomic nucleus, which is an agglomeration of protons and neutrons. The various isotopes of the elements form the set of nuclides.)

According to *Holleman-Wiberg's Inorganic Chemistry* edited by Nils Wiberg, A. F. Holleman, and Egon Wiber, for short-lived elements, the decay constant (or the half-life) can be calculated by the Geiger-Nuttall Rule, which they write as $\ln \lambda = -37.7 + 53.9 \times \log R$ for the uranium decay series. It follows that the decay constant λ of polonium $^{214}_{34}\mathrm{Po}$, for which $R = 6.6$ cm, is about 10^4.

According to the *Dictionary of Material Science and High Energy Physics* edited by Dipak K. Basu, the higher the released energy of alpha decay, the shorter the half-life. Although variations occur, smooth curves can be drawn for nuclei having the same atomic number Z. The explanation for this rule was an early achievement of quantum mechanics and nuclear structure.

The *Wikipedia* online encyclopedia gives this modern formulation for the law: $\ln \lambda = -a_1(z/\sqrt{E}) + a_2$, where Z is the atomic number, E is the total kinetic energy (of the alpha particle and the daughter nucleus), and a_1 and a_2 are constants.

The Einstein-Stark Law of Photon Absorption: Chemistry, 1912

Albert Einstein (1879–1955), German-born American physicist, and ***Johannes Stark (1874–1957),*** Bavarian-born physicist.

This law of photochemistry states that an atom of a molecule undergoing a photochemical process absorbs only a single photon (*Chemical Oceanography* by Mary L Sohn). Another statement of the law: The absorption of one quantum of radiation results in the formation of one photo-excited state (*The Organic Chemistry Problem Solver* by James R. Ogden). A final perspective on the law: The absorption of one photon can cause a change only in one molecule in a strictly photochemical reaction, but in the multistage process of photosynthesis, a number of photons (8–20) of light are needed for each molecule of carbon dioxide reduced (*The Silvicultural Basis for Agroforestry Systems* edited by Mark S. Ashton and Florencia Montagnini).

Interestingly, in 1919, Stark won the Nobel Prize in Physics for his "discovery of the Doppler effect in canal rays and the splitting of spectral lines in electric fields," and he published more than 300 papers during his life. He despised the "Jewish physics" of Albert Einstein and Werner Heisenberg. After Heisenberg defended Einstein's theory of relativity, Stark angrily wrote in the SS newspaper *Das Schwarze Korps* that Heisenberg was a "White Jew." Stark attacked theoretical physics as "Jewish" and wanted to ensure that only non-Jews held scientific positions in Germany. After World War II, a "denazification" court imprisoned Stark for four years.

Leavitt's Luminosity Law: Astronomy, 1912

Henrietta Leavitt (1868–1921), American astronomer.

A Cepheid variable is a star whose brightness varies in a periodic manner. Leavitt's Law states that the periods of Cepheid variable stars (i.e., the time for one cycle of dimness and brightness) are proportional to the luminosity of the stars—the greater the brightness, the greater the period. The luminosity can be used to estimate interstellar and intergalactic distances.

In 1902, Leavitt became a permanent staff member of the Harvard College Observatory, where she studied photographic plates of variable stars in the Magellanic Clouds. In 1904, using a time-consuming process called superposition, she discovered hundreds of variable stars in the Magellanic Clouds. These discoveries led Charles Young of Princeton to write to Harvard College Observatory director E. C. Pickering, "What a variable-star 'fiend' Miss Leavitt is; one can't keep up with the roll of the new discoveries." Sadly, she died young of cancer before her work was complete. Had she lived longer, many speculate that she would have been a strong candidate for the Nobel Prize. (See *The Stars of Heaven* by Clifford Pickover.)

Friedel's Law of X-ray Reflection: Physics, 1913

Georges Friedel (1865–1933), French physicist.

The intensities of reflection for X-rays from opposite sides of a crystal plane are the same. Stated in the language of crystallography, Friedel observed that the intensity distribution in diffraction patterns is centrosymmetric such that $I_{hkl} = I_{-h-k-l}$.

Moseley's Law of X-ray Emission: Physics, 1913

Henry Moseley (1887–1915), English physicist.

This law describes the intensities of regions in X-ray emission spectra of elements. In particular, the most intense short-wavelength line in the X-ray spectrum is related to the element's atomic number Z as $\sqrt{f} = k_1 \cdot (Z + k_2)$. Here, f is the frequency of the main X-ray emission line, and k_1 and k_2 are constants whose values depend on the type of emission line under study. To arrive at his law, Moseley experimented with various metals as X-ray targets and measured the wavelengths of the most intense lines. The lines are useful for confirming the identity of an element and, in fact, can demonstrate that a substance under study is an element. The lines also helped scientists place elements in the proper locations in the periodic table.

Shortly after this famous discovery, Moseley was killed in action in World War I by a sniper at Gallipoli. Because the law was extremely useful in the history of chemistry for assigning atomic numbers to newly discovered elements, many scientists have speculated that Moseley might have won the Nobel Prize if he had not been killed—the Nobel Prize is only awarded to the living. (See *Undergraduate Instrumental Analysis* by James W. Robinson, Eileen M. Skelly Frame, and George M. Frame II.)

Steinmetz's Law of Magnetism: Physics, 1916

Charles Steinmetz (1865–1923), American electrical engineer.

When an external magnetic field is applied to certain materials (e.g., ferromagnets), the material absorbs some of the external field. When the external field is removed, the magnet will retain some field, as part of a phenomenon known as hysteresis. According to this law, we can calculate the work W needed to take a ferromagnetic material around its hysteresis loop: $W = \eta B_m^{1.68}$, where B_m is the maximum value of the induction in the cycle, and η is the Steinmetz coefficient that depends on the material under study.

In order to understand how this law is used practically, James R. Ogden in *Electrical Machines Problem Solver* asks students to use Steinmetz's Law to determine "the ergs loss per cycle in a core of sheet iron having a net volume of 40 cubic centimeters in which the maximum flux density is 8,000 gauss. The value of η for sheet iron is 0.004." To solve, we use the Steinmetz Law and find that W is approximately equal to $0.004 \times (8000)^{1.6} = 7{,}028$ ergs/cm^3 per cycle. The total loss is $7028 \times 40 = 281{,}000$ ergs/cycle or $281{,}000 \times 10^{-7} = 0.0281$ joules/cycle.

Steinmetz, the holder of more than 200 patents, was only four feet tall. Shortly after receiving his Ph.D. in 1888, he was forced to flee Germany after writing a paper criticizing the German government. As we recounted in Bob Fenster's *They Did What!?: The Funny, Weird, Wonderful, Outrageous, and Stupid Things Famous People Had Done*, while working for General Electric in America, Steinmetz once submitted a bill for $10,000 dollars. When his managers asked him for an itemized invoice, he simply wrote:

1. Making chalk mark: $1
2. Knowing where to place it: $9,999

The Bose-Einstein Distribution Law: Physics, 1924

Satyendra Nath Bose (1894–1974), Indian physicist, and ***Albert Einstein (1879–1955),*** German-born American physicist.

The Bose-Einstein Distribution Law gives the average number of identical bosons in a state of energy E. (A boson is a particle, such as the photon, pion, or alpha particle, which has zero or integral spin.) An unlimited number of bosons can be placed in the same state E. According to *Elementary*

Modern Physics by Richard T. Weidner and Robert L. Sells, the Bose-Einstein Distribution Law applies to a system of *identical* particles that are indistinguishable, each having integral spin. A physical system illustrating the Bose-Einstein Distribution Law is that of a blackbody and its radiation.

The Franck-Condon Principle of Electronic Redistribution: Chemistry and Physics, 1925

James Franck (1882–1964), German-born American physicist, and ***Edward Condon (1902–1974),*** American theoretical physicist.

In a molecular system, the electronic redistribution from one energy state to another is sufficiently rapid that the nuclei of atoms involved can be considered to be stationary during the redistribution. (See Jennifer Bothamley's *Dictionary of Theories.*)

This principle describes the transition between two electronic states of a molecule, for example, the photodissociation of a molecule in the visible spectrum. The principle states that the relative positions and momenta of the atoms are preserved during the electronic distribution. (See *Airy Functions and Applications to Physics* by Olivier Vallee and Manuel Soares).

The principle arises from the fact that movement of nuclei is negligible during the time taken by an electronic transition. Thus, the time required for electronic transition is so short that the atoms in a molecule do not have time to change their positions appreciably. (See *Atomic and Molecular Spectroscopy* by M. C. Gupta.)

Pauli's Exclusion Principle: Chemistry and Physics, 1925

Wolfgang Pauli (1900–1958), Austrian-born American theoretical physicist.

No pair of identical particles can simultaneously occupy the same quantum state. For example, electrons occupying the same atomic orbital must have opposite spins. This principle plays a fundamental role in quantum theory and applies to fermions (e.g., electrons, protons, and neutrons) but not to bosons (e.g., photons). (See Jennifer Bothamley's *Dictionary of Theories.*)

According to *Pauli's Exclusion Principle: The Origin and Validation of a Scientific Principle* by Michela Massimi, the Pauli Exclusion Principle (PEP) is a well-tested and commonly accepted results in physics. "From spectroscopy to atomic physics, from quantum field theory to high-energy physics, there is hardly another scientific principle that has more far-reaching implications than PEP." As a result of PEP, one obtains electronic configurations underlying the classification of chemical elements in the periodic table as well as atomic spectra.

According to *The Quantum Quark* by Andrew Watson

> Pauli introduced this principle early in 1925, before the advent of modern quantum theory or the introduction of the idea of electron spin. His motivation was simple: there had to be something to prevent all the electrons in an atom collapsing down to a single lowest state.... So, Pauli's exclusion principle keeps electrons—and other fermions—from invading each other's space.

PEP is certainly one of the most important principles in physics, as it explains why material particles exhibit space-occupying behavior. Thus, PEP might have been placed in the main-entry part of this book. On the other hand, PEP tends to be different from most of the entries in this book because Pauli did not develop the mathematics to go with his empirical principle. Although Pauli formulated his principle to account for certain experimental results, he did so before the flowering of the modern theory of quantum mechanics that was stimulated by the work of such physicists as Werner Heisenberg and Erwin Schrödinger.

The Fermi-Dirac Distribution Law: Physics, 1926

Enrico Fermi (1901–1954), Italian-born American nuclear physicist, and ***Paul Dirac (1902–1984),*** English mathematical physicist.

This law describes the average number of identical fermions (e.g. electrons) in a state of energy E as a function of the Boltzmann constant, the temperature, and α, which is a parameter that depends on temperature and the concentration of fermions.

Fermi in the 1920s was the first to develop a mathematical treatment of how particles such as electrons interact physically. The topic was investigated independently by Paul Dirac. Electrons in metals obey Fermi-Dirac statistics. (See *The History of Science and Technology* by Bryan Bunch.)

The Boltzmann distribution law may be used for the study of systems of interest to chemists at room temperature. In cases where Boltzmann distribution law fails, the Fermi-Dirac or Bose-Einstein distribution law can be applied, depending upon whether the system is composed of bosons or fermions. (See *Statistical Thermodynamics* by M. C. Gupta.)

The Moskowitz-Lombardi Rule of Magnetic Distribution: Physics, 1973

Paul A. Moskowitz (born 1945), American physicist, and ***Maurice Lombardi (born 1942),*** French physicist.

As background, a charged atomic nucleus with nonzero spin produces a magnetic field whose strength may be expressed by the size of its magnetic moment μ. The magnetization is distributed over the volume of the nucleus. The distribution of nuclear magnetization characterizes its deviation from the distribution that would occur for an ideal point nucleus, and is expressed by the parameter ε.

Moskowitz and Lombardi observed that for a series of ten mercury isotopes, a simple relation exists between the magnetic distribution ε and the magnetic moment μ: $\varepsilon = \alpha/\mu$, where α is a constant. The rule has been applied to isotopes of such elements as mercury, iridium, gold, thallium, platinum, tungsten, osmium, and barium. This relationship gives nuclear physicists insight into the complex structure of the atomic nucleus, which may contain more than 200 protons and neutrons.

Moskowitz, a prolific inventor, has since pursued a career in wireless technology at IBM Research. He won more than $50,000 when he appeared on the popular TV game show *Wheel of Fortune*.

Hawking's Black-Hole Laws: Physics, 1970s

Stephen Hawking (born 1942), British astrophysicist.

Many principles that concern black holes have been attributed to Stephen Hawking. Hawking is not given a main entry in this book primarily because his various equations, ideas, and conjectures are often not referred to in an eponymous fashion (as, e.g., are Ohm's Law of Electricity and Newton's Laws of Motion), and the scientific literature often refers to these as equations or theorems rather than laws. Nevertheless, sources do occasionally consider many of the following principles as "Hawking's Laws." Consider, for example, that the rate of evaporation of a Schwarzschild black hole of mass M can be formulated as $\mathrm{d}M/\mathrm{d}t = -C/M^2$, where C is a constant, and the particles are emitted at time t distributed in a thermal spectrum with temperature $1/8\pi M(t)$. (See Stephen Hawking, "Particle Creation by

Black Holes," *Communications in Mathematical Physics* 43(3): 199–220, 1975.)

An observer outside a black hole would observe the hole to be at a finite temperature. Expressed in Planck units, another law of Hawking states the temperature T of a black hole is inversely proportional to its mass: $T = k/m$. Lee Smolin writes in *Three Roads to Quantum Gravity*:

> The constant k is very small in normal units. As a result, astrophysical black holes have temperatures of a very small fraction of a degree. They are therefore much colder than the 2.7 degree Kelvin microwave background. But a black hole of much smaller mass would be correspondingly hotter, even if it were smaller in size. A black hole the mass of Mount Everest would be no larger than a single atomic nucleus, but would glow with a temperature greater than the center of a star.

To understand Hawking's area "law" that follows, the reader should understand the concept of event horizon. Just a few weeks after Albert Einstein published his general relativity theory in 1915, German astronomer Karl Schwarzschild performed exact calculations of what is now called the Schwarzschild radius. This radius defines a sphere or event horizon that surrounds a body of a particular mass. According to classical black hole theory, within the sphere, gravity is so strong that light, matter, or any kind of signal cannot escape. In other words, anything that approaches closer than the Schwarzschild radius will become invisible and lost forever. For a mass equal to the mass of our Sun, the radius is a few kilometers in length. For a mass equal to Earth's, the Schwarzschild radius defines a region of space the size of a walnut. In other words, a black hole with an event horizon the size of a walnut would have a mass equal to the mass of the Earth.

Hawking's 1971 area theorem, sometimes called his Area Law, states that the area of a future event horizon of a black hole can never decrease. For example, Hawking showed that if two black holes unite, the surface area of the final black hole must exceed the sum of the surface areas of the initial black holes. For these reasons, the total black hole portion of the universe is likely to be increasing. Although this law is expected to break down when quantum effects are taken into account, such as when hypothetical black hole evaporation is considered, a suitably generalized form of this law is useful to theoretical astrophysicists who ponder the behavior of black holes.

Physicists Brandon Carter, Stephen Hawking, and James Bardeen have formulated several principles of black hole mechanics, such as the area theorem just discussed, that have analogues in the laws of thermodynamics.

The other laws of black hole mechanics concern such concepts as the "surface gravity" of the event horizon. (See *Current Trends in Relativistic Astrophysics: Theoretical, Numerical, Observational* by Leonardo Fernandez-Jambrina and Luis Gonzalez-Romero. See also J. M. Bardeen, B. Carter, and S. W. Hawking, "The Four Laws of Black Hole Mechanics," *Communications in Mathematical Physics*, 31: 161–170, 1973.)

We may smile at another Hawking "law" (derived from editorial advice given during the writing of *A Brief History of Time*) may be stated as follows: Because every equation halves potential readership of a book, equations within a book should be eliminated or strictly rationed. To be precise, Hawking said in *A Brief History of Time*, "Someone told me that each equation I included in the book would halve the sales. In the end, however, I *did* put in one equation, Einstein's $E = mc^2$. I hope this will not scare off half my potential readers."

In 1974, Hawking determined that black holes should thermally create and emit subatomic particles, a process known as Hawking radiation, and in the same year he was elected as one of the youngest fellows of the Royal Society. He is currently the Lucasian Professor of Mathematics at the University of Cambridge, a post once held by Sir Isaac Newton.

Hawking is disabled by amyotrophic lateral sclerosis, or ALS (commonly known in the United States as Lou Gehrig's disease) and is now almost completely paralyzed. He operates a computer system using an infrared "blink switch" attached to his glasses. Symptoms of his disease started while he was enrolled at Cambridge, and he received a definitive diagnosis at age 21, shortly before his first marriage.

Collaborating with English mathematical physicist Sir Roger Penrose (born 1931), Hawking demonstrated that Einstein's General Theory of Relativity implied that space and time would have a beginning in the Big Bang and an end in black holes. One consequence of such research was that black holes emit radiation and eventually evaporate and disappear. Hawking has also conjectured that the universe has no edge or boundary in imaginary time, which suggests that "the way the universe began was completely determined by the laws of science."

Hawking spent considerable time contemplating scientific laws. In his 1993 book *Black Holes and Baby Universes*, he writes:

> There are well-defined laws that govern how the universe and everything in it develops in time. Although we have not yet found the exact form of all these laws, we already know enough to determine what happens in all but the most extreme situations. Whether we will find the remaining laws in the fairly near future is a mater of

opinion. I'm an optimist: I think there is a fifty-fifty chance that we will find them in the next twenty years.

Hawking goes on to say that even if we do not discover all the laws, a set of laws should exist that completely determines the evolution of the universe from its initial state. "These laws," he says, "may have been ordained by God. But it seems that He (or She) does not intervene in the universe to break laws."

In his article "The Quantum State of the Universe" published in *Nuclear Physics* in 1984, he also writes the following with respect to the laws of the universe:

> Many people would claim that the boundary conditions are not part of physics but belong to metaphysics or religion. They would claim that nature had complete freedom to start the universe off any way it wanted. That may be so, but it could also have made it evolve in a completely arbitrary and random manner. Yet all the evidence is that it evolves in a regular way according to certain laws. It would therefore seem reasonable to suppose that there are also laws governing the boundary conditions.

In the October 17, 1988, *Der Spigel*, Hawking writes:

> What I have done is to show that it is possible for the way the universe began to be determined by the laws of science. In that case, it would not be necessary to appeal to God to decide how the universe began. This doesn't prove that there is no God, only that God is not necessary.

I conclude this last Great Contenders entry with a final quotation from Hawking's *A Brief History of Time* on the topic of scientific theories:

> Any physical theory is always provisional, in the sense that it is only a hypothesis: you can never prove it. No matter how many times the results of experiments agree with some theory, you can never be sure that the next time the result will not contradict the theory. On the other hand, you can disprove a theory by finding even a single observation that disagrees with the predictions of the theory. As philosopher of science Karl Popper has emphasized, a good theory is characterized by the fact that it makes a number of predictions that could in principle be disproved or falsified by observation.

FURTHER READING

Gladstone, John, "Points of Supposed Collision Between the Scriptures and Natural Science" in *Faith and Free Thought* (London: Hodder and Stoughton, 1880).

Benedict, Robert Philip, *Fundamentals of Temperature, Pressure, and Flow Measurements* (Hoboken, N.J.: Wiley, 1984).

Klein, Morris, *Mathematics and the Physical World* (New York: Thomas Y. Crowell, 1959).

Tait, Peter Guthrie, *Scientific Papers*, Volume 2 (Cambridge, U.K.: Cambridge University Press, 1900).

FINAL COMMENTS ON THE BEAUTY OF MATHEMATICS IN SCIENCE

> While the equations represent the discernment of eternal and universal truths, however, the manner in which they are written is strictly, provincially human. That is what makes them so much like poems, wonderfully artful attempts to make infinite realties comprehensible to finite beings.
>
> —Michael Guillen, *Five Equations That Changed the World*

> Before creation, God did just pure mathematics. Then He thought it would be a pleasant change to do some applied.
>
> —John Edensor Littlewood, *A Mathematician's Miscellany*, 1953

THE BEAUTY OF MATHEMATICS

> The meaning of a great scientific equation usually furnishes us with what is called a law of nature.
>
> —Graham Farmelo, *It Must Be Beautiful*

The laws in this book are monuments to the progress of humankind, just as the Apollo spacecrafts were a testament to our quest for the stars, and the discovery of extrasolar planets was a milestone in our endless hunt through the heavens. Although these laws may be with us for as long as humans exist, we will understand and view these laws differently with the passage of time and the increase in our knowledge. For example, the discussion of Fourier's Law of Heat Conduction noted that Fourier wrote and developed his theory in terms of "caloric theory," an incorrect notion that changes in temperature are due to the transfer of an invisible and weightless fluid called caloric. Nevertheless, Fourier's Law *is* correct and in agreement with experiments, even if Fourier's idea of the *nature* of heat was incorrect.

Similarly, physicists today do not consider Maxwell's Equations, discussed under "Faraday's Laws of Induction and Electrolysis," as relating to propagation of waves in ether, an explanation to which Maxwell subscribed. A law can explain how the universe works, even if the researcher who discovered the law is not quite sure *why* it works.

Numbers do seem to rule the universe. Numerical patterns describe the arrangement of florets in a daisy, the reproduction of rabbits, the orbit of the planets, the harmonies of music, and the relationships among elements in a periodic table. Mathematical theories and formulas have predicted phenomena that were confirmed years after the theory was proposed. For example, Maxwell's Equations predicted radio waves. Einstein's Field Equations suggested that gravity would bend light and that the universe is expanding. Russian mathematician Nikolai Lobachevsky (1792–1856) said that "there is no branch of mathematics, however abstract, which may not someday be applied to the phenomena of the real world" (quoted in John dePillis's *777 Mathematical Conversation Starters*).

The British physicist Paul Dirac (1902–1984) once noted that the abstract mathematics we study now gives us a glimpse of physics in the future. In fact, his equations from 1928 that dealt with electron motion predicted the existence of antimatter, which was subsequently discovered. According to the formulas, an electron must have an antiparticle with the same mass but a positive electrical charge. In 1932, U.S. physicist Carl Anderson (1905–1991) observed this new particle experimentally, and it was named the positron. In 1955, the antiproton was produced at the Berkeley Bevatron. In 1995, physicists created the first anti-hydrogen atom at the research facility of CERN, the European Organization for Nuclear Research, which as the largest particle physics laboratory in the world.

A famous incident involving American physicist Murray Gell-Mann (born 1929) and colleagues demonstrates the predictive power of mathematics and symmetry when considering the existence of a subatomic particle known as the omega-minus. Gell-Mann had drawn a symmetric, geometric pattern in which each position in the pattern contained a known particle except for one empty spot. Gell-Mann put his finger on the spot and said with almost mystical insight, "There is a particle." His insight was correct, and experimentalists later found an actual particle with attributes that corresponded to the empty spot in the diagram.

GREAT EQUATIONS OF SCIENCE

> The poetry of science is in some sense embodied in its great equations.... These equations can also be peeled. But their layers represent their attributes and consequences, not their meanings. It is perfectly possible to imagine a universe in which mathematical equations have nothing to do with the workings of nature. Yet the marvelous thing is that they do.
>
> —Graham Farmelo, *It Must Be Beautiful*

Although remarkable equations such as Schrödinger's Wave Equation and Maxwell's Equations are mentioned in this book, some readers may wonder why these formulas are not considered "laws" and mentioned as main entries in this book. Of course, differentiating between laws and "mere" equations is partly the result of historical happenstance. Consider Maxwell's Equations. Scottish physicist James Clerk Maxwell (1831–1879) published *A Dynamical Theory of the Electromagnetic Field*, in which he recast the simple English of Faraday's Law into modern mathematics—as discussed in the Faraday entry. Because Maxwell's Equations have precedent in and comprise Coulomb's Law, Gauss's Law, Ampere's Law, and Faraday's Law (together with the concept that there are no magnetic monopoles), most people do not refer to the Maxwell's Equations as a separate set of laws. Steven Weinberg comments further on these equations in "Sokal's Hoax":

> The equations of electricity and magnetism that are today known as Maxwell's equations are not the equations originally written down by Maxwell; they are equations that physicists settled on after decades of subsequent work by other physicists, notably the English scientist Oliver Heaviside. They are understood today to be an approximation that is valid in a limited context (that of weak, slowly varying electric and magnetic fields), but in this form and in this limited context, they have survived for a century and may be expected to survive indefinitely.

Other important formulations of quantum mechanics, such as de Broglie's Wave Equation, Schrödinger's Wave Equation, Dirac's Equation, and the Klein-Gordon Equation are usually not referred to as laws, perhaps because some of the deep concepts behind the equations had significant overlap. Einstein's famous $E = mc^2$ is usually not called "Einstein's Law," although occasionally it has been called the law of mass-energy conservation. Note that $E = mc^2$ can be derived from Newton's Second Law, $\mathbf{F} = d\mathbf{p}/dt$. (Of course, the precise formulation of $\mathbf{p}$ in special relativity is different from that of $\mathbf{p}$ in classical mechanics.) In 1905, Einstein derived the mass–energy equivalence from the principles of Special Relativity in a short article titled "Does the Inertia of a Body Depend Upon Its Energy Content?"

Professor Clint Sprott of the University of Wisconsin wrote to me about his feelings regarding $E = mc^2$:

> Perhaps $E = mc^2$ is not generally referred to as a law because it is a fairly direct consequence of a deeper truth (special relativity). On the other hand, Kepler's Laws are a consequence of a deeper truth (Newton's Laws), although Newton's Laws came after Kepler's Laws. Perhaps $E = mc^2$ is not a law because it is a statement of equivalence in the sense that gravitational and inertial mass are equivalent. Note that special and general relativity were slow to be accepted—and general relativity is still debated—and thus these formulations were simply called "equations," and the terminology stuck. Finally, because Einstein had so many accomplishments, it would be confusing to talk about "Einstein's Law." Perhaps the lesson to be learned is that if you want something named after you, be careful not to do anything else important after your first big success!

Dr. Daniel Platt of the IBM T.J. Watson Research Center wrote to me:

> Almost no physical principle after 1900 has been called a law. Maxwell unified a set of laws (Ampere's Law, Gauss's Law, and so forth), and found that changing electric fields induced magnetic fields (displacement current) in order to preserve conservation of charge. A growing problem with redundancy began to emerge, such as we see with Gauss's Law versus Coulomb's Law. The early 1900s was a time when mathematical positivism was getting a stronger grip on the scientific community: fundamental physical principles had to be defined according to fundamental terms definable in relation to experiments—what might be called operationalism. This changed the expectation of what "law" meant in the 1900s, and whether it

was appropriate to call such relationships "laws." I reiterate that most "laws" were coined in the 19th century or earlier, with almost no physical law being coined in the 20th century or later.

Schrödinger's Wave Equation, discussed shortly, is usually not called Schrödinger's Law in part because it is more akin to a "definition" and does not describe a physical relationship between observable quantities in a simple or direct manner. Given Schrödinger's Wave Equation, we can calculate the wave function of a particle. Schrödinger's Wave Equation might be considered to be a description of reality that results from an underlying principle that particles behave as localized wave functions.

Obviously, not every important scientific law has been presented in this book, and our knowledge of the universe is continuing to grow at an astonishing rate. My objective was to make the book concise, to focus on laws that are named after a particular lawgiver, and to focus on some of the most influential laws through history, as defined in the introduction. Some principles and laws, although important, were omitted in the interest of space and because they are not definable by simple formulas. One example is Le Châtelier's Principle (1888), named after French chemist Henry Le Châtelier (1850–1936), which can be summarized as follows: If a chemical system at equilibrium experiences a change in concentration, temperature, or total pressure, the equilibrium shifts in order to minimize that change. (The principle does suffer from some notable exceptions.) Some important laws, such as Lenz's Law (1834), named after German physicist Heinrich Lenz (1804–1865), do not have separate entries because they are very closely related, or can be derived from, other laws. In this case, Lenz's Law is closely related to Faraday's Law of Induction and states that any current created by electromagnetic induction flux is in such a direction as to oppose the change in magnetic flux responsible for the induction. The "Great Contenders" section that precedes contains additional interesting laws of nature.

LISTMANIA AND HUMAN ACHIEVEMENT

> Every formula which expresses a law of nature is a hymn of praise to God.
>
> —Maria Mitchell, inscription on her bust in the Bronx Hall of Fame. She wrote the words in 1866.

Charles Murray, in his book *Human Accomplishment*, lists influential people in many fields of human endeavor, covering the time period from

800 B.C. to 1950 A.D. Murray ranks each individual according to a score that is based on the number and nature of written sources in which the individuals are included. Sources include general science books, biographical dictionaries, and reference books. Murray also considers how many pages are devoted to each individual. The following is a list of the top twenty physicists, ranked in order of eminence. Note how many individuals in Murray's book also appear in lawgiver entries in the present book.

1. Isaac Newton
2. Albert Einstein
3. Ernest Rutherford
4. Michael Faraday
5. Galileo Galilei
6. Henry Cavendish
7. Niels Bohr
8. J. J. Thomson
9. James Maxwell
10. Pierre Curie
11. Gustav Kirchhoff
12. Enrico Fermi
13. Werner Heisenberg
14. Marie Curie
15. Paul Dirac
16. James Joule
17. Christiaan Huygens
18. Walter Gilbert
19. Thomas Young
20. Robert Hooke

(Archimedes appears in Murray's "mathematicians" list.)

It is interesting to compare Murray's list of physicists with the end-of-millennium poll published in 1999 by *Physics World* magazine. The survey was conducted among approximately 100 of today's leading physicists. Based on the responses received, *Physics World* determined the following physicists to have made the most important contributions to the field of physics.

1. Albert Einstein
2. Isaac Newton
3. James Clerk Maxwell
4. Niels Bohr
5. Werner Heisenberg
6. Galileo Galilei
7. Richard Feynman
8. Paul Dirac
9. Erwin Schrödinger
10. Ernest Rutherford

Physicist Brian Greene from Columbia University, who participated in the *Physics World* survey, noted that Einstein displaced Newton from the top position and said, "Einstein's special and general theories of relativity completely overturned previous conceptions of a universal, immutable space and time, and replaced them with a startling new framework in which space and time are fluid and malleable." Peter Rodgers, editor of *Physics World*, remarked, "Einstein and Newton were always going to be one and two but what was surprising about the top 10 was that there were seven out and out theorists."

John Galbraith Simmons, in his book *The Scientific 100*, gives the following top 20:

1. Isaac Newton
2. Albert Einstein
3. Niels Bohr
4. Charles Darwin
5. Louis Pasteur
6. Sigmund Freud
7. Galileo Galilei
8. Antoine Lavoisier
9. Johannes Kepler
10. Nicolaus Copernicus
11. Michael Faraday
12. James Clerk Maxwell
13. Claude Bernard
14. Franz Boas Modern
15. Werner Heisenberg
16. Linus Pauling
17. Rudolf Virchow
18. Erwin Schrödinger
19. Ernest Rutherford
20. Paul Dirac

Simmons ranks scientists whose "influence in shaping the contemporary word is pervasive and inescapable." He writes:

> They formulated the laws of motion, discovered how electricity works, and illuminated the structure of the atom. They broke down chemicals into elements and found them in the sun.... And apart from a couple of intellectual discoveries which go back to the Greeks and Babylonians, they accomplished it all in several hundred years.

Michael Guillen, in *Five Equations That Changed the World*, lists the following formulas (in his notation) that he believes led to "the five most powerful and important scientific achievements in history":

1. $F = G \times M \times m \div d^2$ (Newton's Law of Universal Gravitation)
2. $P + \rho \times \frac{1}{2}v^2 = \text{constant}$ (Bernoulli's Law of Hydrodynamic Pressure)
3. $\nabla \times E = -\partial B/\partial T$ (Faraday's Law of Induction)
4. $E = mc^2$ (in addition to Albert Einstein's Special Theory of Relativity)
5. $\Delta S_{\text{universe}} > 0$ (Clausius's Law of Thermodynamics)

"THE GREATEST EQUATIONS EVER"

> Great equations change the way we perceive the world. They reorchestrate the world—transforming and reintegrating our perception by redefining what belongs together with what. Light and waves. Energy and mass.

> Probability and position. And they do so in a way that often seems unexpected and even strange.
> —Robert P. Crease, "The Greatest Equations Ever," *Physics World*

In 2004, Robert Crease conducted a survey of *Physics World* readers who nominated "The Greatest Equations Ever." Maxwell's Equations of Electromagnetism and Euler's identity $e^{i\pi} + 1 = 0$ were the top contenders. Several of Crease's readers wondered about the difference between formulas, theorems, and equations. For Crease, a formula is something that obeys the rules of a "syntax." An equation is a formula that states observed facts and is thus empirically true. Crease says that the equation that describes the Balmer Series of lines in the visible spectrum, $1/\lambda = R_H(1/2^2 - 1/n^2)$, is a good example, as are chemical equations that embody observations about reactions studied in a laboratory. Crease writes:

> However, these distinctions are not really so neat. Many classic physics equations—including $E = mc^2$ and Schrödinger's equation—were not conclusions drawn from statements about observations. Rather, they were conclusions based on reasoning from other equations and information; they are therefore more like theorems. And theorems can be equation-like for their strong empirical content and value.

Crease also notes that a great equation "does more than set out a fundamental property of the universe, delivering information like a signpost, but works hard to wrest something from nature." Michael Berry from Bristol University said in the February 1998 issue of *Physics World*, "Any great physical theory gives back more than is put into it, in the sense that as well as solving the problem that inspired its construction, it explains more and predicts new things."

The following equations relate to descriptions of the physical world and are listed by Crease, using his notation, in order of the number of people who nominated them:

1. Maxwell's Equations $\nabla \cdot \mathbf{D} = \rho$, $\nabla \cdot \mathbf{B} = 0$, $\nabla \times \mathbf{E} = -\frac{\partial \mathbf{B}}{\partial t}$, $\nabla \times \mathbf{H} = \mathbf{J} + \frac{\partial \mathbf{D}}{\partial t}$
2. Newton's Second Law, $F = ma$
3. Schrödinger's Wave Equation, $H\Psi = E\Psi$
4. $E = mc^2$
5. Boltzmann's Equation, $S = k \ln W$
6. Principle of Least Action, $\delta S = 0$
7. De Broglie's Wave Equation, $\lambda = h/mv$

8. Einstein's Field Equations for General Relativity, $G_{\mu\nu} = 8\pi G T_{\mu\nu}$
9. Dirac's Equation, $i\gamma \cdot \partial\psi = m\psi$
10. Hubble's Equation, $v = H_0 d$
11. Ideal Gas Law: $PV = nRT$
12. Balmer Series: $1/\lambda = R(1/n_1^2 - 1/n_2^2)$
13. Planck's Equation: $E = hv$

Graham Farmelo's book *It Must Be Beautiful* presents equations that eleven essayists believe to be the most important equations of twentieth-century science. Six of them are from fundamental physics:

- The Planck-Einstein Equation, $E = hf$, which connects frequency with energy
- Einstein's equation $E = mc^2$
- Einstein's equation that governs general relativity and gravity, $R_{ab} - \frac{1}{2}Rg_{ab} = -8\pi G T_{ab}$
- The Yang-Mills Equation, which describes fundamental particles and their interactions: $\partial \mathbf{f}_{uv}/\partial x_v + 2\varepsilon(\mathbf{b}_v \times \mathbf{f}_{uv}) + \mathbf{J}_\mu = 0$
- Schrödinger's Wave Equation
- Dirac's Equation

These last two I discuss further below. Farmelo's book also includes Drake's Equation that estimates the number of technological civilizations in our galaxy ($N = R^* \times f_p \times n_e \times f_l \times f_i \times f_c \times L$), Shannon's Equations on information theory [$H = -K\sum_{i=1}^{n} p(x_i)\log p(x_i)$], and the logistic mapping that models complicated behavior in the field of chaos theory [$x_{n+1} = rx_n(1 - x_n)$].

Frank Wilczek, one of Farmelo's essayists, includes the following quote by Heinrich Hertz on Maxwell's equations: "One cannot escape the feeling that these mathematical formulae have an independent existence and an intelligence of their own, that they are wiser than we are, wiser even than their discoverers, that we get more out of them than was originally put into them." Steven Weinberg in the book's afterword concludes:

> When an equation is as successful as Dirac's, it is never simply a mistake. It may not be valid for the reason supposed by its author, it may break down in new contexts, and it may not even mean what its author thought it meant. We must continually be open to reinterpretations of these equations. But the great equations of modern physics are a permanent part of scientific knowledge, which may outlast even the beautiful cathedrals of earlier ages.

As I discussed above, the equations of physics can sometimes give birth to ideas or consequences that the equation discoverer did not expect. The power of these kinds of equations can seem magical, according to Wilczek in the essay on Dirac's Equation, which describes the properties of quantum particles. In 1928, Paul Dirac attempted to find a version of Schrödinger's Wave Equation that would be consistent with the principles of Special Relativity. One way that Dirac's Equation can be written is

$$\left(\alpha_0 mc^2 + \sum_{j=1}^{3} \alpha_j p_j c\right) \psi(\mathbf{x}, t) = i\hbar \frac{\partial \psi}{\partial t}(\mathbf{x}, t).$$

The equation describes electrons and other elementary particles in a way that is useful in both quantum mechanics and the Special Theory of Relativity. As I suggested above, the equation predicts the existence of antiparticles and in some sense "foretold" their experimental discovery. This made the discovery of the positron, the antiparticle of the electron, a fine example of the usefulness of mathematics in modern theoretical physics. In this equation, m is the rest mass of the electron, $\hbar$ is the reduced Planck's constant (1.054×10^{-34} J·s), c is the speed of light, p is the momentum operator, $\mathbf{x}$ and t are the space and time coordinates, and $\psi(\mathbf{x}, t)$ is a wave function; α is a linear operator that acts on the wavefunction. Peter Galison writes of Dirac's personality in "The Suppressed Drawing":

> For most of the twentieth century Paul Dirac stood as the theorist's theorist. Though less known to the general public than Albert Einstein, Niels Bohr, or Werner Heisenberg, for physicists, Dirac was revered as the "theorist with the purest soul"... because of Dirac's taciturn and solitary demeanor [and] because he maintained practically no interests outside physics and never feigned engagement with art, literature, music, or politics. Known for the fundamental equation that now bears his name—describing the relativistic electron—Dirac put quantum mechanics into a clear conceptual structure, explored the possibility of magnetic monopoles, generalized the mathematical concept of function, launched the field of quantum electrodynamics, and predicted the existence of antimatter.

Another profound equation in physics, Schrödinger's Wave Equation, describes ultimate reality in terms of wave functions and probabilities:

$$\left[-\frac{\hbar^2}{2m}\nabla^2 + V(\mathbf{r})\right] \psi(\mathbf{r}, t) = i\hbar \frac{\partial \psi}{\partial t}(\mathbf{r}, t)$$

Physicist Freeman Dyson, in his introduction to John Cornwell's *Nature's Imagination*, lauds this formula that represents a stage in humanity's grasp of reality. He writes:

> Sometimes the understanding of a whole field of science is suddenly advanced by the discovery of a single basic equation. Thus it happened that the Schrödinger equation in 1926 and the Dirac equation in 1927 brought a miraculous order into the previously mysterious processes of atomic physics. Bewildering complexities of chemistry and physics were reduced to two lines of algebraic symbols.

With respect to Einstein's General Theory of Relativity and his Field Equations of gravitation, Paul Dirac said that the theory was "probably the greatest scientific discovery ever made." Max Born called it "the greatest feat of human thinking about nature, the most amazing combination of philosophical penetration, physical intuition, and mathematical skill." I briefly discuss the General Theory of Relativity in "Eötvös's Law of Capillarity" in part III.

In the early 1970s, Nicaragua issued ten postage stamps bearing *Las 10 Formulas Matematicas Que Cambiaron La Faz De La Terra* ("The 10 Mathematical Formulas That Changed the Face of the World"). Isn't it admirable that a country so respects mathematics that it devotes a postage stamp series to a set of abstract equations? Have other countries produced a similar series? In addition to scientific merit, perhaps such practical issues as space limitations were considered by the Nicaraguan government so as to avoid long formulas on small stamps.

I conducted my own informal survey as to which formulas should be considered as "The 10 Mathematical Formulas That Changed the Face of the World." The survey was conducted via electronic mail networks, and most respondents were mathematicians and included professors, other professionals, and graduate students. What follows is the answer to this question from approximately fifty interested individuals who gave me their opinions as to the most important and influential equations. The equations are ordered from most influential to least influential, based on the number of different people who listed a formula when they sent their lists to me. For example, $E = mc^2$ received the most votes.

How many of the following formulas can you identify? If you can identify more than five, you are probably more mathematically knowledgeable than 99% of the people on Earth. If you can identify all equations in the Top 10 and all the equations in the Runners-Up list, you are worthy of cavorting with the antediluvian gods. I'll identify these equations in a following section.

According to my survey, here are the ten most influential and important mathematical expressions, listed in order of importance:

1. $E = mc^2$
2. $a^2 + b^2 = c^2$
3. $\varepsilon_0 \oint E \cdot dA = \sum q$
4. $x = \left(-b \pm \sqrt{b^2 - 4ac}\right)/(2a)$
5. $\vec{F} = m\vec{a}$
6. $1 + e^{i\pi} = 0$
7. $c = 2\pi r, \quad a = \pi r^2$
8. $F = Gm_1 m_2/r^2$
9. $f(x) = \sum c_n e^{in\pi x/L}$
10. $e^{i\theta} = \cos\theta + \sin\theta$, tied with $a^n + b^n \neq c^n$, $n \geq 2$

Other mathematical expressions that did not score high enough to be included in the Top 10 but that scored favorably were

1. $f(x) = f(a) + f'(a)(x-a) + f''(a)(x-a)^2/2! \ldots$
2. $s = vt + at^2/2$
3. $V = IR$
4. $z \rightarrow z^2 + \mu$
5. $e = \lim_{n\rightarrow\infty}(1 + 1/n)^n$
6. $c^2 = a^2 + b^2 - 2ab\cos C$
7. $\int KdA = 2\pi \times x$
8. $\mathrm{d}/\mathrm{d}x \int_a^x f(t)\mathrm{d}t = f(x)$
9. $1/(2\pi i) \oint f(z)/(z-a)dz = f(a)$
10. $dy/dx = \lim_{h\rightarrow 0}(f(x+h) - f(x))/h$
11. $\partial^2\psi/\partial x^2 = -[8\pi^2 m/h^2(E - V)]\psi$

How many do you recognize? I identify a number of these in the next section.

NICARAGUA POSTAGE STAMP LIST

Here is a list of Nicaragua's postage stamp equations for *Las 10 Formulas Matematicas Que Cambiaron La Faz De La Terra.* Note how many of these formulas agree with the "Top Ten" list based on my own informal survey.

1. $1 + 1 = 2$
2. $F = Gm_1 m_2/r^2$
3. $E = mc^2$

4. $e^{\text{In}N} = N$
5. $a^2 + b^2 = c^2$
6. $S = k \log W$
7. $V = V_e \ln m_0/m_1$
8. $\lambda = h/mv$
9. $\nabla^2 E = (Ku/c^2)(\partial^2 E/\partial t^2)$
10. $F_1 x_1 = F_2 x_2$

Do you recognize several of these formulas? Study them before reading further.

Here are the solutions for the Nicaragua stamp list:

1. Basic addition formula.
2. Newton's Law of Universal Gravitation. If the two masses m_1 and m_2 are separated by a distance, r, the force exerted by one mass on the other is F, and G is a constant of nature.
3. Einstein's formula for the conversion of matter to energy.
4. John Napier's logarithm formula. This allows one to perform multiplication and division simply by adding or subtracting the logarithms of numbers.
5. Pythagorean Theorem relating the lengths of sides of a right triangle.
6. Bolzmann's Equation for the behavior of gases.
7. Konstantin Tsiolkovsky's rocket equation, which gives the speed of a rocket as it burns the weight of its fuel.
8. de Broglie's Wave Equation, relating the mass, velocity, and wavelength of a wave-particle. h is Planck's constant. Louis de Broglie (1892–1987) postulated that the electron has wave properties and that material particles have associated wavelengths.
9. Equation describing electromagnetic radiation, derived from Maxwell's Equations, which form the basis for all computations involving electromagnetic waves, including radio, radar, light, ultraviolet waves, heat radiation and X-rays.
10. Archimedes' lever formula.

The following are explanations for some of the formulas in my own survey: Number 3 is one of Maxwell's Equations for electromagnetism. Number 4 is the quadratic formula for solving equations of the form $ax^2 + bx + c = 0$. Number 5 is Newton's Second Law, relating force, mass, and acceleration. Number 7 gives the circumference and area of a circle. Number 9 represents a Fourier series (complicated wave disturbances may be represented as the sum of a group of sinusoidal-like waves). In Number 10, the first formula is Euler's identity relating exponential and trigonometric functions, and the second formula represents Fermat's Last Theorem. In the runners-up list, number 7 is the Gauss-Bonnet formula, where x is the Euler characteristic, and number 9 is Cauchy's integral formula in complex analysis. (The Gauss-Bonnet formula occurs in the field of differential geometry and concerns the curvature of surfaces.)

A few respondents suggested that Fermat's Last Theorem be included on the list of the ten influential mathematical expressions because a significant amount of research and mathematics have been a direct result of attempts to prove the theorem. This theorem by Pierre de Fermat (1601–1665) states that there are no whole numbers a, b, and c such that $a^n + b^n = c^n$ for $n > 2$. In 1995, English-American mathematician Andrew Wiles (born 1953) published a famous paper in the *Annals of Mathematics* that finally proved Fermat's Last Theorem. In 1769, Leonard Euler stated that he thought the related formula $a^4 + b^4 + c^4 = d^4$ had no possible integral solutions. Two centuries later, Noam Elkies of Harvard University discovered the first solution $a = 2,682,440$, $b = 15,365,639$, $c = 18,796,760$, and $d = 20,516,673$.

PHYSICS AND RELIGION

> I'm sorry to say that the subject I most disliked was mathematics. I have thought about it. I think the reason was that mathematics leaves no room for argument. If you made a mistake, that was all therc was to it.
>
> —Malcolm X, *Mascot*

As highlighted in this book, many important physicists were quite religious. In some ways, the mathematical quest to understand the universe parallels mystical attempts to understand God. Both religion and mathematics struggle to express relationships among humans, the universe, and infinity. Both have arcane symbols and rituals and seemingly impenetrable language. Both exercise the deep recesses of our minds and stimulate our imagination. Mathematicians and theoretical physicists, like priests, sometimes seek "ideal," immutable, nonmaterial truths and then often venture to apply these truths in the real world. Are mathematics and religion the most powerful evidence of the inventive genius of the human race? Edward Rothstein notes in "Reason and Faith, Eternally Bound" that faith was the inspiration for Newton and Kepler as well as for numerous scientific and mathematical triumphs. Rothstein writes, "The conviction that there is an order to things, that the mind can comprehend that order and that this order is not infinitely malleable, those scientific beliefs may include elements of faith."

Of course, many differences exist between mathematics and religion. While various religions differ in their beliefs, remarkable agreement usually exists among mathematicians.

FURTHER READING

Berry, Michael, "Paul Dirac: the Purest Soul in Physics," *Physics World*, February 1, 1998; see physicsworld.com/cws/article/print/1705.

Crease, Robert P., "The Greatest Equations Ever," *Physics World*, October 2004; see physicsweb.org/articles/world/17/10/2.

dePillis, John, *777 Mathematical Conversation Starters* (Washington, D.C.: The Mathematical Association of America, 2002).

Durrani, Matin, "Physics: Past, Present, Future," *Physics World*, December 1999; see physicsweb.org/articles/world/12/12/14.

Dyson, Freeman, "Introduction," in John Cornwell's *Nature's Imagination* (New York: Oxford University Press, 1995).

Elkies, Noam, "On $a^4 + b^4 + c^4 = d^4$," *Mathematics of Computation,* 51(184): 825–835, 1988.

Farmelo, Graham, *It Must Be Beautiful: Great Equations of Modern Science* (New York: Granata Books, 2003).

Galison, Peter, "The Suppressed Drawing: Paul Dirac's Hidden Geometry," *Representations*, 72: 145–166, 2000.

Guillen, Michael, *Five Equations That Changed the World* (New York: Hyperion, 1995).

Murray, Charles, *Human Accomplishment: The Pursuit of Excellence in the Arts and Sciences, 800 B.C. to 1950* (New York: Harper Perennial, 2004).

Pickover, Clifford, *The Loom of God* (New York: Plenum, 1997).

Pickover, Clifford, *A Passion for Mathematics* (Hackensack, N.J.: Wiley, 2005).

Pickover, Clifford, *Wonders of Numbers* (New York: Oxford, 2001).

Rothstein, Edward, "Reason and Faith, Eternally Bound," *New York Times*, B7, p. 7, December 20, 2003.

Simmons, John Galbraith, *The Scientific 100: A Ranking of the Most Influential Scientists, Past and Present* (New York: Citadel Press, 2000).

Weinberg, Steven, "Sokal's Hoax," *New York Review of Books*, 43(13): 11–15, August 8, 1996.

INTERLUDE: CONVERSATION STARTERS

None of the laws of physics known today (with the possible exception of the general principles of quantum mechanics) are exactly and universally valid. Nevertheless, many of them have settled down to a final form, valid in certain known circumstances.

—Steven Weinberg, "Sokal's Hoax," *The New York Review of Books*, August 8, 1996

The operating system...governs the flow of information through a computer just as an eternal law of nature is thought to guide physics. But...there could be other kinds of architectures and operating systems that themselves evolve in time.

—Lee Smolin, "Never Say Always," *New Scientist*, September 23, 2006

Describing the physical laws without reference to geometry is similar to describing our thoughts without words.

—Albert Einstein, 1922 Kyoto lecture

REFERENCES

> It is commonly assumed in science that underneath all the appearance of irregularity there are some quite simple and quite general laws governing what happens. Thus objects in the world exhibit a vast variety of motions. But Newtonian mechanics managed to show that all the motions can be shown to obey a small number of perfectly general laws of motion. Bodies behave in a large variety of ways when heated; but in thermodynamics it can be shown that all these happenings obey a few, quite general, laws of thermodynamics.
>
> —Robert Nola, "Laws of Nature"

I've compiled a list of references that identify much of the material I used to research this book. Many additional references are throughout the book. I include Internet Web sites in addition to books and journals. As many readers are aware, Internet Web sites come and go. Sometimes they change addresses or completely disappear. The Web site addresses listed in this book provided valuable background information when this book was written. You can, of course, find numerous other Web sites relating to the laws discussed in this book by using standard Web search tools.

If I have overlooked an important principle—either a famous law or one that you feel has never been fully appreciated—please let me know. Just visit my Web site, www.pickover.com, and send me an e-mail explaining the law and how you feel it influences our lives today or the history of science.

Arons, Arnold, *Development of Concepts of Physics* (Reading, Mass.: Addison-Wesley, 1965).

Atiyah, Michael, "Pulling the Strings," *Nature*, 438: 1081–1082, December 22, 2005.

Bothamley, Jennifer, *Dictionary of Theories* (Washington, D.C.: Gale Research International Ltd., 1993).

Bryson, Bill, *A Short History of Nearly Everything* (New York: Random House, 2003).

Bueche, Frederick, *Introduction to Physics for Scientists and Engineers* (New York: McGraw-Hill, 1975).

Carroll, John W., "Laws of Nature," in *Stanford Encyclopedia of Philosophy*; see plato.stanford.edu/entries/laws-of-nature/.

Casti, John, *Paradigms Lost* (New York: William Morrow, 1989).

Considine, Douglas, managing editor, *Van Nostrand's Scientific Encyclopedia*, 7th edition (New York: Van Nostrand Reinhold, 1989).

Cropper, William, *Great Physicists* (New York: Oxford University Press, 2001).

Durbin, Paul, *Dictionary of Concepts in the Philosophy of Science* (New York: Greenwood Press, 1988).

Encyclopædia Britannica, www.britannica.com/.

Farmelo, Graham, *It Must Be Beautiful: Great Equations of Modern Science* (New York: Granta Books, 2002).

Feynman, Richard, *The Character of Physical Law* (New York: Modern Library, 1994).

Francis, Erik Max, "The Laws List"; see www.alcyone.com/max/physics/laws/.

Frayn, Michael, *The Human Touch* (New York: Metropolitan Books, 2006).

Gardner, Martin, *Order and Surprise* (Amherst, N.Y.: Prometheus, 1983), chapter 4.

Gillispie, Charles C., editor-in-chief, *Dictionary of Scientific Biography* (New York: Charles Scribner's Sons, 1970).

Grun, Bernard, *The Timetables of History* (New York: Touchstone, 1975).

Guillen, Michael, *Five Equations That Changed the World* (New York: Hyperion, 1995).

Hall, Carl W., *Laws and Models* (Boca Raton, Fla.: CRC Press, 1999).

Halliday, David, and Robert Resnick, *Physics* (New York: John Wiley & Sons, 1966).

Hart, Michael, *The 100: A Ranking of the Most Influential Persons in History* (New York: Hart Publishing Company, 1978).

Hawking, Stephen, *Black Holes and Baby Universes* (New York: Bantam, 1993).

Kaku, Michio, "Parallel Universes, the Matrix, and Superintelligence," KurzweilAI.net, June 26, 2003; see www.kurzweilai.net/meme/frame.html?main=/articles/art0585.html.

Krauss, Lawrence, *Fear of Physics* (New York: Basic Books, 1993).

Krebs, Robert, *Scientific Laws, Principles, and Theories* (Westport, Conn.: Greenwood Press, 2001).

Merton, Robert K., *The Sociology of Science* (Chicago: University of Chicago Press, 1973).

Nave, Carl R. (Rod), "HyperPhysics," Department of Physics and Astronomy, Georgia State University; see hyperphysics.phy-astr.gsu.edu/HBASE/hframe.html.

O'Connor, John J., and Edmund F. Robertson, *The MacTutor History of Mathematics Archives*, School of Mathematics and Statistics, University of St. Andrews, Scotland; see www-history.mcs.st-andrews.ac.uk/.

Parker, Sylvia, editor-in-chief, *McGraw-Hill Encyclopedia of Science and Technology*, 8th edition (New York: McGraw-Hill, 1997).

Penrose, Roger, *The Road to Reality* (New York: Knopf, 2005).

Peterson, Ivars, "MathTrek Archives"; see www.sciencenews.org/pages/sn_weekly/math_arc.asp.

Pickover, Clifford, *A Passion for Mathematics* (New York: Wiley, 2005).

Smolin, Lee, "Never Say Always," *New Scientist*, 191(2570): 30–35, September 23, 2006.

Tipler, Paul, *Physics* (New York: Worth Publishers, 1976).

Trefil, James, *The Nature of Science* (New York: Houghton Mifflin Company, 2003).

Weisstein, Eric, "Eric Weisstein's World of Physics"; see scienceworld.wolfram.com/physics.

Weisstein, Eric, "Eric Weisstein's World of Science"; see scienceworld.wolfram.com/search/.

Wikipedia: The Free Encyclopedia; see en.wikipedia.org/.

Wikipedia, "List of Laws in Science"; see en.wikipedia.org/wiki/ List_of_laws_in_science.

Wikipedia, "Scientific Laws Named after People"; see en.wikipedia.org/wiki/Scientific_laws_named_after_people.

Wilson, Jerry, "Scientific Laws, Hypotheses, and Theories"; see wilstar.com/theories.htm.

Our world is filled with the statues of great generals, atop of prancing horses, leading their cheering soldiers to glorious victory. Here and there, a modest slab of marble announces that a man of science has found his final resting place. A thousand years from now we shall probably do these things differently, and the children of that happy generation shall know of the splendid courage and the almost inconceivable devotion to duty of the men who were the pioneers of that abstract knowledge, which alone has made our modern world a practical possibility.
—Hendrik Willem van Loon, *The Story of Mankind*

ABOUT THE AUTHOR

Clifford A. Pickover received his Ph.D. from Yale University's Department of Molecular Biophysics and Biochemistry. He graduated first in his class from Franklin and Marshall College, after completing the four-year undergraduate program in three years. His many books have been translated into Italian, French, Greek, German, Japanese, Chinese, Korean, Portuguese, Spanish, Turkish, Serbian, Romanian, and Polish.

One of the most prolific and eclectic authors of our time, Pickover is author of the popular books *The Heaven Virus* (Lulu, 2007), *A Beginner's Guide to Immortality* (Thunder's Mouth Press, 2007), *The Möbius Strip* (Thunder's Mouth Press, 2006), *Sex, Drugs, Einstein, and Elves* (Smart Publications, 2005), *A Passion for Mathematics* (Wiley, 2004), *Calculus and Pizza* (Wiley, 2003), *The Paradox of God and the Science of Omniscience* (Palgrave/St. Martin's Press, 2002), *The Stars of Heaven* (Oxford University Press, 2001), *The Zen of Magic Squares, Circles, and Stars* (Princeton University Press, 2001), *Dreaming the Future* (Prometheus, 2001), *Wonders of Numbers* (Oxford University Press, 2000), *The Girl Who Gave Birth to Rabbits* (Prometheus, 2000), *Surfing Through Hyperspace* (Oxford University Press, 1999), *The Science of Aliens* (Basic Books, 1998), *Time: A Traveler's Guide* (Oxford University Press, 1998), *Strange Brains and Genius: The Secret Lives of Eccentric Scientists and Madmen* (Plenum, 1998), *The Alien IQ Test* (Basic Books, 1997), *The Loom of God* (Plenum, 1997), *Black Holes: A Traveler's Guide* (Wiley, 1996), and *Keys to Infinity* (Wiley, 1995). He is also author of numerous other books—including *Chaos in Wonderland: Visual Adventures in a*

Fractal World (1994), *Mazes for the Mind: Computers and the Unexpected* (1992), *Computers and the Imagination* (1991), and *Computers, Pattern, Chaos, and Beauty* (1990), all published by St. Martin's Press—and the author of more than 200 articles on topics in science, art, and mathematics. He is also coauthor, with Piers Anthony, of *Spider Legs* (TOR, 1998). Pickover is currently an associate editor for the scientific journal *Computers and Graphics* and is an editorial board member for *Odyssey*, *Leonardo*, and *YLEM*.

Editor of the books *Chaos and Fractals: A Computer Graphical Journey* (Elsevier, 1998), *The Pattern Book: Fractals, Art, and Nature* (World Scientific, 1995), *Visions of the Future: Art, Technology, and Computing in the Next Century* (St. Martin's Press, 1993), *Future Health* (St. Martin's Press, 1995), *Fractal Horizons* (St. Martin's Press, 1996), and *Visualizing Biological Information* (World Scientific, 1995), and coeditor of the books *Spiral Symmetry* (World Scientific, 1992) and *Frontiers in Scientific Visualization* (Wiley, 1994), Dr. Pickover's primary interest is finding new ways to continually expand creativity by melding art, science, mathematics, and other seemingly disparate areas of human endeavor.

The *New York Times* has proclaimed, "Pickover contemplates realms beyond our known reality." The *Los Angeles Times* writes, "Pickover has published nearly a book a year in which he stretches the limits of computers, art and thought." Pickover received first prize in the Institute of Physics "Beauty of Physics Photographic Competition." His computer graphics have been featured on the cover of many popular magazines, and his research has recently received considerable attention by the press—including CNN's "Science and Technology Week," The Discovery Channel, the *New York Times*, *Science News*, the *Washington Post*, *WIRED*, and the *Christian Science Monitor*—and also in international exhibitions and museums. *OMNI* magazine described him as "Van Leeuwenhoek's twentieth century equivalent." *Scientific American* several times featured his graphic work, calling it "strange and beautiful, stunningly realistic." *WIRED* magazine wrote, "Bucky Fuller thought big, Arthur C. Clarke thinks big, but Cliff Pickover outdoes them both." Pickover holds more than forty U.S. patents, mostly concerned with novel features for computers.

For many years, Dr. Pickover was the lead columnist for *Discover* magazine's "Brain-Boggler" column, and he currently writes the "Brain-Strain" column for *Odyssey*. His calendar and card sets, *Mind-Bending Visual Puzzles*, have been among his most popular creations.

Dr. Pickover's hobbies include the practice of Ch'ang-Shih Tai-Chi Ch'uan, Shaolin Kung Fu, and piano playing. He owns a 110-gallon aquarium filled with Lima shovelnose catfishes. These bizarre creatures resemble sharks with ultratiny, alien eyes. He advises readers to maintain a shovelnose tank in order to foster a sense of mystery in their lives. Look into the fish's eudaemonic eyes, dream of Elysian Fields, and soar.

Visit Dr. Pickover's Web site, www.pickover.com, which has received more than a million visits. He can be reached at this Web page or at P.O. Box 549, Millwood, NY 10546-0549 USA.

INDEX